Holography MarketPlace

8th Text Edition

The reference text and sourcebook for holography worldwide.

The original version of this book contained holograms from the vendors and many of them are no longer in business. Therefore this version of the book contains everything that was in the original version but it has no holograms.

Edited by Alan Rhody and Franz Ross

Copyright © 1999 Ross Books

Library Catalogue Information
Holography MarketPlace - Eighth Text Edition
Bibliography:
1. Includes Index.
2. Holography
3. Directories - Holography Industry
4. Photography
5. Physics
ISBN 978-0-89496-101-4

Printed in Canada

ABOUT THE HOLOGRAM ON THE COVER OF THIS BOOK

Recommended Viewing Instructions:

1) Hold this book at arm's length in front of you underneath a single spotlight, under any lamp that has a clear lightbulb in it, or in direct sunlight. Any overhead light source that makes distinct shadows (such as a flashlight) will work well. *If you view this hologram under a fluorescent light, with more than one light, or under cloudy skies you will see a blurry image. Move to a better spot!*

2) Tilt the book up and down until the hologram image is viewable. You should see "Parrots in the Jungle".

3) Tilt the book from side to side and up and down until the image appears brightest. Notice the variety of vibrant colors displayed and the depth of this 3D image!

*****

You are looking at the first fully dimensional, full-color photopolymer hologram ever published on the cover of a book. In fact, to our knowledge, it is the first time this type of mass produced hologram has been presented to a world-wide audience (outside a holography science conference). It was made by Dai Nippon Printing Co. Ltd. (of Japan) using a panchromatic holographic recording material developed by DuPont.

To date, Dai Nippon Printing (DNP) is the only company in the world that has demonstrated the willingness and the ability to combine all the components (panchromatic film, multiple lasers, specially designed replication equipment) needed to mass-produce bright, high quality holograms that are fully dimensional and realistically colored. You will find more detailed information about the process used to make this particular hologram in Chapter Six. Other companies that are pusuing related work with color photopolymer holograms are also profiled.

It is a great honor for Ross Books to feature one of these magnificent new DNP holograms on the cover of the Holography MarketPlace 8th Ed. It is a first for the holography industry and the publishing industry as well. We are proud to commemorate the first 50 years of holography with such a hologram and sincerely believe that it represents an important and invigorating development in our industry.

Foreword

Welcome to the Eighth (1999) edition of Holography MarketPlace. This edition includes information that will benefit professionals that work within the holography industry as well as those in other fields that need to know more about holograms and their applications. HMP 8 also provides basic knowledge that will appeal to the curious businessperson, student or casual reader. We hope that you enjoy reading this publication and find it a very useful resource and a valuable addition to your bookshelf.

For your convenience, HMP 8 has been divided into four sections. The first section explains what a hologram is and covers the basics of hologram production, including design and origination. Major commercial applications are reviewed. It also includes an in-depth discussion about large holograms used for exhibition and a description of the existing holographic giftware distribution network.

Section Two features new and emerging holographic tech- nologies that we believe will have widespread commercial applications in the future: panchromatic photopolymers (for realistic looking 3D images), holograms directly em- bossed onto fabric (an environmentally-safe coloring method), "instant" holographic printers that make easily viewable, fully dimensional, animated images from digi- tal input (a computer peripheral), and new methods for producing real-time dimensional images (for entertainment and educational displays).

The third section reviews production equipment and materials that are relevant to hobbyists and professionals, alike. The fundamentals of laser theory are introduced and the types of lasers most commonly used by holographers are described. The newest lasers available from major manu- factures (CW and pulsed) are featured. For holographers working with a more limited budget, we've included im- portant information about refurbishing their existing la- sers and about buying surplus equipment.

Substantial space in this section is devoted to the newest silver halide recording emulsions available and related processing methods. Photopolymer, photoresist, and DCG recording materials are also described. This section concludes with a review of existing dot matrix production equipment, as these automated hologram origination machines are being increasingly employed by large manufacturers.

Section Four contains an updated international business directory of the holography industry and easy to use cross-index tables. Please notify us of all relevant changes.

Acknowledgments: We thank all the companies that included holograms in their advertisements. Additionally, we'd like to express our gratitude to all the authors that submitted articles to us. We look forward to including contributions that could not be included in this particular edition in a future publication. We encourage all of our readers to submit information to us at any time that they think will benefit our industry.

Alan Rhody and Franz Ross
Ross Books WInter 1999.

Index to Advertisers

SECTION 1
The Basics

Chapter 1 - Introduction to Holography

Chapter 2 - Design and Artwork

Chapter 3 - In the Holography Studio

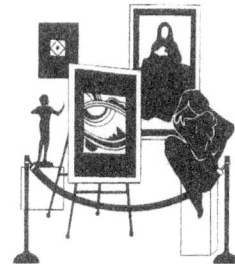

Chapter 4 - Exhibition Holograms

Chapter 5 - Holograms as Giftware

RARE & COLLECTABLE EDITIONS

The following titles are now available in limited quantities. Each volume of HMP contains pertinent reference material, historically significant information and rare, collectable holograms!

Now only $60.00 for the entire set of HMP 2, 3, 4, 5, 6 & 7!

HOLOGRAPHY MARKETPLACE 2nd EDITION (1990)

Includes the hologram: *Statue of Liberty (American Bank Note) embossed (2"x3")* Plus: Introduction; Making Embossed Holograms; Computer Generated Holograms; Non Destructive Testing; Holographic Optical Elements; also includes 300 literature references. 220 pages.

HOLOGRAPHY MARKETPLACE 3rd EDITION (1991)

Includes these holograms: *Brain Skull (Polaroid) two channel image photopolymer (3"x3"); Earth/Space/Grid (American Bank Note) embossed (3"x3"); Floating Alphabet (American Bank Note) embossed (3"x4"); Magic Wizard (American Bank Note) embossed (1"x1"); Woman/Fruit/Flowers (Light Impressions) embossed true color still life (4"x4"); Prehistoric Man (Bridgestone Graphic Technologies) embossed (3"dia.); Space Shuttle in Orbit (Archeozoic/Polaroid) photopolymer (1"x1").* Plus: Introduction to Holography; Varieties of Holograms; Recording Materials; Lasers; HOEs; Non Destructive Testing; Computer Generated Holograms; Holography in Education; Embossed Holograms; Business of Holography; Businesses by Category; Database of Individuals; Bibliography; Glossary. 174 pages.

HOLOGRAPHY MARKETPLACE 4th EDITION (1993)

Includes these holograms: *Transamerica Pyramid (Polaroid) photopolymer (2.5"x3"); Earth/Grid (AD2000 Inc./ABN) embossed (3"x3"); Disney Characters (Hologramas de Mexico) color embossed (2"x3"); Ghostbusters (American Bank Note) 2 channel embossed (4"x5"); Butterfly (Holography Presses On) embossed diffraction (.5"); Egyptian King (Holopress) embossed (2"x2"); 4 Image Montage (The Diffraction Company) embossed (4"x4"); Inaugural Invitations (CFC Applied Holographics) embossed (1.5"x1.5"); Bouquet (CFC Applied Holographics) full color embossed (2.5"x3.5"); Folding Package (Transfer Print Foils/Light Impressions) embossed (3"x5"); Earth/Lab (Global Images/Chromagem) embossed (5"x5").* Plus: Sales and Distribution; Direct Mail Marketing; Model Making; Holography Basics; Advanced Principles; HOEs; Heads Up Display; Computer Generated Holograms; Holographic Non Destructive Testing; Embossed Holograms; Lasers; Recording Materials; Main Business Listings; Bibliography; Glossary. 192 pages.

HOLOGRAPHY MARKETPLACE 5th EDITION (1995)

This edition features: *TV- a limited edition photopolymer hologram which was produced using a newly developed process that incorporates digital, optical, and holographic technologies to create a truly amazing image!* Plus: *Harry 4"x5", Chinese Lion Dancers 3.5"x5" (The Lasersmith); Mount Rushmore 6"x9", Rock Solid (TPF); Space Shuttle, Flag, Eagle, Fireworks 5.5"x11" (Crown Roll Leaf); Map (Hologramas de Mexico); Initials (HPO); Wizard 2"x3" polymer (Lazer Wizardry); Butterfly 3"x3", Valid (CFC/Applied); Tiger (Krystal) Holographics); Matrix (Dimensional Arts)* Plus: Sales and Distribution; Holographic Stereograms; Embossed Holograms; Lasers; Recording Materials; Business Listings, Appendix and more. 160 pages.

HOLOGRAPHY MARKETPLACE 6th EDITION (1997)

This edition includes: *17 rare, state-of-the-art holograms from the world's best manufacturers. Featuring a large, digitally-generated, full-color embossed stereogram on the front cover!* Plus: Making holograms at home on a budget; Color Reflection Holograms; Using Digital Tools to Originate Artwork; Digital Holo-Printers; 3D Modelmaking; New CW Lasers; New Pulsed Lasers; Buying Surplus Lasers; Large Format Holography; Recording Materials; Industrial Applications; Sales & Distribution; Business Listings and more! 192 pages.

HOLOGRAPHY MARKETPLACE 7th EDITION (1998)

This edition features: *20 rare, state-of-the-art holograms from the world's best manufacturers. Featuring a large, digitally-modeled, photopolymer hologram of the Mona Lisa on the front cover! 3 more photopolymer holograms and high quality embossed holograms for security, packaging, etc., inside!* Plus: Introduction; Artwork and Mastering; Color Holography: Silver-halide emulsions (New materials from Russia, Germany, China); Other Recording Materials (Photopolymer, DCG, Photoresist, Photo-thermoplastics); Optics and Equipment (Fringe Stabilization); New CW Lasers; New Pulsed Laser systems; Sales and Distribution; Business Directory and more! 240 pages.

Ordering Information

Continental USA:	Single $30	**Entire Set 2-7 $75**	Prices **include** shipping and handling.
Alaska, Hawaii, Canada, Mexico:	Single $35	**Entire Set 2-7 $85**	VISA, MC, AMEX, check, money order
All other overseas countries:	Single $45	**Entire Set 2-7 $110**	accepted ($) payable to "Ross Books".

Ross Books: P.O. Box 4340 Berkeley, CA 94704
Phone: 800-367-0930 (toll-free in USA) or 510-841-2474 Fax: 510-841-2695
Email: sales@rossbooks.com **PREVIEW THESE BOOKS ON OUR WEB SITE:** www.rossbooks.com

Introduction to Holography

This chapter assumes that you have no prior knowledge of holography and its uses. It begins with a discussion of the unique attributes of this fascinating visual medium, provides a simple explanation of what holograms are, and describes how basic holograms are made. The major commercial applications of holography are also described.

HOLOGRAMS ARE EVERYWHERE

You wake up in the morning and grab a new tube of toothpaste. The box sparkles with prismatic colors. You go out to breakfast. The restaurant is giving away hologram trading cards with each meal. You grab one for your son because you know he collects them. You finish your breakfast and head off to your job. Not paying close attention to the road at this early hour, you exceed the posted speed limit. The friendly Highway Patrol Officer who pulls you over takes your driver's license and examines it closely in the sunlight. He is looking for a hologram laminate that indicates your license is authentic.

At work, you open a new package of computer software. The box is sealed with a hologram label and a holographic logo is "etched" on the CD-ROM. On the way home from work you stop at the music store to buy a present for your . daughter. A 3D holographic portrait of her favorite group is on the cover of the CD you pick out and, even more amazing, the band members appear to dance a bit as you tilt the box back and forth! You pick out a greeting card to go with the gift. It has colorful holographic foil decorating the front. Then you pay for your purchases with a credit card that has a hologram on it. Once home, you pour yourself a martini. The bottle has a holographic tax stamp across the cap. The video you pick out to watch that evening has a hologram on the box.

This scenario is an example of the frequency with which we see holograms in our everyday life. Unlike a decade or more ago, holographic images and holographic materials have entered the mainstream and are commonly used for a variety of commercial applications. This development is partly due to the standardization of manufacturing methods and the inevitable maturation of the entire industry. More important is the fact that the clients who commission holographic originations are achieving the results they desire.

Holography has proven that it is a viable technology that can be successfully integrated into a wide range of commercial endeavors, including advertising campaigns, marketing promotions, security programs, and retail sales. The medium has established a track record of deterring counterfeiters, attracting shoppers, and adding value to products . As the technology evolves further and potential users become more informed about the holography marketplace, a host of other applications will certainly arise.

Judging from some of the articles included in this book, before too long you may use a holographic printer at your office to output 3D pictures from your computer, you may see a holographic billboard at the airport, you may have your own holographic portrait taken at the mall, and you may even wear clothes made with holographically colored fabrics!

MAJOR ATTRIBUTES OF HOLOGRAMS

Although often compared with photography, holography is really a completely different medium. Holography is based on optical principles that are different from photography's and holograms have different physical attributes than photographs. They are comparable only because both are ways of recording an image onto a piece of photosensitive material (film), and because, at times, similar equipment and materials are used in making them.

The most obvious difference between a photograph and a hologram is image dimension and image depth. For instance, when we look at a photograph and move it from side to side we are unable to see "around" the scene or perceive any depth. We cannot see over or under the image. We only see a flat (two-dimensional) picture displayed on the surface of the film. The picture is actually only a collection of light and dark shapes that we recognize as a particular subject. Our memory might remind us that the subject really has dimension and depth, but this visual information is not recorded in the photograph.

Parallax, Dimension, and Depth

A hologram is also flat, but the picture "on it" is not. When we look at a hologram and move it from side to side we can see many different views of the scene. We can also look behind foreground elements to see things in the back- ground. This property is called parallax and it is closely tied to the process of visual perception. The Random House Dictionary defines parallax as "the apparent displacement of an observed object due to the difference between two points of view." Our two eyes see slightly different things. Our brain automatically uses these multiple views to create image dimension and image depth.

The relevant point here is that if we see at least two views of the same object, we can perceive a three-dimensional image. A hologram records and displays many views of the same object-a photograph is limited to only one. In fact, the word hologram, coined by physicist Dennis Gabor in 1948, is commonly defined as "whole picture" based on its Greek roots. The terms holography and holographic are typically used to discuss anything related to holograms, though the word "holograph" actually has another unrelated meaning.

To better illustrate the difference between the two media, imagine looking out through a small window at a particular view. If the window represented a photograph, you would be frozen in front of it in one position. Consequently, you would only see one perspective of that scene and would have no way of perceiving depth. However, if the window were a hologram, you could move around in front of the window and see different parts of the scene from many different viewing angles. The scene would look three dimensional and display depth. To further elaborate on this analogy, if this window could retain the scene, and you could carry it around and show it to someone else later- they would be seeing a hologram of that scene.

Conventional Stereo Imaging

It is possible to create an image with depth using two similar photographs taken from slightly different angles if these pictures are presented to our eyes in just the right way. The two pictures are called "stereo pairs." Perhaps you are familiar with stereogram posters, "Viewmaster TM" binocular viewers, or virtual reality helmets. These artificial methods work, but they require people either to look at things unnaturally or to employ special viewing equipment. For many commercial applications it is more

practical to use holography when dimensional imaging is required. Methods have even been developed that incorporate stereo imaging with holography. These techniques will be explained later in the book.

Images That "Project" Out of the Hologram

Another difference between holograms and photographs is that holographers can position their images to "project off' or "float over" the surface of the film. One popular holographic image is of a water faucet that projects a foot or so out of the picture frame. A viewer can reach out and put his hand right through this apparently solid image. Other holographic images can be positioned to appear some distance "behind" the picture frame. Still others straddle the surface of the film (called the image plane). No photograph can do that!

Other Properties and Considerations

Holograms do have two properties which should be addressed, as they affect commercial applications. One is viewing angle. Unlike a normal picture hanging on the wall or in a magazine, holographic images can only be seen within certain viewing parameters. If a viewer moves too far off-center, the holographic image will disappear.

Another problem is lighting. All holograms need to be properly illuminated in order to be seen. The lighting conditions that exist in many display environments are not the best for viewing holograms. Potential users must anticipate how their holograms are going to be displayed before starting production. There are ways to solve these problems using good design practices, as well as technological approaches which are currently being developed by researchers worldwide. Both these issues are covered in greater detail throughout this book.

HOW HOLOGRAMS ARE RECORDED

The degree to which you can look around an object and the distance the image forms in front of or behind the film depends on how the hologram was made. This brings up another major difference between the two media-the procedures used to record imagery. We are all familiar with photography; we need a camera, film and an adequate amount of light. The light reflects off our subject, passes through the camera lens and exposes the film. Bright subjects expose the film a lot, darker subjects expose the film less. The picture we see is composed of varying tones.

Holography is different. It records an image in an entirely distinct way. Film is still employed; however, conventional cameras and ordinary lighting are not. Instead, to make a hologram the film needs to be exposed by a beam of coherent light-that is, light which is composed of light waves that have identical frequencies and which are vibrating together "in phase." Light from the sun and from light bulbs will not work. These sources emit light of varying frequencies with randomly varying phase.

Lasers do emit such light, and are therefore utilized. Since most of these lasers are not portable, and many delicate optical components are used in the process to further manipulate the laser beam, most holographic recordings are

Major Commercial Applications of Holography

Security/Authentication Holograms are difficult to produce and/or duplicate by the average criminal and therefore have become integrated into many government and commercial security programs. They have been attached to official documents, currency, tax stamps, event tickets, ID cards, credit cards, and product labels. To further deter professional counterfeiters, hologram manufacturers have added sequential numbering, tamper-evident materials, and other covert features to their security holograms.

Packaging Attracting attention is the name of the game and holograms and holographic materials help differentiate one product on the shelf from another. The unique dimensional images and/or colors inherent to the medium catch a shopper's eye. Billions of square inches of holograms have been generated for this purpose. Many packaging companies (especially those associated with the labeling and hot stamping industries) have successfully integrated holographic materials into their existing production lines. It is common to see holographic designs on foil wrapping, plastic films, cardboard containers, and other paper products. Holograms are often combined with more traditional printing methods to achieve the best effect.

Advertising/Promotion In a world swamped by merchandise displays and marketing gimmicks, holograms still attract attention long enough to convey a message-the goal of any advertiser. Holograms have been used in a variety of promotional campaigns including direct-mail, P.O.P., and even billboards. They have appeared on flyers, posters, magazine inserts, ad premiums, T-shirts, and executive gifts. Ad-industry trade journals have repeatedly reported successful and measurable results from companies using holograms in their advertising programs.

Value-Added Decoration Textiles, paper products, and even candy have been embellished with holograms. Many companies use holograms to spruce up their stationery, annual reports, and product catalogs. Others decorate their products with attractive holographic materials. Still others produce actual holograms to increase the value of items, such as collectible trading cards and figurines.

One of the fastest growing decorative applications is in the fashion industry. For instance: one company has developed threads of holographic materials that can be woven into fabrics using household sewing machines; another company has developed techniques to hot stamp holographic accents on bolts of cloth in large quantities at the factory, and still another is involved in marketing holographic iron-on transfers which can be washed and dried repeatedly. One researcher, Dr. Munzer Makansi, has even developed a method by which diffraction gratings and holographic images can be embossed directly on synthetic materials, such as nylon.

Signs From indoor point-of-purchase displays to roadside billboards, holograms of all sizes and types are being used to deliver information to the viewer. Fully dimensional pictures, animated holographic images, and holographic portraiture are being employed to complement, and even replace, printed signs, photographs, and transparencies. A major automotive manufacturer has recently announced the production of a full-color hologram measuring 40 square feet!

Illustration By providing dimensional information on a flat surface, holograms can illustrate books, catalogs, and magazines in new ways. Although the educational uses have been mainly limited to children's books, the potential exists for instructional applications in medical, scientific, and industrial publications.

Giftware Over the past decade, a sizable retail market has developed for hologram pictures (wall decor), jewelry, watches, and related optical novelties. Once confined to museum shops and specialty stores, holograms have gone mainstream. Although the sales boom may have slowed as the initial novelty wore off, unique hologram products are constantly being introduced, as the giftware industry requires new merchandise each year.

Trade Show, Museum, and Lobby Displays Over the years, numerous holograms have been produced for corporate clients and museum exhibitions. They are usually required to be large, measuring several square feet or more. When properly displayed, the effects are indeed remarkable. Images can project up to several feet in front of, and behind, the hologram. Fully detailed, multicolor images can be produced. Animated effects are possible as well.

Since holograms capture microscopic details, museums have recorded archival images to exhibit in place of the actual objects, which are safely stored away. Objects too impractical or too heavy to move (like industrial equipment) can be recorded on a hologram, rolled up in a tube, and displayed in full detail at the next trade show.

Art Some of the most distinctive and spellbinding art in the past thirty years has been produced by holographers. Still in its infancy compared to photography, and seemingly surpassed by the proliferation of digital technologies, holography still can do things other media can not. As the technology becomes even more accessible, more artists may start utilizing the unique attributes of holography to express their ideas.

Holographic Optical Elements (HOEs) One attribute of holograms is that they can direct light in a desired path. Therefore, industrial designers are using HOEs to replace bulky and breakable optical devices, such as glass lenses and mirrors. A flat HOE can be produced on lightweight plastic that duplicates the properties of the glass elements. These holograms are much less expensive to manufacture and service. They are now being used to enhance the viewability of LCD screens, watch faces, and other displays. HOE solar collectors are being researched.

made in a darkened holographic studio, where conditions can be precisely controlled.

THE HOLOGRAPHIC PROCESS

The process used to produce commercial holograms consists of three main parts:

1) recording the image,

2) regenerating the image so that we can see it, and

3) replicating it, if it looks good.

This process is somewhat analogous to the process used in making an audio recording. (Obviously, holography deals with visual images rather than sound, but the basic production steps can be compared.) We begin by recording the performer during a session in a recording studio using specialized equipment. Next, we "play back" the original recording, using related methods. If we like what we hear, we can duplicate the original recording (stored on a "master tape") onto cassettes, CDs, et cetera for sale to the public.

Let's consider the first step in the holographic process-recording the image. (This step is discussed in greater detail elsewhere in this book, but for beginners the following explanation should suffice.) As mentioned, holographers must use the coherent light from a laser to illuminate the subject, or object being recorded. The recording is captured on a photosensitive material, which is typically a sheet of special, high-resolution black-and-white film.

In most holographic recording setups, the single beam of light leaving the laser is immediately divided into two smaller beams, of equal length. One of these smaller beams travels directly to the film. The other beam reflects off the object and back onto the same film. As the two beams of light converge on the film they interact and combine to form a complex pattern. This pattern is called an interference pattern. An interference pattern created in this way and recorded by any means is called a hologram.

The interference pattern is recorded onto the film during an exposure that lasts from fractions of a second to minutes, depending on the film's sensitivity and the amount of laser light reaching it. The film is then developed and processed in order to permanently store the interference pattern: i.e., a hologram is produced. But if we looked at it, we would only see an indecipherable microscopic pattern of closely spaced overlapping lines. No recognizable image is seen, just a recording of the interference pattern.

To see the image which was recorded, we perform a second procedure-image "playback." By properly illuminating the film (in this case, with the laser light that originally exposed it), a viewable image will appear. The image that results is commonly referred to as a hologram, even though the phrase "holographic image" is more correct.

The image that is produced is a replica of the unique set of light waves which bounced off our subject and exposed the film during the recording process. How does this work? The interference pattern stored on the film interacts with the incoming laser light to reconstruct an image of our subject. Whatever the film "saw" during the recording process, a viewer looking at the finished hologram would

see. If we made a hologram of a physical object, we would see a three-dimensional image of that object. Since it is impractical to use lasers to play back all holograms, methods were developed that allowed holograms to be seen and enjoyed under ordinary lighting conditions.

The third step of the process is replication. If the holographic recording process was successful, the original hologram can then be mass-replicated in a variety of ways for commercial applications (usually on plastic, paper, or glass).

This simplifies the process quite a bit, but the basic facts to learn are that: a laser is needed to illuminate our subject and thereby expose the film; an interference pattern is recorded onto the film (not a visible picture); once this film is developed and illuminated correctly, a three-dimensional image of the subject is generated.

Two Types of Lasers Can Be Used

We will discuss lasers in depth later in this publication, but it is important that we touch on the subject in this introductory chapter, too. The type of laser you use affects what subjects you can record.

There are two kinds of lasers used by holographers; the continuous wave (CW) laser and the pulsed laser. The CW laser emits a steady wave of laser light, whereas the pulsed laser emits laser light in bursts. The CW laser is by far the most common laser used in holography. The power of a CW laser is typically measured in watts (W). In holography labs, most of these lasers fall in the 5 to 50 milliwatt (mW)range.

Remember that what we are recording on the plate are two laser beams converging (or interfering) with each other at the plate. If the object moves even a microscopic amount (on the order of a fraction of a wavelength) from one moment to the next, we will record two different interference patterns and the holographic image will look blurry or will not even appear.

An exposure with a CW laser can take from less than a second to several minutes. Because there cannot be any motion at all during the exposure, we need to eliminate any vibration coming from the ground. To do this we make or buy a vibration isolation table on which to put our laser, optics, and objects. Since it is absolutely critical that we have no motion at all, the subjects that we holograph with CW lasers have to be "dead" or immobile objects.

Pulsed lasers, quite the opposite of CW, emit extremely quick bursts of very powerful laser light. The output is measured in joules. Consequently, the exposure time is much shorter than a CW laser. Exposures can be made in nanoseconds (one nanosecond is one billionth of a second). You do not need a vibration isolation table for the pulsed laser. What can you shoot? Anything you want. You can shoot people, splashing water, animals. Why such freedom? Because your subject cannot move significantly in a nanosecond.

What are the drawbacks of pulsed lasers? Why doesn't everyone use one? The answer is expense. They typically cost tens of thousands of dollars or more , and require a lot of extra overhead and care. Lasers don't last forever and when a pulsed laser bums out it is costly to fix. Holographers are anxiously awaiting a low-cost, easily maintained pulsed laser. Progress is being made in this area.

Recording Materials

Although we have discussed how holograms are made, we have not discussed the photosensitive materials on which they are recorded. In photography, the most common item used to capture images is a silver-halide emulsion coated onto a film base. In holography, there are a number of materials used to record your image. The most common recording media are:

1) silver-halide

2) photoresist

3) photopolymer

4) dichromated gelatin

Note that we use the phrase "recording materials" instead of emulsions. This is because not all the items used to capture holographic images are emulsions. In an embossed hologram, for example, the holographic image is literally stamped into clear plastic or foil using a mechanical, rather than an optical, process. Holograms have even been recorded on chocolate candies and lollipops. A discussion of the different recording materials requires a chapter of its own, which you will find later in this book.

Artwork Origination

Another important topic that we should cover in this introductory chapter is what kinds of subject matter you can make a hologram of.

As with any creative art form there are many choices available, and much depends on what you want to accomplish. To simplify matters, we list some of the most common things that are used for holographic subject matter.

1) 3D objects (sculptures, miniature models, or actual objects),

2) 2D objects (flat graphics, illustrations, photos, etc.),

3) Stereographic composites (specially shot motion pictures, video, or computer graphic files that are arranged to produce 3D imagery).

Although some of the above topics are covered in more detail in the next chapter, we will give an overview of each now.

3D objects These are the most common subjects recorded in holograms. What your model is, of course, depends on the type of laser you are going to use for your exposure. With a pulsed laser, as we have already mentioned, you can shoot live subjects and just about anything you wish. Most holograms, however, are made with a CW laser and immobile objects are required as subjects. Holographers will frequently commission sculptors to create highly detailed miniatures of things which can't fit in the holography studio.

2D objects You will see flat artwork being used with great abundance in embossed holograms. Camera-ready art for 2D holograms is created much the same way you

Lighting Holograms Properly

To fully grasp what holograms look like, and to understand how much their look depends upon proper illumination and viewing angle, one should examine the holograms in this book under various light sources. (All the holograms in this book are designed to be lit with a light source positioned above and on the reader's side of the hologram. Hold the book in front of you and tilt it back and forth until the holographic images appear brightest and most distinct.)

Use a Single, Bright"Point" Source of Light

An overhead, single beam of bright light (such as direct sunlight) works best. However, sunlight is not usually the most practical light to use. There is a whole range of other light sources with which to view a hologram; some are better than others. Good light sources cast sharp shadows. Ideally, holograms require a light source that mimics the laser light that originally made the hologram - the beam should contain the original exposure wavelength, should have enough intensity to replay the hologram, and should emanate from a single "point" source (such as a very small light bulb). Fluorescent tube lights that are found in most offices do not have all these attributes.

Thus, whenever you go into a shop that specializes in holograms, one usually finds that the, shop has subdued overhead lighting with a single spotlight focused on each hologram, or group of holograms. This serves the dual pur- pose of creating a pleasant lighting environment as well as providing a proper illumination for holograms. People who display holograms in their homes find that an inexpensive way to illuminate them is by using a clear (unfrosted) light bulb with a single, small filament inside. Bulbs with vertical filaments often work better than bulbs with horizontal filaments. Halogen spotlights work very well, too. These bulbs are available at any shop with a large selection of lamps and lighting supplies. Put them in a ceiling lamp or lighting arm.

Avoid Multiple Lights and Diffuse Lighting

When a hologram is illuminated with light coming from different places (at different angles), each light source makes a separate image. This mixture of multiple images makes the hologram look blurry. Try to block off every source but the one positioned directly overhead. Then adjust the illumination angle. Diffuse light sources (such as fluorescent lights, frosted light bulbs, light from behind lamp shades and clouds) also create blurry, indistinct images. Avoid using them.

Whether or not a hologram can be seen well in ordinary room light also depends on how that hologram was designed. If the designer knows the hologram will end up in a room with lots of lights or under fluorescent lights, he will use flat graphics as artwork or record a 3D subject with very little depth. The multiple light sources still will create multiple images, but the flat, shallow images will overlap more and the image still will be viewable. These shallow images are also more recognizable under diffuse light.

would create art for a conventional printer, except that the graphics can be positioned in layers to create an image with depth.

2D/3D objects You also can have a combination of photos, line art and 3D objects in your final hologram, although the depth of the 3D object is often limited due to practical considerations.

Stereographic composites The holographic stereogram is one of the most exciting compositions in holography today. It allows artists to incorporate a wide range of visual effects in their images--especially animation and dimension. Today's computer technology is making the production process more accessible and the finished products more refined. We should point out that several techniques are used to make holographic stereograms. Sorting out the jargon can become confusing. Some of the names you will hear that refer to holographic stereograms are:

Holographic Stereograms This is probably the most common, safe, and inclusive name. It is used to name any hologram that belongs to the group of holograms that are designed to achieve their effect by utilizing a matched pair of images and a human's capacity for stereo vision.

Integral Holograms In general, this is a term that refers to a finished image that is constructed from many discrete units.

Multiplex Hologram This phrase describes a hologram produced using a system developed by Lloyd Cross and refined by the Multiplex Company that utilizes stereographic and integral techniques. This is probably the first commercial holographic stereogram process developed and it involves filming a subject on a rotating stage.

Incorporating LCDs

A major advance in holographic recording systems was the introduction of the LCD (liquid crystal display). It did not take long for holographers making stereograms to see the benefits of using the LCD as a source for the image being recorded. LCD origination substitutes graphics displayed on a liquid crystal display screen for cinematic footage, thereby allowing digitized images (with all their advantages) to be easily incorporated into a hologram. Using this method, a wide variety of cinematic, video, and still images can be scanned in a computer, manipulated, and displayed electronically using LCD technology. The picture on the LCD is the object that is recorded on the hologram. Recording a sequence of related pictures can produce a 3D image and/or an animated one. Computer assisted design and origination of artwork will be discussed further in the next chapter.

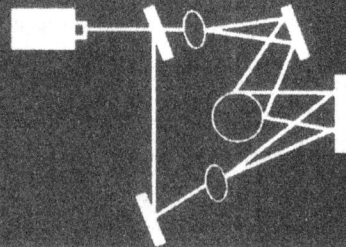

2
Design and Artwork

The first step in producing a holographic image is designing and preparing suitable artwork. This is a crucial step, as the type and quality of the subject matter recorded in the hologram will determine how the final image will look. This chapter discusses the different kinds matter that can be supplied to the holographer and ex- plains how each type is incorporated into the production process.

ARTWORK ORIGINATION

In the early years of holography, specially designed works of art or sculptures were commonly used as subject matter for holograms. These items are still used today. However, it is now possible for computer graphic artists using readily accessible hardware and software programs to electronically generate "camera-ready" artwork that holographers can assemble into images that display dimension, depth, projection, and motion. Holographers are substituting this digitally originated artwork in place of the time-consuming drawings, hard-to-record physical objects, and expensive cinematic shoots that they traditionally utilized. They are using the computer to increase flexibility and versatility in the design and production processes, as well as to cut production costs. This merger between electronic imaging systems and optical-based ones is resulting in new and profitable opportunities for all those involved, especially the artists and designers that are able to best utilize both media to achieve their clients' goals.

In most instances, the "camera ready" artwork prepared by the design team is output as a series of computer graphics files which is sent to a hologram origination facility. These computer files correspond to various graphic elements of the holographic image being produced. In brief, the holographer uses these graphics files to generate a "master" hologram which is recorded on a high-resolution photosensitive material using a laser and specialized optics. The master hologram can then be mass-replicated in a manner suitable for commercial applications.

In this chapter, we discuss how artwork for your hologram can be produced and explain how it is integrated into the manufacturing process.

COMMON WAYS OF MAKING A MASTER

There are several different methods that can be used to make your master hologram in the production studio. A clear understanding of these methods will go a long way toward helping you understand the manufacturing process and helping you plan your project. Therefore, we list the four most common methods of making a master hologram and discuss each method.

1) 3D Artwork A physical object (or person) is used as the subject to be recorded in the studio.

2) 2D and 2D/3D Artwork A computer image (such as a PhotoshopTM or Illustrator fileTM) is created, output to film or paper, and used in the studio shot. You can use several images to create the feeling of depth if you wish.

3) LCD Artwork and HOPs The computer image in method 2 can be made and, instead of outputting to film, the image can be illuminated on an LCD (liquid crystal display) screen. The LCD image becomes the object you make a hologram of. In the advanced form of this method you can create a series of images (a cinema) which is mechanically exposed in sequence onto your master hologram creating a mini cinema for the viewer. This method is very popular and the machinery used to expose these optical images is a "holographic optical printer" (HOP).

4) Dot Matrix Machines Another method is to skip the LCD altogether and simply bum the image you want directly into the photosensitive emulsion. Typically, am image is converted into pixels and each pixel is recreated by a diffraction grating. The machines that perform this are referred to as "Dot Matrix" hologram machines and they are becoming more and more popular in embossed hologram production.

As you can imagine, there are pluses and minuses to each of the above methods, and the artwork used in some methods will not work in others. We will now discuss how each of the above methods works, and the type of artwork that should be supplied when using it.

Method 1: 3D Model Making

Although live subjects can be recorded using specialized equipment, most holographers shoot inanimate models.

In thinking about models, it is important to note that the "depth of field" in your final hologram is closely tied to what type of hologram you intend to mass-produce and how the hologram will be illuminated. Embossed and dichromate holograms generally reconstruct fairly shallow images (one inch or less), while photopolymer films can replay images of several inches in depth. Holograms produced on silver-halide glass plates can reconstruct images several feet deep under proper illumination. On the other hand, the unit cost for silver-halide and photopolymer production runs is much higher than embossed runs. You obviously need to discuss carefully with your holographer the specifications of your sculpture before creating it.

Basically, a hologram can be thought of as a window through which you view a scene. One of the first steps in 3D model making is to pretend you are standing directly in front of a window which is the size of your final hologram. This impresses on you the limits of the viewing space. Imagine yourself moving from side to side-which allows you to see around a given object and into areas of the room which would otherwise be hidden. Designers should predetermine viewing angles with the holographer and communicate these measurements to the model maker in order to maximize the usable image area. Designers also need to take into account that a holographic image has "volume"- it can have depth (an object can be behind the window) and/or projection (an object can appear in front of the window). Images should be designed to best utilize this front- to-back dimension. Many model makers make the most important visual components of their images focus right on the image plane (right on the surface of the hologram), which stays in focus under less than ideal lighting conditions.

Execution

After the design process comes execution, a step which utilizes a model maker's craftsmanship and technical proficiency. One craftsman who is quite experienced in this field is George Sivy of Richmond Development Group (formerly Gray Scale Studios). Sivy has been a holography model maker for twelve years. He worked for Polaroid during its first years in holography, making models

utilized for custom and stock images (such as the popular "Brain! Skull" which appeared on the cover of the Holography Marketplace 3rd Ed.). In addition, he has collaborated on numerous commercial projects, ranging from embossed security holograms to photopolymer holograms designed for the giftware market.

In terms of materials, Sivy uses "what will solve a given problem." Although some materials and combinations that he uses are proprietary, he did mention the readily available Sculpey, a synthetic clay that can be baked. This ensures a stable model, which is important since most holographic mastering uses continuous wave lasers that require absolutely no motion during exposure periods. Long exposure times might require even more stable materials, and Sivy may use Sculpey first, then make a mold which is used to generate an even more stable sculpture.

After the model is created, Sivy says that it is often painted to create contrasting areas of light and dark. Since most, but not all, holograms are intended to be reproduced as monochromatic images, the "coloring" process is quite different from that in ordinary model making. Less experienced model makers will often work under a safelight which duplicates the laser light which will illuminate the model during exposure. Good model makers will also utilize textures, shading, and special effects to maximize the hologram's visual impact. Again, this requires communication among the design team, the model maker, and the holographer.

There are many other professional model makers who work in the field of holography. If you are interested, an interview with some of them appears in Holography MarketPlace 6th Ed.; they discuss how they produce their work in more detail.

Method 2: 2D/3D Artwork

A computer image (such as a Photoshop™ or Illustrator file™) can be created, output to film, and used in the studio shot. You can use several images to give the illusion of depth if you wish.

The production steps used by one business that does a lot of 2D/3D work are listed on the next page to give you an idea of what is involved in making a master. The details involved vary from business to business.

The 2D/3D hologram method is used in making a master for photopolymer, dichromate, silver-halide, or embossed holograms but we will restrict the examples in our discussion to the embossed hologram master because it illustrates the process best.

The creation of 2D images obviously needs to be done with care. Imagine breaking a conventional print ad into three levels of related graphics-a foreground, a middle ground, and a background. Picture each element to be a separate flat graphic (a 2D) or a photographic transparency. (For example, the foreground image might be a corporate logo, the middle image a picture of a product, and the background image a landscape.) Arrange the three elements front to back, yet separated from each other by a 114 inch or so of space (3D). This array represents the multi-level imagery associated with a typical 2D/3D hologram.

Production Steps-Creating an Embossed Multilevel (2D/3D) Hologram

Image Design

1. Designer consults with holographer regarding job-specific design requirements.

2. Designer prepares client's artwork for holographic reproduction. Creates a multilevel image "on paper" consisting of black-and-white line art drawings and/or photographic images.

3. Drawings, photos, and graphics that comprise the image are scanned into the computer using Adobe's Photoshop. If the desired imagery must be copied from existing corporate artwork, Photoshop tools can be used to extract the desired graphics.

Digital Image Assembly

4. Digitized image is assembled and checked. Line art is cleaned up. (Black outlines should separate image components and all lines must be unbroken.)

5. Bit map images are converted to postscript using Adobe StreamlineTM (now included with Adobe IllustratorTM).

6. Postscript images imported into Illustrator. Image is broken into multiple levels - one file created per image plane (i.e., an image with a foreground, a middle ground and a background requires three files).

7. Designer assigns appropriate colors to image components on each level.

8. Completed files sent to holography studio by diskette, SyQuest™, DAT, optical disk, e-mail, etc.

Creating a Production Tool

9. These files are reviewed and imported into new Illustrator™ "master" file standardized for that particular holography studio. Images are ganged if necessary; they are sized to fit production equipment; and pre-designed cut guides, registration marks, and TMsymbols are added.

10. Adobe SeparatorTM is used to generate color separation "sub-files" for imagery levels that are multicolor, and composite "sub-files" for imagery levels of single color. These "sub-files" will be used to make masks that the holographer will use when exposing the recording material.

Photoshop in its RGB mode generates necessary separations if non-primary colors are being copied from existing artwork.

11. Output "sub-file" separations on paper to check color assignment.

12. Take Illustrator "sub-files" (on SyQuest drive) to image setter for output.

13. Image setter outputs film positive transparencies: one per "sub file." Colored areas are now solid black.

14. Holographer uses these film positives to make glass negatives on Kodak HRP (color areas are now clear to allow laser light to pass through and expose plate at specified angle per selected color).

Holographic Recording

15. Holographer shoots one layer at a time, one color at a time. Each time masks are positioned to block off portions of the recording material that should not be exposed. Every time another color is required, the holographer must adjust the optical setup to change the reference beam's angle.

16. Holographer repeats the exposure process for each level of imagery until the master hologram is complete.

Holographic Replication

17. Finished hologram is checked for flaws and prepared for electroplating.

18. The hologram is metallized and stamping dies are created that reproduce the microscopic patterns on the original hologram.

19. These stamping dies are used to emboss the hologram on rolls of foil or plastic.

20. Holograms are die cut, finished, and sent off for application (hot stamping, packaging, etc.).

(Editor's note - Special thanks to MikeGrogan of Holographic Dimensions Inc. for information provided.)

Digital Design Tools

These "levels" of artwork can be easily generated using digital tools familiar to most computer graphics artists. Adobe's Illustrator™, Corel's Draw™, and MacroMedia's Freehand™ can be used to create original drawings. Adobe's Photoshop can be used to import and touch up a client's existing artwork to make it suitable for holographic reproduction. Adobe's Postscript™ or Microsoft's TruetypeTM font collections are often utilized to create logos. Kai's Power ToolsTM and Bryce™ programs can be used to further enhance imagery. In short, a variety of software programs are capable of doing the job.

Once the artwork is finished, the digital files are sent to the hologram production studio either electronically or physically. There, an in-house designer will typically use Adobe's Separator™, Photoshop™, or Quark's Express™ to break the image into the appropriate component layers, if the original artist has not already done so. These files will be sent to an image-setter (a machine that outputs film) which will generate film transparencies corresponding to the different levels of imagery. These transparencies will be copied onto rigid glass plates. These glass plates will then be stacked in a sequential array. This array will constitute the physical object to be recorded.

Creating Dimension and Depth

Designing an image for a 2D/3D hologram is similar, but not identical, to designing an image for print. Since the artist is designing a multilevel image, subject matter obviously should be positioned to take advantage of these unique dimensional properties. For clarity, the most important elements of the scene are usually placed directly on the image plane. Foreground elements intended to float "above" the image plane should be easily recognizable, as they will blur out under less than ideal lighting. The same applies to background images with great depth. Drop shadows, textures, and shadings are often incorporated to exaggerate dimensional effects.

Designers commonly arrange the different image levels' in one of two ways:

1) So that in the finished hologram the foreground image appears to "float" slightly above the surface of the embossed material, the primary image is on the hologram's surface (called the "image plane"), and the background is behind the other two.

2) Or, the foreground image might be positioned directly on the image plane, with the middle ground and background images underneath, which further exaggerates the apparent depth of the image.

Design Guidelines-Parallax

Designers must also consider taking advantage of parallax-the ability to look around the sides of an image. It is important to note that due to standardized holographic production methods, most embossed holograms only display horizontal parallax, i.e., a side-to-side view, rather than an over-and-under one. This is usually adequate as it mimics the way we ordinarily look at the world.

To use parallax effectively, foreground images positioned to float off the hologram's surface need to be sized correctly so edges do not "cut-off' prematurely if a viewer moves off center. Since background imagery will be in sight when a viewer looks "behind" the foreground elements, the graphics for the background should extend completely from one side of the hologram to the other. For these reasons, and for finishing purposes, foreground and background artwork should be oversized in relation to the final size of the hologram. Only the imagery that is planned to appear on the hologram's surface should be actual size.

Although parallax considerably expands the viewing zone of the image, this attribute does have certain restrictions. If the viewer moves too far off center, the image will disappear. It is wise to consult with the holographer to determine the viewing parameters of a particular manufacturing process before starting design work.

Design Guidelines-Size

Although embossed holograms can be produced in various sizes, cost considerations, manufacturing equipment, and marketing requirements generally favor making holograms of 6" x 6" or less. It can be quite a challenge to achieve the visual impact required by a client while working on such a small canvas. To save production time and expense, it is common to gang a number of smaller images on one "master" hologram. "4-ups," "9-ups," and "36-ups" are common arrangements. The entire set of images is replicated together, and then each separate image is die cut out during the finishing process.

Design Guidelines-Color

Also the designer always should consult with the holographer beforehand to determine what colors are best utilized in a particular type of hologram. Embossed holograms create different colors by bending light to varying degrees. It is like creating a customized prism that will direct the light according to the designer's wishes. Therefore, the graphic artist needs only to assign colors to specific image areas in the artwork - the holographer then "colors it in" during the exposure process by adjusting his optical set up.

Another unique property of embossed holograms is that they do not display permanent colors-that is, as the viewer moves up and down in relation to the hologram, the colors will shift through the entire rainbow. In practice, it is common to take advantage of these unnatural color shifts to emulate movement and increase visual impact. Clients usually desire the brilliant, dynamic effects which result.

The designer only needs to specify the colors that are intended to play back when the finished hologram is viewed directly at eye level under proper illumination. Unlike standard four-color printing that utilizes combinations of cyan, magenta, yellow, and black (CMYK), the primary colors utilized in embossed holography are red, green, and blue (RGB). These three colors can be combined holographically to create yellow, white, and a range of secondary colors. Complex colors are a result of halftone screen mixing of RGB; different densities result in different colors. In holography, black results from unexposed areas of the film.

Some hologram mastering facilities are able to recreate "true color" using proprietary techniques based on pixilated renditions of the artwork. In some cases the studio will provide a palette of colors that designers can work from. Other studios claim that any color can be holographically duplicated. Again, we advise that designers consult with the production house to determine the most appropriate way to proceed. Ask to see samples that have been produced for other clients.

Experience dictates that lots of black and white image areas are visually unappealing: matte white is commonly reserved for registration marks and trademark information. Drab colors should be avoided in favor of the bright colors inherent to this process-red, yellow, blue, and green. Colors should be arranged to best contrast imagery and accent dimensional effects. Each area of the image should be assigned a color separate from its neighbor, and different from the areas that may layover or under it. Colors in the center of the color spectrum are usually recommended for major image components, as they are the brightest and will be the last to disappear as the viewer shifts position.

At the Holography Studio

Once the graphics work has been done, files should be organized, labeled, and shipped to the holography studio for review. Most studios are capable of accepting a variety of file formats including diskette, DAT, SyQuest™ disks, and optical disks. At this point the designer's work is finished.

Once the holographer records the image, the resulting master hologram can be proofed by the designer and/or client. If it is acceptable, it is sent to a production facility to be replicated. After replication, the holograms are finished in accordance with the client's wishes. Most embossed holograms are backed with an adhesive for hot stamping to paper stock or delivered on rolls for "peel and stick" applications. They work best when attached to a rigid materials that have flat and smooth surfaces, such as magazine covers, bank cards, and cardboard packaging.

Method 3: A Series of Pictures

Instead of outputting the computer generated image used in method 2 to film, the image can be illuminated on an LCD screen. The LCD image becomes the object you make a hologram of By using a series of pictures and special equipment, you can produce animated holographic images. The hardware which mechanically records a sequence of images onto the master hologram is called a holographic optical printer (HOP). Such devices are gaining popularity among holographers due to the unique effects they are capable of producing.

In some of the first HOP devices, a frame of movie film was used for the object. (See figure 1) Each frame of movie footage was sequentially projected through a slit placed in front of the hologram film. As the movie film was advanced one frame , the roll of holographic recording film was similarly advanced. In this way, an entire mini-movie was recorded onto the holographic mm.

Today, the movie projector is usually replaced by an LCD screen that is attached to a computer. This eliminates the cost of film and allows for easier artistic editing.

The phrase "holographic optical printer" comes from the fact that we are taking an optical image, as opposed to an object such as a sculpture or 2D artwork, and "printing" it on the film. If you are illuminating a sequence of images, and you film the sequence correctly, your final result will be a stereoscopic cinema. This is called a holographic stereogram and it is how many "moving" holograms are produced.

The holographic stereogram does differ from a traditional stereogram. As Steve Larson of Chromagem Inc. points out: "Similar in some respects to the old Viewmaster™

concept (a binocular-like 3D slide viewer that displayed two near identical pictures taken from slightly different angles), a holographic stereogram differs in that the number of stereographic pairs can be in the hundreds, instead of the single pair that was provided with the Viewmaster. What that means to the holographer/artist, is that now, motion and time can be captured and displayed holographically."

Filming

There are different ways to shoot your film for holographic stereograms, and professionals that do this usually have special stages or tools to do the filming. It is important to control the motion between frames because the viewer will be seeing two different images at a time and the brain will be rendering the two images into one to visualize depth. Obviously, you need to control the amount of motion between frames or the brain will not be able to render the two images into one 3D image. One common way to obtain a large angle of view around your actors in the cinema and also control the motion between the frames, is to have your actors on a stage which you can rotate in front of the camera (while the actors are acting). Another way is to have some variation on a railroad track design in front of your actors, and move the camera along the track. In both cases, the intent is to create a horizontal "angle of view" in your final film, with a controlled amount of motion between frames .

Digital Modeling and Rendering

Most recently, affordable digital design, modeling, and layout programs have been used to create realistic looking images using no physical objects as subject matter. Hundreds of different perspectives of a "cyber-scene" can be rendered, combined, and output to create fully dimensional holograms.

As holographer Steve Smith states, "Early on, pricy work-stations such as the first series of Silicon Graphics machines were required to run expensive image modeling and rendering programs; these cost upwards of $20,000, generally out of the range of professional holographers. With the release of powerful yet lower priced graphics work-stations, such as the Pentium™ and the PowerMac™, and with more capable yet affordable image modeling and rendering software, a new realm of imaging techniques are setting the stage to become the way to create holographic stereograms."

In brief, the computer graphic artist first models a real or an imaginary object or scene "on screen." Lightwave 3- D™, Byte by Byte's Sculpt4D™ or Alias/Wavefront's Power Animator™ are good modeling programs to use. The computer is then used to render an appropriate sequence of graphics files which correspond to the various visual perspectives which the designer wants to appear in the finished hologram. Pentiums and PowerMacs can be used to render simple rotations. DeskStation Technologies' Raptor™ rendering engine or SGI hardware have been used to render more complex imagery, i.e., hundreds of different perspectives of one image.

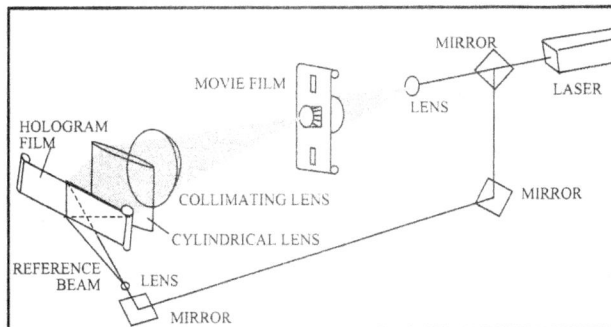

Figure 1. Diagram of an early HOP that used movie footage.

George Sivy, a model maker who specializes in holographic design, elaborates on the aforementioned process. "I use an accelerated Amiga 3000 Tower computer running VideoToaster™ to import and mix existing imagery. I use LightWave to model, light, surface, and manipulate single objects or even entire scenes. In layout mode, I animate the object, and animate a virtual camera in accordance to the holographer's requirements. (I typically either rotate the object in relation to a stationary camera or pan the scene in a smooth horizontal path.) I use a Raptor rendering engine provided by DeskStation Technologies to generate 180 frames. This translates to six seconds of real-time animation which may be incorporated into the final hologram. Next, the rendered frames are edited further using the VideoFlyer™ and copied onto S-VHS tape for proofing. Once approved, the files are downloaded onto an appropriate storage medium and shipped to the holographer."

Clients who want to generate their own imagery are advised to consult closely with the holographer before originating and delivering any artwork. Most holography studios are capable of accepting a variety of file formats which you can ship to them electronically or by diskettes, DAT, SyQuest disks, or optical disks. Since there are currently only a few computer graphics artists familiar with the procedures required to create animated stereograms, the holographer's in-house artists will probably need to review. your files, clean them up and re-sequence them.

Editor: See the covers of Holography MarketPlace Editions 5, 6 and 7 for examples of digitally modeled and rendered holograms. Editions 5 and 7 have photopolymer holograms on the cover; Edition 6 has an embossed one.)

Method 4: Dot Matrix Machines

This method does not use a physical object as subject matter and simply records a pattern of diffraction gratings directly into a photosensitive recording material. The collection of gratings produces an image the same way the dots in a photograph or pixels on a television screen do. The equipment that translates your artwork into these gratings is called a "Dot Matrix" hologram machine. They are designed to make the mastering process for embossed holograms easier.

The great benefit of the dot matrix method is that it is very cheap (no film at all) and totally digital. The drawback is that you will only be working with embossed holograms and the resulting image's depth of field is very limited. To produce holograms, 2D artwork is furnished to facilities that have one of these machines. A surface-relief master hologram results. It can be replicated by various embossing methods.

Many dot matrix holograms are made for security applications as they can display very complex images. Since the image is constructed from tiny pixels, many covert features can be included in a small hologram label. In the past year, companies have developed methods for producing poster-size dot matrix holograms suitable for "seamless" packaging applications. The kinetic images and prismatic colors produced by the thousands of arranged diffraction gratings create very eye-catching effects. Progress is also being made on producing more dimensional and more realistic looking images. Because of the gaining popularity of the dot matrix machines, another article in Chapter Eleven describes dot matrix technology in more detail.

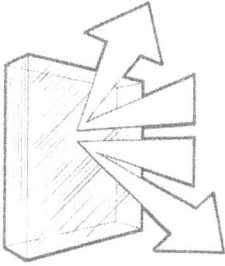

3

In the Holography Studio

The following information is provided in order to intro-duce the reader to what is involved in making ordinary holograms in a holography studio. It is by no means a complete description of what occurs; however, it should prove useful to beginners and potential users o f the tech-nology. More detailed instructions are available in the Holography Handbook-Making Holograms the Easy Way, also published by Ross Books.

BASIC PRODUCTION TECHNIQUES

Suppose one enters a holography studio where a simple hologram is about to be made. The first thing you will notice is a special vibration-free table in the room called an **isolation table.** During a typical exposure (which can last seconds to minutes), a movement smaller than a wavelength of light can ruin days of work. Therefore, the table is designed to isolate the holography setup from the smallest vibrations in the surrounding environment. It is usually quite massive and built to float on a cushion of air.

On the table is a laser, some mirrors and a piece of photo-sensitive material positioned in a **plate holder**. This pho-tosensitive material is typically a silver-halide emulsion coated onto a glass plate, and thus will be referred to as the **recording plate**. Since the recording plate is sensitive to light, the studio must be darkened when the plate is out of its box. Sometimes certain safelights can be used so that the holographer can see, but extreme care must be taken not to pre-expose the recording plate.

Everything on the table is arranged in a carefully mea-sured manner. The object to be holographically recorded is positioned in front of the plate holder in the middle of the table.

As mentioned, to record a hologram we use a laser that emits a single beam of light at one wavelength. We cannot use the sun or just any light as our source because the light from common light bulbs or daylight contains many con-stantly changing wavelengths: it is not coherent. If we make the exposure using incoherent light, the changing

wavelengths would create a multitude of interference pat-terns and the resulting holographic image would be com-pletely blurred and useless.

TWO MAJOR CATEGORIES OF HOLOGRAMS

The terms "transmission" and "reflection" are two of the primary ways to categorize holograms. Very simply, the terms refer to how the hologram is illuminated during the viewing process.

Transmission holograms require that the illuminating beam of light pass through the hologram in order for an image to be seen. Therefore, these holograms must be backlit. Sometimes positioning a light behind the holo-gram is difficult.

Conversely, **reflection holograms** require that the illu-minating beam emanates from a source on the viewer's side of the hologram. The light reflects off the hologram back to the viewer's eyes. Reflection holograms are often favored by the general public because they can be hung on easily a wall and illuminated with a ceiling light or a lighting arm.

Most of the embossed holograms in this book seem to be reflection holograms; however, they are actually trans-mission holograms with a mirror attached to the back. The mirror sends the light back through the recording material and creates an image. Very thin coatings of metal are used in place of glass mirrors. This is a clever and practical way to solve the rear lighting problem while retaining the advantages of using transmission holograms.

Within the two major divisions of holograms (reflection and transmission), there are many variations. Like any other specialized field, holography has its own lingo, and in some cases the same hologram can be described using more than one name.

MAKING A TRANSMISSION HOLOGRAM

To make a transmission hologram, first we turn on the laser and aim the beam at Mirror 1. Due to the fact that Mirror 1 is only partially reflective, part of the beam is reflected toward Mirror 3, and the other part passes through Mirror 1 to Mirror 2. Because the beam is split, Mirror 1 is referred to as a beamsplitter. (See figure 1.)

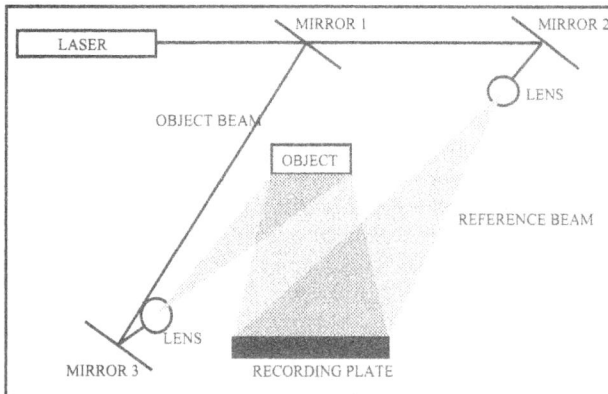

Figure 1 - Transmission Hologram

The beam that passes through Mirror 1 to Mirror 2 is called the reference beam. The reference beam's path always ends at the recording plate without ever illuminating the object. For instance, in this setup after the reference beam strikes Mirror 2, it is reflected through a lens toward the recording plate. The lens' function is to spread the beam so that it will cover the entire plate (in some cases, the lens is placed in front of Mirror 2; in either case its function is the same-to spread the beam). At the same time, the other beam, which we call the object beam, reflects off Mirror 3 and also passes through a lens. This lens spreads the beam out so that it illuminates the entire object. The laser light reflects off the object (hence the name "object beam") and strikes the photographic recording plate. *The two beams must travel exactly the same distance so that when they recombine at the recording plate they will be in sync with each other, and an interference pattern will be formed.*

After exposure, the photosensitive plate is developed using the appropriate chemistry and standard darkroom procedures (i.e., developer, bleach, fixer, washing, drying, etc.) The resulting developed plate is the hologram.

Holding the developed plate up to light, we see that the plate is semitransparent. On closer inspection we see that the darkness of the plate is caused by developed emulsion. The plate seems to have countless swirls of threadlike developed emulsion which are called fringes. The fringes look like the swirls that make up your fingerprints or the boundaries on topographic maps. There appears to be no order to the swirls. In fact, the fringe pattern is a recording of the interaction between the reference beam and the object beam.

Viewing the Image Using a Laser

To see the image, we put the recording plate back in the plate holder on the table in exactly the same place it was for the exposure. Then we remove the object and Mirrors 1 and 3 from the table. Now, when the laser is turned on again only the reference beam illuminates the plate. When you look through the plate, an image is seen of the original object, in its original place and at its original depth. (See figure 2.) This reconstructed image is indistinguishable from what you would see if the object was not removed! The first time holography students see this happen they are quite amazed!

Why We See an Image: A Simple Explanation

A detailed explanation of why this happens would occupy many pages. A simple explanation might go like this: the two beams strike the photosensitive recording material at the same time. Since they both originated from the same laser beam, and traveled equal distances, they are precisely in sync with each other. When two such waves of light recombine, their interaction produces an interference pattern.

This pattern is recorded on the photosensitive material during the exposure step. When we develop and process the recording plate, the interference pattern is stored on the plate as the fringes that we see. Because we are recording the interaction of light waves (which are quite small), the fringe patterns are microscopic (on the order of 1,000 or more fringes per millimeter).

After development, if we aim only the reference beam at the plate (at exactly the same angle that originally exposed the plate) the interference pattern which was recorded causes the light waves passing by to change direction. This phenomenon is called diffraction. It occurs whenever light passes through small apertures, like the space between fringes. This diffracted light has exactly the same form as the light waves which were originally reflected from the object.

In other words, a properly illuminated hologram regenerates or recreates the way light was reflected from the object to the recording plate. You can see an image of the object without that object even being there. Under laser light, it looks exactly like the original thing being recorded.

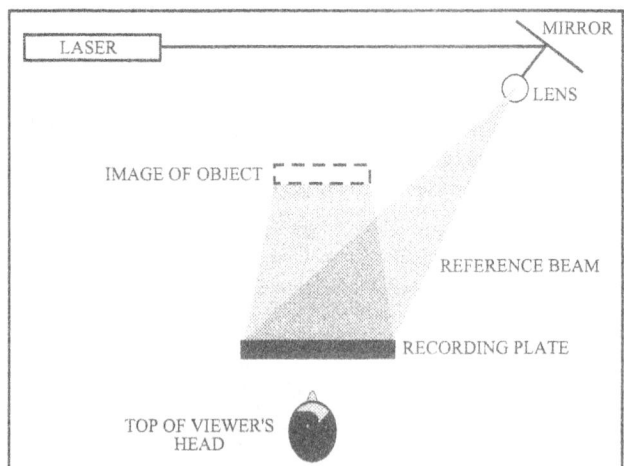

Figure 2. Viewing a transmission hologram

MAKING A REFLECTION HOLOGRAM

If we start with the basic setup previously depicted, but transfer the reference beam around with mirrors so it illuminates the recording plate from the side opposite the object beam, we create a reflection hologram. Remember that reflection holograms are meant to be illuminated from the "front," so changing the position of the reference beam will achieve this. (See figure 3.)

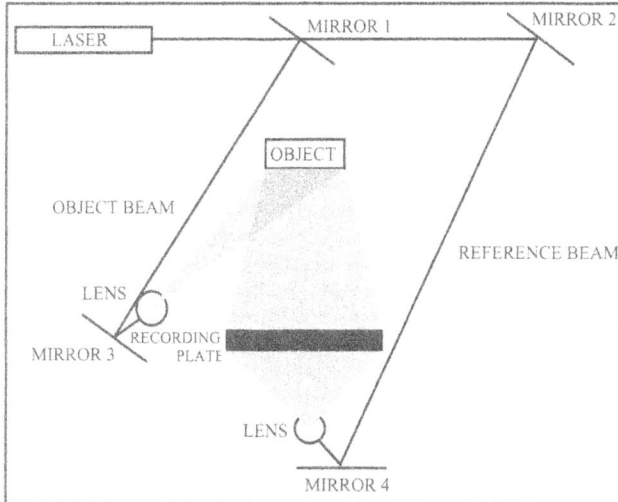

Figure 3. Reflection Hologram

POSITIONING THE HOLOGRAPHIC IMAGE

It is important that we cover the topic of the H1 and a master hologram in this introduction because it is a fundamental procedure in the making of almost every commercial hologram. H1 stands for "hologram one," which simply means it is the first hologram you make on the path to your desired final hologram. Sometimes the H1 is the master hologram from which you make multiple copies. Frequently, though, there is more than one hologram that needs to be made before you get the finished hologram from which you will make copies. If this is the case, the next hologram in the sequence is called the H2, then H3, and so forth.

A question that immediately comes to mind is, "Why would anyone want to make an H2?" Well, historically one of the big problems that holographers had was placing the subject exactly where they wanted it. Suppose, for example, you want the object in the final hologram to appear half in front and half behind the recording plate. How would you do it? You obviously can't do it on your first shot because the object would have to be going right through your photographic plate.

This problem was solved by the following procedure:

a) First, make an H1 transmission hologram. Since the H1 hologram creates an image of the object, why not use the image (generated by our HI) as our subject, and make another hologram (an H2) of it? In other words...

b) then make a hologram of a hologram! This H2 hologram can be a transmission or reflection hologram, depending on your need. It sounds strange, because you are making a hologram of an image and not an

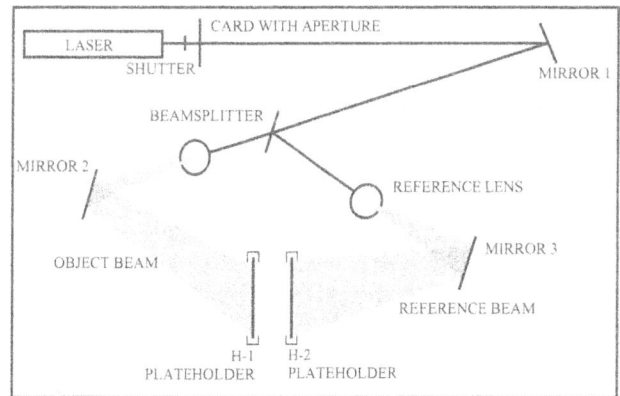

Fig. 4 – Reflection H-2 being made from H-1

object. But it works. (See figure 4.)

c) Now, since you can make a hologram of the HI's image, take time to move the image around to wherever you want it positioned. In this case, adjust the H2 recording plate so that the image of the object is half in front and half behind the plate and then make your H2. The problem of getting half of the object in front of the plate, and half behind, is solved.

In short, there are at least three good reasons why an H2 should be made:

1. The H2 allows you to reposition the image of your subject. When you reposition your image from the HI, you may make your subject focus in front of the recording plate, behind the plate, or anywhere within the limits of your equipment (you are usually limited by the laser's ability and the quality of the optics). The creative potential here is enormous because you are able to move solid objects around as if they are ghosts. You can have two objects occupying the same space, etc. The process of moving the image around to make the H2 is called image planing.

2. It gives the holographer a chance to brighten up the image. Since you may move your image anywhere, you can focus the image right at the recording plate. This concentrates the light directly on the recording material and brightens up the image considerably. This is commonly done with silver-halide reflection holograms.

3. It saves time on remakes. If you develop the H2 and decide you don't like the position of your subject astride the recording plate, you don't have to find the original subject and set it up again. This can be important if there are large costs in arranging the H1 shot.

Going through the pains of making HI, H2, etc. to produce a master for commercial replication is usually required. It is technically possible to get results from the first shot-but most professional holographers shoot a series of holograms in order to end up with a suitable production tool.

ALWAYS FOLLOW PROPER SAFETY PROCEDURES IN YOUR HOLOGRAPHY LAB.

IMAGE PROJECTION OF HOLOGRAMS

Although transmission holograms seem to be naturally designed to create a hologram with considerable projection, one can also make reflection holograms that have a great deal of projection. In fact, reflection holograms with considerable projection are a favorite among artistic holographers and the buying public. They are favored because they can be hung on the wall and illuminated just like a painting, whereas transmission holograms need to be lit from behind, often requiring a much larger viewing area.

Laser-viewable transmission holograms can demonstrate amazing depth and projection when the correct equipment is used to make and display them. It should be noted that the depth of the holographic image is not so much a function of the power of the laser as it is of the coherence length of laser light. Theoretically, the maximum image projection in front of the hologram plate can be as great as the projection in back of the plate (depth of the image). Unfortunately, it is difficult for our brains to make sense of greatly projected images. Because of this, and the fact that usually there are optical distortions created in the image planing process, projected distances in transmission holography usually are kept under four feet.

Laser transmission holograms have the widest parallax and display deep images best. There are laser transmission holograms, for example, of people and objects in a 4,000 cubic foot room, made by pulsed lasers. Not surprisingly, projected hologram images like this generate some of the highest shock and thrill responses from viewers.

We mentioned earlier that although it is necessary to use a laser to make a transmission hologram, it is not always necessary to use a laser to see a transmission hologram. In fact, most transmission holograms can be seen in sunlight. This may seem confusing, because we have said that in order to see a holographic image you have to shine on it the laser beam that made the hologram. This is true, but sunlight contains a multitude of wavelengths of light, including the one from the laser that we used to make our exposure. Also, the sun is such a great distance from earth that it appears to be a single beam of light shining on our plate. It would seem that we have only to position the plate at the proper angle, and we should see our image.

But it also stands to reason that if sunlight passed through a transmission hologram we would also get images being formed by all of the other wavelengths that are somewhat close to the wavelength of the reference beam. These other frequencies of light would diffract at a somewhat different angle than the original reference beam. The result would be a multitude of images forming right next to each other, creating a blur instead of a clear, crisp image.

That's exactly what does happen and it took a while for a solution to be developed. Around 1969, Dr. Steve Benton came up with a solution. The resulting hologram is sometimes referred to as a "Benton" hologram, or more frequently, a rainbow hologram.

RAINBOW HOLOGRAMS

Benton reasoned that since our problem is too much imagery at the point of reconstruction for our object, why not block off some of it? In other words, suppose we put up an opaque mask against the transmission hologram, with a long, narrow horizontal slit through which we view our transmission hologram. This would certainly clean out a lot of the annoying secondary images that are blurring the primary image's reconstruction. (See figure 5.)

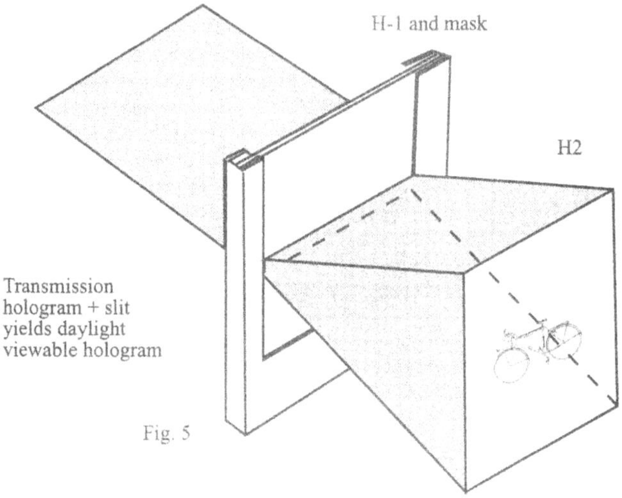

Fig. 5

This "cleaning" comes at a price, however, because the mask causes loss of vertical parallax (the ability to be able to see over and under our object). We would, however, still have our horizontal parallax (ability to see side-to-side around the object). Humans, with feet fixed on the ground and eyes on a horizontal plane, are actually more accustomed to horizontal parallax than vertical.

The procedure to produce this masked hologram is as follows:

a) First a normal transmission hologram is made.

b) Next, an H2 copy of the transmission hologram is made, but an opaque card with a horizontal slit is placed between H1 and H2.

To see the resulting holographic image (using the laser light we used to make it or light of an identical frequency), our eyes must be positioned at the "real image" of the slit. This viewing geometry will be apparent in the studio, but what happens when we want to see the hologram outside?

Imagine viewing this H2 hologram in two different colors (two different frequencies) of light. A hologram of the image made through the slit will be played back, but each of the two wavelengths of light will diffract through the hologram fringes at a slightly different angle. There will be two different images of the object, each a different color and each at a slightly different vertical position.

Next, think of the image in white light or sunlight. All of the wavelengths present will reconstruct their own image, all slightly displaced vertically with respect to one another. As you move up and down in front of the plate the color of the image will shift through all of the colors of the rainbow (hence the name "rainbow hologram").

As you move from side to side you will have horizontal parallax because nothing has been done to destroy it. By careful planning, the image can play back any desired color at the correct viewing angle, or even a combination of colors (a multicolor rainbow hologram).

In effect, the hologram is filtering the white light, while all that is sacrificed is vertical parallax, which, as we mentioned, our two horizontally positioned eyes usually don't miss anyway. Also, these rainbow images are often extremely bright, because all of the frequencies in white light are being used to form the image.

So the rainbow hologram technique is a way of making a transmission hologram sunlight-viewable. Other names for this are "daylight-viewable" or "white-light viewable" holograms. They all mean the same thing-a hologram you can see without needing a laser.

MAKING A HOLOGRAPHIC STEREOGRAM

Most historians credit Lloyd Cross and his cohorts in San Francisco with the development of a process that resulted in the first reliable method for producing a holographic stereogram - it resulted in a three-dimensional cinematic image that appeared to "float in space." Their method, developed in the early 1970s, allowed live subjects, life-size models, and special visual effects to be incorporated into their holograms in a practical and affordable way, as expensive pulsed lasers were unobtainable.

In order to commercialize the endeavor, Cross and his colleagues manufactured a motorized display unit for their freestanding 360-degree version. They also developed a stationary wall-mounted unit that displayed 120 degrees of viewing angle as the person moved around it. The idea of creating a self-contained holographic display device was quite revolutionary and very admirable. The complete units, which incorporated an inexpensive light source (an unfrosted light bulb with a vertical filament) along with the hologram, sold for several hundred dollars. The Multiplex Company has been producing units based on this process for more than twenty years.

Here is a simplified description of the process:

1. Make a rotating stage.

2. Place an object or a scene with live actors on the stage.

3. Set up a stationary movie camera in front of the stage.

4. Film the subject as the stage rotates 360 degrees, making sure to shoot at least three frames for each degree of rotation. In addition to the stage moving, the subject is allowed to move slightly in a manner that will result in a smooth animated sequence. Rapid or uneven motion, however, will create undesired "blurring" effects .

5. Develop the movie footage in a normal manner.

6. HOP Transfer: We now want to make a hologram of each frame o f the movie footage . These holograms will be sequentially exposed onto a sheet of film using a holography setup whose elements are collectively referred to as the "holographic optical printer" (see earlier text). The HOP setup illuminates each individual frame of movie footage with laser light. Another laser beam meets the beam that went through our movie frame at the emulsion by another path to create the hologram. Each frame is optically "condensed" into a narrow strip on the film using lenses and a mechanical slit aperture that restricts the image to one, narrow, vertical slit. The film is advanced and the process is repeated. A series of vertical slit holograms, running the length of the film, results. (See figure 6.)

7. After the process is complete, you will have a length of film with hundreds of thin vertical holograms on it. Once processed, you can take the film and wrap it into a cylindrical shape. When the film is illuminated from inside the cylinder (behind the film) with an appropriate light source, the viewer will see an apparently solid image floating in space inside the cylinder! As the cylinder rotates, or the viewer walks around it, the image looks fully dimensional and appears to move!

These dramatic effects result from the fact that each eye sees a slightly different image at the same time. Our brain then combines these images to give us a "stereogram" effect. One limitation to Cross' approach is that this technique creates images that display horizontal parallax only (i.e., you cannot see above and below the image). This is very adequate in most situations because we generally inspect images by looking side-to-side and not over-and-under.

Subsequently, holographers produced variations of Cross's concept. Some made stereograms with different degrees of view, commonly 60 or 90 degrees. Others began shooting the sequence of frames by moving the camera along a track (instead of moving the stage). They went on to flattening out the cylinder, which allowed the holographic stereogram to be produced and handled more easily. Eventually, researchers embossed these holograms onto mirror-backed plastics or produced copies which allowed front lighting (which is more practical in most situations).

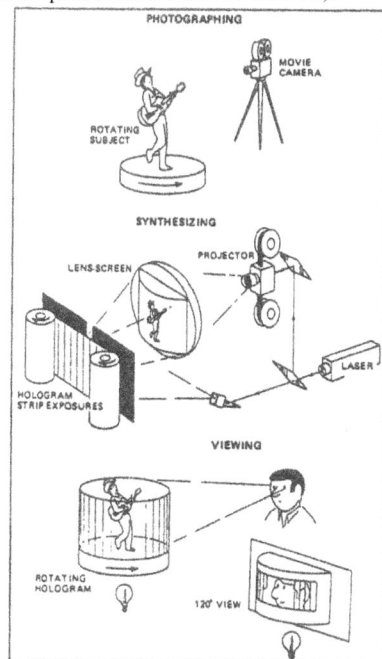

Fig. 6. Making a holographic stereogram using film footage

Exhibition Holograms

4

This chapter will discuss the kind of holograms that most people notice-large pictorial holograms. They are often called "display holograms", as most are intended for installation in public spaces as advertising, educational exhibitions, signage and art. Since the images are frequently life-size and the hologram itself is measured in feet or meters, this catagory of holograms is also refered to as "largeformat".

INTRODUCTION

In our last (7th) edition of Holography MarketPlace we included an article by John Perry called "Display Holography-Common Misconceptions, Myths and Reality." Perry's company, Holographics North, Inc., specializes in the production of "large format" holograms for display applications. The holograms he makes are typically incorporated in museum exhibits, trade show displays, corporate lobbies, and storefronts. In his 15 years of experience, he has encountered a number of misconceptions regarding holograms intended for exhibition related purposes.

COMMON MISCONCEPTIONS

The myths he addressed and debunked in his article included:

1) Holograms project images into empty space, to be seen from all sides;

2) A holographic installation requires a lot of equipment;

3) A laser is needed for display;

4) A dark environment is preferred when showing and viewing a hologram;

5) Those 3D pictures with ribbed surfaces are holograms;

6) Holograms of any size can be mass-produced;

7) Large holograms can be displayed in shallow light boxes.

The following article was contributed by Ron Olson of Laser Reflections, a company that also produces holograms for display applications. Olson's company specializes in making highly detailed, fully dimensional, deep-image holograms that are recorded with the use' of a pulsed laser- a type of laser that fires a powerful burst of laser light so fast that long exposure times are unnecessary. Therefore, immobile subject matter and elaborate vibration con- trol systems are not required, allowing live subjects (people, animals), moving objects, and even special effects (smoke, splashing liquids) to be recorded "instantaneously" in the studio.

Employing a pulsed laser not only speeds the origination process, it frees the holographer from many of the creative constraints imposed by the use of CW lasers, and, hopefully, makes the customer more comfortable with holographic imaging technology. The holographic recording session becomes more analogous to what happens in a traditional photography studio-a client comes in with a human model; the model poses with a product while the holographer shoots film; the film is processed; proofs are generated; and once proofs are approved, copies are mass-produced.

Over the years, Olson has directed a considerable amount of effort to promoting his services to the advertising, signage, and exhibition industries. Like John Perry, he continually finds himself confronting misconceptions and myths about holography. Here he relates a frustrating experience he had with a roomful of potential customers which inspired him to contribute a sequel to Perry's article.

Perry has contributed another article for this edition of HMP that explains his work in more detail. We hope that the information presented will inspire more artists, businesses and educational facilities to use the type of holograms that these holographers and their associates make.

EXHIBITION HOLOGRAMS: COMMON MISCONCEPTIONS, MYTHS AND REALITY

by Ron Olson (Laser Reflections)

I was invited to speak at a luncheon gathering of exhibit design professionals and I used the occasion to probe into the reasons for the obvious scarcity (bordering on blatant disregard) of holography within the museum and trade show environments. I began by asking the question, "Why don't you list holography among your palette of visual display tools along with photography, videography; dioramas, et cetera?"

The primary reasons I got were repetitious and worth repeating. All were misconceptions that I had heard before, but to hear them from professional exhibit designers seemed a bit much. In short, not one exhibit designer in attendance had ever had occasion to learn about this noble, but completely misunderstood, visual display technique. Collectively, their responses gave me a good idea as to why exhibit design houses weren't returning my phone calls and instilled in me an awareness of the magnitude of my sales and marketing problem. I asked to contribute this article in the spirit that perhaps by presenting the "facts" in print, in a publication of this type, it will spare me, and other serious display holographers, such ignominy in the future.

Myth #1: It costs too much. Most of the design professionals whom I questioned believed that creating an original hologram of decent proportions would cost anywhere from $5,000 to '$10,000.

At Laser Reflections we charge $1,500 for a single custom image measuring 16 by 12 inches, which includes a reflection copy (viewable with an attached halogen frame-light) and up to three transmission masters (viewable in laser light). We charge $2,450 for the same thing in a larger (24 by 14 inches) format. Additional copies range from $650 to $995 as a function of size and quantity ordered. Discounts apply to customers who commission more than a single image (assuming they can be shot during the same day's session).

These numbers compare rather favorably with commercial studio photographers' half-day rates and are typically dwarfed by the costs for on-location shoots as favored by many free-spending advertising directors.

In addition, interested exhibit designers operating on a very tight budget can put head-turning holography to work starting at under $600, by using stock images. Collections of stock images, with attached lighting, make excellent temporary exhibits.

There was also a perception among the designers that holography should be getting cheaper (a la ink jet printing). The facts are as follows: the hardware for the laser imaging system I built for our studio (a Neodymium:Y AG/ Glass laser) costs more today than it did twenty years ago (almost $200,000). Over the last five years, the cost of holographic plates has doubled to a current level of $45 per plate (for 16 by 12 inches).

The breakthrough in image pricing will come with high volume purchases-we are currently quoting $295 for a self-contained holographic beverage sign as an alternative to neon signs (in quantities greater than 500); $595 for a retail product display case (in quantities greater than 10). These prices are realistic and competitive.

Myth #2: It takes too long. I got answers ranging from weeks to months regarding lead times.

From scheduled shoots (or sittings) we can deliver ready-to-hang copies in as little as 24 hours-with standard processing we guarantee delivery of the first copy within five working days of the shoot. Unlike many other holography companies, we are not constrained by the need to design and produce models-we work from real objects/subjects as supplied by customers. Master images are created and processed on-site at the rate of one per hour. When studio time is of the essence (recent examples being a wildlife shoot of a Peregrine falcon and an Atlanta Falcon) we can shoot up to six images per hour and delay processing until later.

Myth #3: Subject matter is limited. I was told that because of the need for extremely long exposures, holographic compositions were limited to lifeless images of coins, Star Trek™ toys and printed circuit cards-things that can be rigidly mounted to a massive optical table.

Our studio uses a pulsed (Q-switched) laser with exposures of ten nanoseconds to effectively freeze motion. We routinely image people, animals (on a recent Saturday afternoon we imaged a mountain lion and a lynx), tanks of fish, etc. nearly anything which can be carried, led, or coerced into our downtown San Francisco studio for the required 60-90 minutes shoot.

Myth #4. Holograms are hard to light properly - improper lighting leading to less than optimal viewing.

All of the finished holograms we produce at Laser Reflections include lighting. We designed a low-voltage halogen spotlight which attaches directly to the frame of our wall art; and we include integrated lighting in all of our display cases. In trade show or retail environments where overhead lighting can be particularly brutal (and can interfere with image reconstruction), we offer plexiglass shields which address the problem quite elegantly.

Recently we have begun offering transmission holographs which can be backlit with inexpensive, eye-safe lasers. The payoff for laser lit imagery is that scene depths can be extended from 24 inches (pretty much our standard using halogen spotlights) to more than six feet of crystal-clear scene depth and in-your-face projection exceeding three feet!

Myth #5. The work we've seen isn't of sufficient quality for use in our (high-profile) environment. The consensus was that holograms were little more than novelty items-typically the images were of children's playthings or ghoulish monsters-out of place in areas not accessible by skateboard.

Our experience suggests otherwise. Our portrait of Mayor Willie Brown hangs proudly in two prestigious locations: at San Francisco's Moscone Center and the Main Public Library. The Moscone's general manager, Dick Schaff, responds: "The dimensional, lifelike rendering of the Mayor is engaging and blends with our building's state-of-the-art design." Of a 23-piece showing we had at the Technology Museum of Innovation in San Jose, museum President Peter Giles said, "The holographic displays have captivated visitors with their three-dimensional beauty and realism. The visitors' response to the exhibit is beyond anything we've experienced."

Myth #6: They're always green. Many of the shrink-wrapped variety commercial holograms are green because green is an easy color to see and cheap to produce in high quantities. All of the images produced at Laser Reflections are monochrome-however, the single color can be specified as red, orange, yellow, green, or blue-with trade-offs in visibility at the rainbow extremes. Because the images are a single color, we advise customers away from custom images with a critical color component (a multicolored soda can or a deluxe burger, for example) and towards subject matter like people, animals, or fine works of sculpture, whose spectral shortcomings often go undetected.

In addition, by mixing true color reality and single color virtuality, we further minimize "the color issue." To achieve this, we created our "now you see it/now you don't" display case for use with real objects which are time-shared with holographic images. By oscillating two built-in halogen lights, we effectively move the viewer's attention from the holograph (located on the front panel) to the product or artifact (located inside the display box). This concept- which we call holographic timesharing-works in museum cases, in trade show exhibits, even on vivariums or aquariums. (Whoever first said that a picture is worth a thousand words anticipated my problem in describing this phenomenon-you'll have to trust me: the effect is extremely engaging.)

Myth #7: It's gratuitous, stagnant technology. Several designers felt that holography contained no new visual information compared to photography and that the 3D aspect was little more than a gimmick. Not one exhibit designer among those I questioned was aware that a holograph has approximately 1,000 times the resolution of a photograph (we can resolve single strands of hair). This property makes it possible to exhibit highly detailed images of fragile or rare objects that otewise would remain unseen in a museum's vaults. None were aware that images could be animated via an array of playback light sources; and lastly - not one was aware that a holograph can be turned on and off with the flip of a light switch (now you see a high-resolution three-dimensional image/now you see through a transparent piece of glass). All useful and impressive visual effects for educational and entertainment applications.

For potential users who require images to fill vast spaces, we can create mixed-media packages combining conventional 2D digital print graphics with holographic elements - but we remind would-be users that the optimum response to our work comes when viewers can get really upclose

and personal. At close range the interactive nature of our unique style of holography kicks into high gear: inviting the viewer to peer around objects, compelling them to study otherwise hidden details, and then to wonder aloud in compulsory fashion, "howdotheydothat?"

Furthermore, it was implied that holography had already tried and failed to make a mark within exhibit display - and that nothing had changed. It's ironic that these are the same people that upgrade their computers every six months but they somehow believe that within laser engineering - everything stands still. All of the images to which the museum exhibit designers referred (i.e., those on sale in their gift shops) were produced using laser technology dating back two or three decades.

CONCLUSION Holography is a cost-effective high-impact tool for use within exhibit design. It commands far more attention than lightboxes and creates a very high quality first impression (essential in getting viewers to return with their friends and associates). At shows of our work we routinely observe people standing attentively in front of images for more than a minute - holography's ability to get and maintain attention puts it in a league apart from any competitive visual display technique.

The bottom line regarding holography in exhibit display is that it works. With a new millennium waiting in the wings; a "been there / done that" look to much of what's supposed to be new and exciting in exhibit design; and a retail environment which increasingly begs for compelling visuals - holography's time to gain acceptance would seem to be at hand.

For further information, contact:
Laser Reflections, Inc
589 Howard Street
San Francisco, CA 94105
Phone: (415) 896-5958
Fax: (415) 896-517
Email: hologram@laser-reflections.com
Web site: www.laser-reflections.com/

Photography by Brendan Beirne

Live "wild"cat being recorded in darkened pulsed laser studio.

LARGE FORMAT HOLOGRAPHY

A MAJOR PRODUCER OF LARGE FORMAT DISPLAY HOLOGRAMS EXPLAINS HIS TRADE

an interview with John Perry (Holographics North)

Although security and packaging applications currently account for the majority of business in the holography industry, large format display holograms remain the most impressive (and, perhaps, the most inspiring) works produced by holographers. No other type of hologram has the same visual impact as a big, colorful 3D image. Unlike many small, embossed holograms that display shallow images that go mostly unnoticed, a meter-square hologram with projecting and/or receding images commands attention.

In addition, no other type of hologram affects the public's perception of what our industry is capable of producing in the same way. The first hologram most people ever paid attention to was probably a large hologram hanging in a museum, trade show booth, or corporate lobby. These large format display holograms continue to be both popular and effective. We encourage potential users to see a high quality, large format display hologram themselves if they are unfamiliar with them. It is a visual experience that words can not adequately describe.

In an attempt to further educate potential users about this exciting format, Holography MarketPlace (HMP) is proud to present an interview with the most prolific and experienced large format holographer working in the U.S. today, John Perry (JP) of Holographics North. His company has been producing noteworthy holograms for the past 15 years.

THE BASICS

HMP: How do you define "large format" holography? What size holograms are included in the large format category?

JP: I consider "large format" anything larger than the "standard" commercially available plates or sheet films- so larger than 50 by 60 cm (about 20 by 24 inches). This definition comes more from the labs doing the work than anything else.

HMP: Are there many holography companies making these large format holograms?

There are perhaps three dozen labs that have done up to 50 by 60 cm, but only about six to eight that have made bigger holograms. Most of those are now gone. I think we are the only North American lab now producing meter scale or larger holograms.

HMP: Please list the basic production steps involved in producing a typical large format hologram.

JP: For a traditional hologram, we start with a three-dimensional physical subject of some kind. The object being recorded must be rigid and the same size as the desired image. We place it directly on the holographic vibration isolation table, illuminate it with laser light, and make a hologram of it (as described elsewhere in this volume). Viewable only by laser light, this hologram is called the "master."

We then illuminate the master with our laser to form a focused image in space. We place a fresh piece of film where this image is located and make a new hologram, the "transfer" of it. This transfer hologram is the final product and is white light viewable. It is, literally, a hologram of a hologram .

For holographic stereograms (which are also described in more detail elsewhere in the book), we first shoot 180 35- mm photographs of the subject as it rotates and animates. The appropriate pictures can also be generated from a rotating image on a computer monitor screen. We then make the master hologram from this series of two-dimensional photographs (or from the digitally generated views): essentially building a "3D looking" object up from left to right.

HMP: What should a client bring to you? (An idea, a physical model, a computer file?)

JP: There are lots of answers here. We can work from a physical model or subject, including people and places. We can also work from a computer model sent to us on a

EARLY MAN 11" x 15" (28 x 38 cm)
Evolution of man's head in five views. Models by the American Museum of Natural History. Reflection hologram, Holographics North.

disk, or FTP'd directly to us. Or we can create and animate a computer model from scratch.

HMP: Please address the design and production issues concerning color, image depth, and motion.

JP: Color-I have always recommended multicolor transmission holograms to our large format clients. Very strong subjects do well in one color, such as the T-Rex skull we made for the St. Louis Zoo in 1989. But usually the mixture of colors adds a lot of sizzle.

Full color is possible, and we have done three for commercial clients. We do the color separations ourselves, and it is about twice the work of a "pseudo-color" three-color shot. I would say the difficulty of registering the color components increases about as the square of the linear dimension of the piece; so that what is demanding in, say, 8 by 10 inches, is nearly impossible in 32 by 40 inches.

Color is controlled by the geometry of the setup in transmission work, not by the use of different colors of laser light. And the problems stem from chromatic aberration, which means that the blue light focuses to a different point than the red. This leads to a chain of corrections that must be made; each one requiring another. In the end, the question really is how much mis-registration the image can allow, and still look good. Each one of these we have done has taught us a lot, and I always think the next one will be perfect.

Depth-Image depth depends on the type of hologram. We usually produce white light transmission holograms, since images can be up to four meters deep. We never recommend laser transmission holograms anymore, because of the difficulty, cost, and danger of laser illumination.

Reflection holograms are beautiful, but are not really practical in very large formats. First, they are hard to produce, since the master must be very big and must be open aperture (as opposed to the narrow slit master used to make transmission holograms). Also, the tolerance to movement in the system is much less, since the fringe spacing is only about 112 wave, as opposed to several waves in transmission.

Then there is the limited depth of field. Usually our clients want the depth of the image to roughly match the size of the film, or exceed it. With reflections, a meter-square hologram cannot have any more sharp image depth than a 4-by-5-inch one - which is only about eight inches. And then they are, of course, never as bright.

Motion-The subject of motion only became a concern for us six years ago, when we built our stereogram printer. I like to see the image animation slow enough that the difference from one frame to the next is barely noticeable. This usually equates to an image point moving about halfway across the hologram width in the whole series of frames. But we have exceeded this rule many times for clients who really want to see things screaming by. We warn them about time smear, and try to keep the fast-moving stuff small and central.

GOLD MINER
Miner pans for gold and searches for a nugget in this animated and highly detailed image produced for a museum. Hologram by Holographies North.

For example, our "Gold Miner" piece is very popular. It was originally made for the Placer County Museum in northern California. There is one point in the animation where the prospector lifts his right hand and puts it into the pan of silt. The motion here is way too fast, so that his hand broke up into what looked like a lot of little fish jumping into the pan. We doctored our software so that the critical four frames of rapid movement were each repeated, to slow it down. This should, of course, mess up the depth, but this turned out not to be noticeable.

I try always to have the same computer modelers do our work, but many of our clients want to do it themselves. Almost invariably, they animate the subject too fast the first time around.

HMP: Based on your experiences, where are large format holograms most effective?

JP: In science centers, holograms are great for punctuating an exhibition, and waking up an audience about to doze off or walk out. In trade shows, they can suck people in from fifty yards away, and hold them for quite a while so the staff can do their job. People are asked what they remembered in the show- "that great hologram in the XYZ booth."

For fine art, I think the medium adds a completely different and fresh form of image making, with its own qualities and possibilities. This is a difference in kind, and not just in degree, as it may be in the other applications.

And like early photography, to repeat that tired old analogy, it seems to be taking half a century to develop artists that know what to do with it. This is not at all to take away from the efforts of the artists now working in holograms, but rather to applaud how important that foundation work is, in developing a vocabulary for the medium as art.

I believe it takes time, like half a century, no matter who is out there, to really start creating meaningful work in any new medium. I guess I'm saying that I think fine art (as opposed to commercial art) is the brightest future for display holography of all sizes.

HMP: Do you still work with artists interested in producing large format holograms?

JP: We have now worked with at least thirty-two artists, and some of these many different times, like Setsuko Ishii from Tokyo and Marie Christiane Mathieu from Montreal. Artists usually come here and stay for a number of weeks, working with us as we produce their pieces. Setsuko has left with as many as thirty 3-by-5-foot holograms after a month in residence.

HMP: What types of work do the artists try to produce?

JP: Setsuko, for instance, sometimes paints with inks on rolls of clear acrylic, which we backlight and make masters of. She positions them carefully in space, one behind another, curving and rolling across the film area. Then we combine these into three- and four-color transfers, usually about 3-by-5-feet in size. The result is a spectacularly colored world of tenuous shadow forms, defining huge volumes in space.

HMP: How long does a large format hologram take to produce?

JP: Most jobs take about two weeks, but we like to have at least three weeks notice before a due date. We have never missed a deadline in fifteen years.

HMP: What do you charge for your work? How do potential customers react to the costs of production?

JP: We have not changed our prices since we started. Streamlining has kept pace with increases in overhead and supplies. A large format custom piece can cost anywhere from $3,000 to $25,000, depending on size, colors, and image details. Our average job is probably about $10,000. Stock images run about one-third the cost of custom work.

It is rare that we get a job that does not involve something new, that we basically have to figure out how to do. I sometimes quote the job when there are still some questions, but I'm sure to have at least one fallback plan, before taking any job.

We get totally mixed reactions to our prices. Some callers say, "Oh, my heavens, you've got to be kidding." And others say, "What a relief. I thought it would be in the hundreds of thousands." Still others say, "But we've been quoted four times that to get the same holograms made." (One wonders where they are actually going to be made, and where the rest of the money is going.) And occasionally we get the very disturbing response, "But we've been quoted half of that to get the work done." (Back when there were other labs doing this sort of work, this was just part of life in business. Now, it seems more sinister.)

HMP: How do you insure that the holograms you make are transported and displayed properly?

JP: We build our own shipping crates, and have never lost a hologram in any of them. We handle the installation on only about 25% of our work. We have a pretty tight installation diagram form , and a cleaning instruction sheet. But some installations still get botched. So we are putting a lot of red ink on the cover sheet.

We do encourage our clients to pay us a visit at some point, if they can. Once they have seen the showroom here, and we have shown them the details of lighting, they usually do a good job on their own.

CURRENT WORK
HMP: What type of jobs are you currently working on?

JP: Two weeks ago (August 1998), veteran holographer Jody Bums and I went to Virginia and shot the film footage for a life-size stereogram o f a skeleton they discovered two years ago near the Jamestown Fort excavation. This young man died of a musketball wound in the right knee-and it's still in there, after almost four hundred years. This was a tough dolly shot, since the camera had to be up more than ten feet above the floor, and roll five feet. We could not tilt the guy up, or he would fall apart. We rented a ton

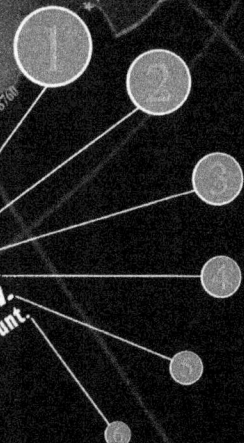

The Holographic revolution starts here

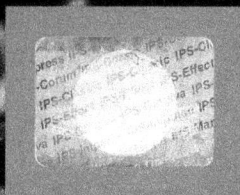

This hologram contains different levels of security, red laser hidden image and various linear or micro text.

We have great experience for all type of holograms:
- 2D
- 3D
- 2D/3D
- Colorgrams
- Stereograms
- DOVD's and OVD's
- Electro Beam
- Holomatrix™ from 50 to 2000 dpi
- 3D Dot matrix
- 3D Computer generated, and any combination of the above.

HoloWebs, LLC located in San Diego, CA is the corporate headquarters for our group of vertically integrated companies. Conceived in 1984, we have developed into a multinational force in the holographic industry. For over 14 years HoloWebs, LLC and its affiliated divisions have brought to their customers innovative holographic labeling and packaging ideas and state-of-the-art HSE (high-speed embossing) technology and equipment.

This technology and equipment made available by HoloWebs, LLC includes; lasers, electroforming, high and low resolution Holomatrix™ Dot Matrix, HSE (High Speed Embossing) equipment, both wide and narrow web, which allows for high speed production of both holographic and non-holographic substrates. HoloWebs, LLC finishing services include kiss-cutting, flexographic overprinting, laminating, hot-stamping and labeling application to continuous forms and other security formats.

HoloWebs, LLC and it's international sales department are responsible for the development and sales of our holographic manufacturing systems as well as holographic products. Producing every step of the holographic process in-house, HoloWebs, LLC provides embossing services to clients worldwide, in a minimum of manufacturing time and at an accessible cost.

HoloWebs, LLC

9475 Chesapeake Drive, Suite A,
San Diego, CA, USA 92123
Tel. (619) 576 1778 Fax. (619) 576 2181
Email holowebs@holowebs.com   Home Page http://www.holowebs.com

FLEXIBLE & RIGID HOLOGRAPHIC PACKAGING MATERIALS

Holographic material with an infinite designs and incredible visual effects is highly suitable for application to rigid and flexible packaging. The technology offered today by HoloWebs, LLC is state o the art HSE (high speed embossing). This technology allows companies to use holography as a permanent feature in their packaging and advertisements, rather than as a special promotion.

HOLOGRAPHY AS A PROMOTIONAL AID

Without a doubt holograms can today be considered the most highly enticing of visual aids, owing to their combination of kinetic colors and unique visual effects. Since holography can be combined with other graphic processes such as printing, laminating, hot-stamping, etc., spectacular promotional articles can be created, such as stickers, labels, collector cards and displays. Absolutely anything the imagination can think of!

SECURITY SOLUTIONS AND MATERIALS

HoloWebs, LLC takes the security application further by adding additional nearly invisible authentication markings revealed only to the client and due to our advanced technology, we are able to provide these proven defense measures cost effectively.

Customized imagery design, tamper evident labels, hot-stamping, bar codes, serial number, sequential numbering of certificates of authenticity, credit cards, high security documents such as stocks, bonds, share certificates and currency... the applications are limitless.

TECHNOLOGY TRANSFERS

HoloWebs, LLC is responsible for all technology transfers and "know-how", providing services to a global clientele. Technology Transfer training is proudly provided by HoloWebs, LLC in the areas of Computer Graphics Equipment, Photoresist Technology, the Electroforming process and the creation of holograms in our Optics Laboratory, as well as in the Embossing, Laminating, Ink Jet and Kiss Cutting areas of finishing.

THE SOURCE OF QUALITY HOLOGRAMS

IHMA
INTERNATIONAL HOLOGRAM MANUFACTURERS ASSOCIATION

Holograms now play an essential part in the production of documents, displays, packaging, gifts and other products throughout the world

Obtaining your customised hologram from an IHMA member gives you many advantages

Their work benefits from the most advanced and highest quality production techniques

Members commit themselves to the Association's Code of Practice, offering their clients assured security and reliable business ethics

And only IHMA members can record your hologram on the unique Hologram Image Register – a worldwide database of holograms, which helps reduce the risk of fraudulent manufacture and copyright infringement of holograms.

For further information on the **IHMA IN EUROPE** IHMA Runnymede Malthouse, Runnymede Road, Egham Surrey TW20 9BD England Phone: +44 (0)1784 497008 Fax: +44 (0)1784 497001 E-mail: Ian_Lancaster@CompuServe.Com	**For further information on the** **IHMA IN THE USA** IHMA PO Box 887 Englewood CO 80151 USA Phone: 800 741 6552 E-mail: ReconnUSA@aol.com

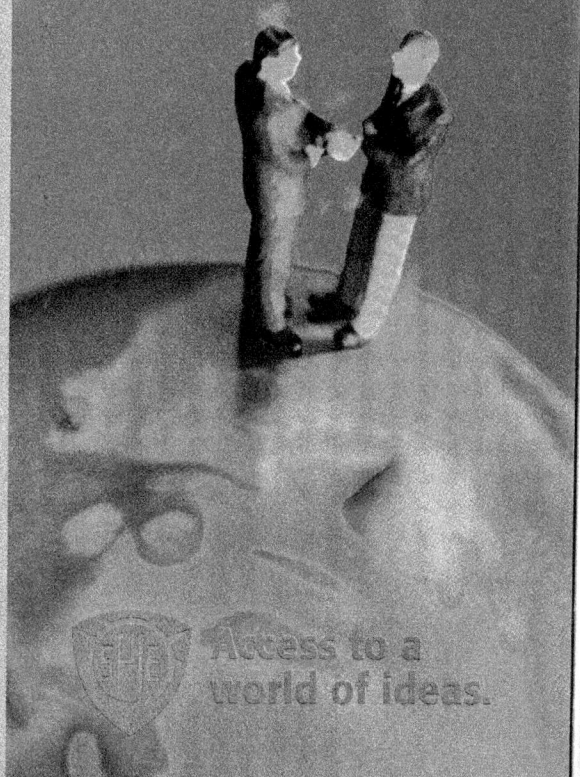

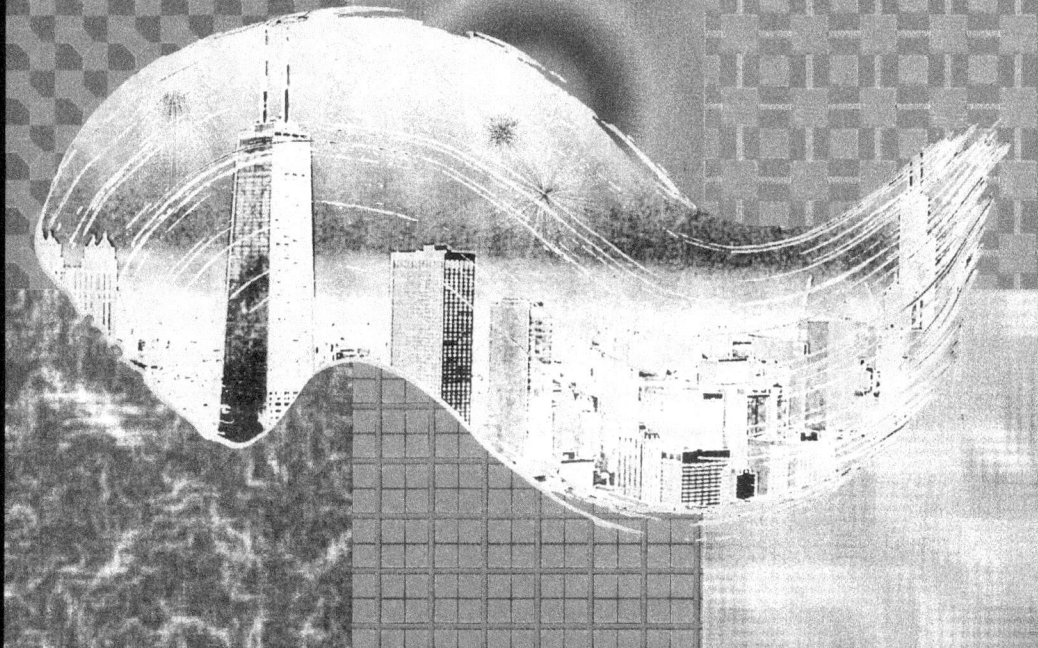

Holographic Masterpieces...

...Rock Solid Security

Transfer Print Foils, Inc.
9 Cotters Lane, P.O. Box 538, East Brunswick, NJ 08816, USA
(908) 238-1800, (800) 235-FOIL, Fax: (908) 238-7936

A HoloPak Technologies, Inc. Company

Holographic Masterpieces

As the nation's leading manufacturer of embossed holography, Transfer Print Foils, Inc. offers the widest range of holographic technique. Perhaps it is the reason that everyone, from people beginning their first holography project to industry experts, relies on TPF to see their project through to completion.

Three dimensional images, 2D/3D images and TransFraction™ (prismatic) patterns are all available from TPF. We also proudly offer "pixel" images and patterns as well as an in-house Design Center to help turn your concepts into reality.

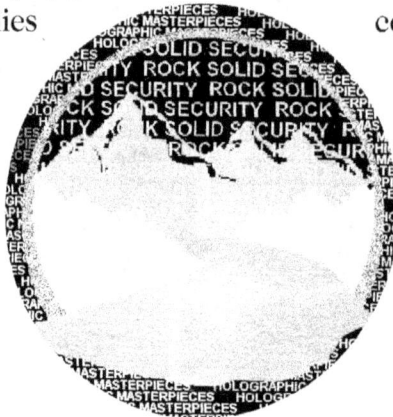

Rock Solid Security

High-tech security products such as our exclusive HoloClear™ provide the highest level of Rock Solid Security available today! Counterfeiters will run when they see a document or ID card with the HoloClear™ overlay. Call us for samples of this fraud-stopper immediately!

With a security production facility, advanced research and development and "state-of-the-art" technology, we provide unique images that are used for security on drivers licenses, international ID cards, trading cards... the list goes on and on!

Transfer Print Foils, Inc.

9 Cotters Lane, P.O. Box 538, East Brunswick, NJ 08816, USA
(908) 238-1800, (800) 235-FOIL, Fax: (908) 238-7936

A HoloPak Technologies, Inc. Company

of dolly equipment in Baltimore, crossed our fingers, and the hologram turned out great.

HMP: How many pieces have you produced this past year?

JP: This year we have produced six giant, animated holograms of underwater scenes for the Grand Princess cruise ship; a full-color 3D rendition of the Renaissance painting "Rape of Europa" for a Boston collector (with the computer help of Ryder Nesbitt at MIT); the Jamestown Fort skeleton hologram that I mentioned earlier; six computer-generated trade show pieces for various companies; and five huge pulsed master transfers (originally pulsed at Fringe Research) based around the concept o f "A Midsummer Night's Dream" for the Canadian artist Evergon. Most of this work has been really exciting, with lots of challenges and great results.

HMP: Provide an example of a challenging commercial job.

JP: We just made a hologram of three computer-gener- ated concentric rings, rotating in different directions about their common center. The client wanted them to move fast. They are about thirty-four inches in diameter, and the trade show booth only allows a six-foot light path for illumination.

The problem here is that the master slit, or viewing slit, is pushed way back by the short light path, with the result that when viewed from about 8 to 10 feet away, the left and right side of the rings are seen in very different time frames, and they look like pretzels.

The piece is mirror-mounted for front lighting, which we usually discourage in trade shows, because of ambient light problems. But here it is fortunate. The booth was canopied, first of all, so no ambient light problems. And the short light path can be somewhat corrected in the horizontal axis by slightly curving the acrylic-mirror-backed hologram into a concave arc. We figured a way to do this easily on the booth wall, and it worked out very well.

HMP: How do you currently market and sell your work?

JP: Our market is now equally split between trade show, fine art and science center applications. We have a stock list of about twenty-five images for sale off the shelf, which is designed mostly for the science centers. The trade show work is almost always custom, and of course the artists come with their own concepts.

Through the years, we have had many people, both inside and outside the holography community, marketing for us. Some have been very helpful, some not. But we have a strict policy of never accepting work directly from their contacts. "End runs" are one sure way of losing the respect of your colleagues, which generally leads to a failed business. I am a believer in keeping the respect of not only your colleagues, but also your clients. I would rather take a financial hit on a job than have a disgruntled client. And I would rather help out a competitor, and hope for the same some day, than make an enemy.

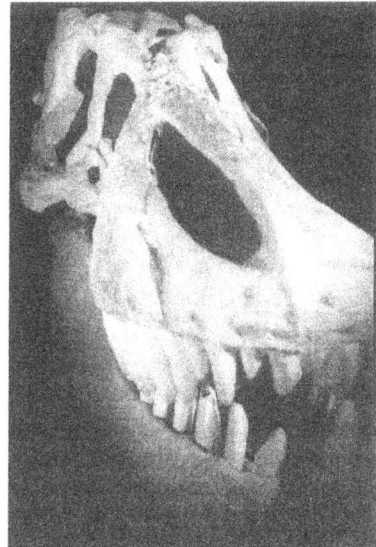

T-REX SKULL
30" x 42" (75 x 105 em) Life-size (4' long) dinosaur skull sports fantastic teeth two feet in front of the hologram. Fossil cast by the American Museum of Natural History. Transmission hologram by Holographies North.

Evolving Techniques

HMP: What type of lasers, recording materials, optics, and studio equipment did you work with when you started out? Where did you get your supplies? How much did your basic materials/equipment cost?

JP: Lasers-My company, Holographics North, Inc. (HNI), started making large format holograms in 1984 with a new 25 milliwatt HeNe, without an etalon. I hired two of my recently graduated university students and we produced deep subject images only in laser transmission mode, by using several object beams, staggered in path length.

I could see no way to make large white light viewable transfers without more coherence length, which would require an etalon. The reference length just varies too much over the height of the film, so that the object and reference beams cannot be mutually coherent. This laser transmission work included our first job, for Pratt and Whitney, of three turbo airplane engines. These are still some of my favorite pieces.

The next year, 1985, we squeezed a series of twelve 22-by-30-inch white light transmission transfers out of this laser, (still without an etalon). These were for "Expo 86" in Vancouver, for the pavilion of artwork by a Canadian artist, Michael Snow. Later that year we got an old 4-watt Argon laser with an etalon, but we needed to generate our own three-phase power, since it does not exist in our neighborhood. I hated this laser, which needed constant attention, and consumed massive amounts of power and water.

In 1987 we got an old Spectra-Physics 125 HeNe and had Don Gillespie (from El Don Engineering) retrofit an etalon into it. We now have three of these, and I love them. We generally get about 70 milliwatts through the etalon, and have about twenty second exposures for meter-square shots. Don has consistently done a good job for us. Each refurbished laser cost about ten thousand dollars.

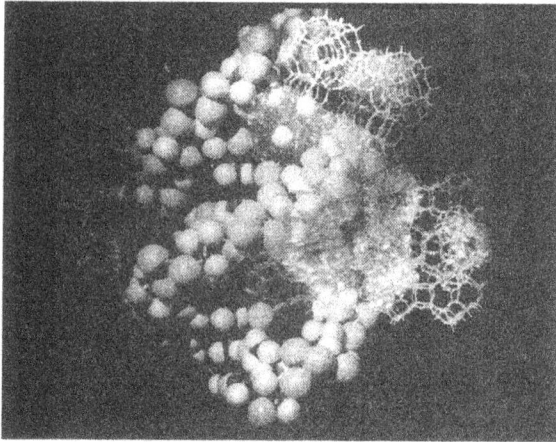

MOLECULES 36" x 30" (90 x 75 cm)
Full-spiral DNA model seen from the left, joined by a multicolor group of Zeolite molecules as viewer moves to the right. Four-channel, 3-color, transmission hologram by Holographics North.

Recording materials-We have always shot on Agfa HD films. Originally it was the polyester backed stuff, and since about '87 the acetate backing, with a big increase in emulsion quality. We started out paying $425 for a 10-by-1.15- meter roll. Our last order cost us $1,740 a roll.

Optics-Most ofour early optical and mechanical equipment came from a local junk yard that would get the throw-outs from the big IBM and Simmonds Precision plants nearby. We would pay eleven cents a pound for steel, and thirty cents for aluminum-three-dimensional translation stages (4 lbs.), mercury vapor lamphouses with f/1 lenses (8 lbs.), and so forth. Unfortunately, they're gone now, though much of the equipment we buy "new" is often surplus or discounted for some reason or other.

Film holder-We use a lot of gravity, which is cheap in Vermont. Our first meter-square film holder was glass, but the second was a 4-by-4-foot, two-inch-thick slab of native Vermont granite. That became my desk, and we now load on a 6-by-6-foot vertical slab of steel.

Isolation table-We had shot the 32-by-42-inch Pratt and Whitney pieces, and many others, on our original 5- by-la-foot concrete table. We poured our 12-by-22-foot floor-level concrete optical table in spring of 1985, to do the Expo job on.

HMP: What applications were you making holograms for in your early years?

JP: Our early work, like the Pratt and Whitney pieces, was almost all for trade show applications. Our second year saw the "Expo ' 86" pieces, which were really fine art display, and then a meter-by-meter piece for the artist Frank Stella in 1987.

HMP: What type of subject matter were you typically recording?

JP: For about our first six years, we were confined by the technology to physical things- products, machines, models-the usual holography problem. We expanded that quite a bit when we started shooting some masters with Fringe Research's pulsed ruby laser in Toronto. I was a partial investor in that laser, which was acquired in 1989.

So now we could do people and other shaky subjects. And the quality of pulsed images is, of course, gorgeous. The stereogram technology has now, of course, really opened up the spectrum of subjects.

HMP: What technological obstacles needed to be overcome?

JP: The technological problems were enormous. In the '80s we went through staggering amounts of film, trying to get all the bugs out of the systems. We sometimes spent two whole rolls with no success. But luckily film was still cheap. It was truly grueling, especially considering how physically difficult the procedure is, with preparing, loading, and processing giant pieces of film.

Holding thefilm still-One of the most challenging problems we encountered was how to hold a 45-by-72-inch piece of film for one minute without allowing it to move a millionth of an inch. Large format work is especially demanding because cost and difficulty of mounting various components increases in proportion to the hologram's linear dimension squared. And so does the exposure time! But the tolerance to movement is exactly the same as if you were making a 4-by-5-inch shot.

We have always used the method we developed at the University fifteen years ago--printer's ink rolled onto the back of the film acts as both adhesive and antihalation backing. It's slow, but very clean and consistent.

We once experimented with the concept of a liquid gate film holder, about 4-by-4-feet, which can produce very clean results. I calculated the pressures involved but I couldn't believe they would be so great. So we tried to build a holder out of two-inch granite on the back and 3/8 inch glass on the front. The glass was bowed out about three inches and the gate only half full of alcohol, three liters, when it broke. (Every now and then the calculations actually work!) So we could easily have made l2-by-16-inchers with it, but nowhere close to meter-square.

Vibration isolation-Movement control during exposures took about ten years to iron out, and is still mischievous. For the first ten years or so, we religiously used an electronic fringe locker to overcome instabilities in our twenty-two-foot-long concrete table. We don't use it any more since we found that the table is solid, after all.

Many other problems required custom solutions: film processing and drying, film storage and cutting, geometry for various types of shots within the framework of our 12- by-22-foot table, packing and shipping large flat holograms, etc.

HMP: Do you have any patents on the techniques that you developed?

JP: First of all, having looked into the patent situation, and done some searches, I long ago decided that they may be useful for getting funding and selling your business, neither of which I wanted to do. But they really don't do any good for protecting a small company like mine against unfair competition. The best way to protect procedures is to keep coming up with new ones. So we don't bother with patents, and we don't really keep secrets, anyway.

Many experienced holographers have visited the lab and worked here alongside the staff, and have seen it all. Our table and procedures have been copied several times, but these always ended in failure, for one reason or another.

HMP: How did your basic production equipment and materials change throughout the last fifteen years?

JP: There was a lot of equipment jockeying during the first six years of business. For instance, we went through several variations of our processing tank (which evolved to look a lot like a tanning bed). But once we settled on things like HeNe's with etalons, our curved, 6-by-6-foot steel film holder (mounted to a vertical 4-by-6-foot optical table for abase), and a number of other successful tools, very little has changed.

One major breakthrough came when we built our large format stereogram printer in 1992. Basically, the system allows us to make a hologram from a series of 2D photographs. It took about six months to design and build, and then about another two years to get it really working as well as it does today. Now we can computer-generate and animate our images, shoot on location, and change image scale as much as we want.

For instance, for one trade show piece we took a slice of rabbit tibia, 114 inch around, with a sample of artificial bone implanted into it, and blew it up to twenty inches across in a 34-by-44-inch hologram. You could clearly see the microstructure of the implant material.

Probably the best indication I can give of the improvements made is that ten years ago we were getting about one good shot out ofthree in, say, 42-by-42-inch size. Now we get about nine good shots in ten. And "good" is, if anything, a lot better than it used to be. It is rare that we lose a shot; and I always make my apologies to the holography gods when I say that. (They hate arrogance.) But this is a statistic I am truly proud of, especially in these days of fixed film supplies.

HMP: Have you integrated computers into your production systems?

JP: Not much. We have a little, ancient thing that runs the stereogram system, and does it really well. And now, of course, a lot of our images are computer-generated, which is absolutely wonderful. Our stereogram system requires 180 rendered images of the model, rotated 0.3 degrees per frame. But, as mentioned, we farm all of that modeling out either to the clients or to two terrific digital modeling guys we have nearby.

HMP: What problems did you encounter related to your sales and marketing efforts?

JP: Pricing was a problem at the beginning. I suppose it always is whenever there is little competition. I never felt comfortable asking our few competitors what they charge, even though we are pretty good friends and sometimes even work together on different phases o f the same job. So it was pretty much a matter of figuring what we needed to get, and what the market would bear. Somebody once said pricing is the key to any business. We must have hit it

BALLOON BLOW-UP 38" x 28" (95 x 70 em)
Woman blowing up a balloon which disappears as viewer moves to the right. Two-color, 2-channel. transmission hologram by HN

pretty well at the beginning, since we haven't changed them and we're still here.

HMP: Please list some significant holograms you produced over the last fifteen years.

JP: I looked through our pile of eighty to ninety backup copies oflarge format pieces from over the years, and every one ofthem has a great story. One ofmy favorites is the one we made in 1988 for HUE legwear.

The two women who started HUE, and experienced meteoric success, wanted something different to dress up a boutique area in the Bloorningdales flagship store in Manhattan. I suggested a many-channeled image oflegs, feet, stockings, shoes, ribbons, bubbles, and other legwear related stuff, all changing rapidly as you walk by the hologram. "Staccato" was the word that sold the job.

We met in Toronto, picked a leg model, and shot fourteen masters with the pulsed laser at Fringe Research with the help of holographer Michael Sowdon. Back here at HNI, we cut the masters up and arranged the pieces into three slits, for a three-color transfer, so that the image in each color would change at different times. We had a total of twenty-two channels of imagery.

The result is a New Year's Eve sort ofimage, with confetti, ribbons and bubbles, legs, feet, and torsos filling big volumes ofspace, draped in all sorts ofgreat stuff. One image is of the lower 3/4 of a female body, wearing a filmy polka-dot skirt and socks, gyrating and standing on a Toronto phone book. You can even see the "416" area code. When we needed another leg, the HUE owners would stand in. The piece was a great success.

Another favorite was the "Teleglobe" image, created for a trade show. It was before the days of stereogram. We started with a four-foot-diameter foam sphere, and had a local model maker put continents on it. We applied glitter to the urban areas to catch the laser light, and made a master of this for one color of the transfer. The second color was a small satellite model, with two dish antennae, orbiting the earth, about three feet out in front. And the third color was a series ofcommunication lines, first connecting Ottawa and Amsterdam to the dishes of the satellite, and then one

changes to a line coming from the satellite straight out ten feet toward the viewer. I have seen people actually grabbing their friends by the shoulder to keep them from getting impaled on this line.

THE FUTURE

HMP: How do you see large format holography evolving in the next decade?

JP: Sadly, not at all if the film situation does not improve (Agfa stopped manufacturing holographic film in 1997). At HNI, we can go about five years on our Agfa stock. Then we need another source, or another technology.

I understand there is an effort to mosaic photopolymer film into large format images, but it's being kept under wraps. In Lithuania, Slavich claims it is soon going to have meter-wide rolls of film, but not yet, and it may be fairly slow.

The market for big work is very limited anyway, largely because of the price, which is largely because of the labor. Lighting and limited viewing field are also factors. Point-of-purchase (POP) displays are not a viable option because of the "point-source" lighting requirements, and theatrical ap- plications won't usually work because of the viewing field.

The stereogram/computer image advancement, and several minor technical improvements we have come up with recently, have widened our playing field quite a bit in terms of subject matter and brightness. But large format work is really a showpiece, and not a great medium for conveying a lot of detailed information. Sometimes that is what is needed, and other media may do it better.

For further information, contact:
Holographies North, Inc.
444 south Union St.
Burlington, VT 05401
Phone: 802-658-2275
Fax: 802-658-5471
Email: jp@holonorth.com
Web site: www.holonorth.com

5

Holograms as Giftware

This chapter discusses the commercial development of artistic holography, especially the sale and distribution of stock holograms and related products by the giftware industry. The major categories of merchandise are described. Tips on opening your own hologram store are included.

THE COMMERCIALIZATION OF ARTISTIC HOLOGRAPHY

In its early stages holography remained unseen by the general public. Only scientists and researchers had access to the lasers and other specialized equipment that were needed to create and view a hologram. When methods were developed in the late 1960s that enabled a hologram to be created and viewed in more practical ways, holography slowly left the laboratory and began a journey that has resulted in a multimillion-dollar, worldwide industry. A great portion of this industry deals with "artistic" holography (i.e., three-dimensional images of things) which is also commonly referred to as "pictorial" or "display" holography.

During the 1970s and early '80s holograms were made by individual holographic "artists" on a one-by-one basis. The process was labor-intensive and time-consuming. Production techniques were developed through trial and error. Raw materials such as film emulsions were scarce, equipment was often homemade, and production quality often inconsistent. Unfortunately, the individuals and small companies that were capable of making high-quality holograms generally did not have the money or marketing expertise needed to get their work into widespread distribution, so holograms were still out of view of the public-at-large.

The handful of galleries and stores that did show holograms proved that the public was fascinated by this emerging medium. Although most holograms were treated as futuristic art work or novelty items and were relatively expensive, the public constantly asked for more affordable ones. Enterprising gallery owners, retailers, and hologra-

phers recognized this demand for holographic merchandise, and a small industry slowly evolved. Holographers and their entrepreneurial partners began to create products rather than art work; well-connected retailers began to distribute holograms to other retailers; and hologram aficionados became customers. Everyone recognized the potential of this new industry, yet manufacturing and display lighting problems still needed to be overcome before holograms and related products could enter the mainstream marketplace. Limited production runs kept prices high.

Over the next decade technological advances enabled holograms to be mass-produced in a variety of ways. This made it feasible for artists, technicians, and businessmen to join together to create facilities dedicated solely to producing large runs of affordable, high-quality holograms. These holograms were intended for a variety of commercial applications including security, packaging, and advertising-as well as products for the giftware industry.

Art holographers copied their most popular images onto film (which was less expensive and easier to handle than holograms produced on sheets of glass) and began to use assembly-line production methods in their labs. Whole catalogues of images soon became available, intended for sale as wall decor. Retail price points dropped considerably. Other holographers perfected methods of mass-producing very bright dichromate holograms for use as jewelry. Still others concentrated on developing high-speed automated replication technologies capable of embossing holograms on very inexpensive foils and plastics-perfect for use on toys, optical novelties, and paper products.

Once reliable supply lines were established, it became feasible for other companies to package and market these holograms in a variety o f ways and integrate them into the nonnal chain of giftware distribution. Businesses in the United States and England quickly grew into major distributors. Film holograms were matted and/or framed and marketed as high-tech art. Holographic fashion accessories (including watch faces, pendants, and earrings) were developed. Executive gifts and desktop accessories were created. Rolls and rolls of kids' stickers were produced. New toys were invented. Holograms started to appear at national gift shows, in giftware catalogs, and in the media.

Savvy retailers soon realized that holograms and related products were very popular with the buying public, and if displayed correctly, could prove quite profitable. A good display of holograms drew a crowd, generated customer excitement, and more importantly, generated dollars! (Most giftware items sold in stores are priced at twice their cost.) Holographic merchandise spread from science museum gift shops to mainstream outlets, and even included a number of specialty stores set up to sell only hologram products.

Increased visibility created greater public awareness of the product and demand for new and better holograms. Artists added color and motion to their images. Manufacturers automated further and invented materials especially suited for holographic applications. Holograms became brighter and easier to see under typical viewing conditions. Distributors created new product lines by integrating holograms into existing merchandise. Packaging was brought up to commercial standards. Wholesalers adopted more sophisticated marketing techniques, while retailers offered a wider selection of goods.

Today, a variety of holograms are manufactured around the world for the giftware industry, with the highest concentration of factories located in North America and Europe. English holographers have traditionally dominated the silver-halide film replication business. American holographers are actively developing photopolymer replication factories. The production of dichromate holograms seems to have slowed, while facilities capable of producing embossed holograms have multiplied significantly, especially throughout Asia. Surprisingly, very few holograms are exported from Japan.

The number of distributors and wholesalers dealing exclusively in holograms has dwindled as the market has diversified. To stay profitable several major distributors have developed their own custom images in order to target specific consumer groups. One major U.S. distributor has developed a very successful product line based on the ever popular hologram eyeglasses with stock and custom photopolymer images as lenses. Another has developed close working relationships with the product development departments at several major retail chains, thereby ensuring long-term sales. The sale of "licensed" holographic images featuring popular sports figures, cartoon characters, and movie scenes has grown steadily, while the sale of more mundane images has stagnated.

There are fewer holography specialty shops in business now than a few years ago. Those that continue to do well have increased the variety of goods that they carry and often include related optical novelties in their product mix.

Sales of holographic artwork are practically nonexistent. However, more stores than ever before are carrying some sort of hologram-related product. It is not uncommon to find an inexpensive hologram item at the comer store.

THE CHAIN OF DISTRIBUTION

Let's examine the chain of distribution as it typically exists for a holographic product.

The Copyright Holder

The distribution process starts with the copyright holder. Any unique work of art, including a painting, photograph, or computer-generated graphic, can be protected from unauthorized duplication (in most countries) by registering the image in the appropriate manner. In the case of holography, the original work of art is either a model, a graphic, or a computer program designed to generate a holographic image. Whoever creates the unique work of art that later becomes a hologram is considered the copyright holder of the image. It is also possible to copyright the finished hologram itself as a unique work of art, provided that none of the components that appear in the holographic image belong to another party. There are, however, statutory limits stating that after a number of years, a piece of art can become public domain and may be used freely.

Every hologram, if properly copyrighted, has only one owner, the copyright holder. The copyright holder therefore controls all subsequent distribution and is positioned at the top of the distribution chain. The copyright holder can be an individual or a group of people such as a business. Most commonly, a business commissions an artist to make a model or to design graphics and the artist turns over all copyright privileges to the business as part of the arrangement. This is legally known as "work for hire" and each party's responsibilities and rights must be documented to avoid problems concerning ownership. Holograms that are not copyrighted can be copied by whomever owns the "master" hologram.

The Manufacturer

Different companies specialize in manufacturing specific types of holograms. The manufacturing process generally involves three stages- mastering (creating the original hologram), reproduction (producing some quantity of copies), and finishing (lamination, cutting, sorting, etc.). Some companies do everything; others subcontract out some part of the job. Often a company that manufactures holograms also owns copyrights in order to have a selection of stock images to offer its customers. A few manufacturers bypass the nonnal chain of distribution and sell directly to retailers.

Typical costs, margins, and profits for hologram with retail price of $20:

Business	Pays	Sells	<Retail	Markup
Copyrt Holder	$4	$6	80%	50%
Distributor	$6	$8	70%	33%
Wholesaler	$8	$10	60%	25%
Retail Shop	$10	**$20**	50%	100%
Customer	**$20**	N/A	N/A	N/A

The copyright holder needs to know the exact cost of each unit produced, since the manufacturer's charge will obviously influence the final price billed to the end user. To figure the unit cost, one would take the total bill from the manufacturer (including any additional shipping and handling charges) and divide it by the number of usable copies actually delivered. As in most manufacturing businesses, prices decrease as quantity increases. In order to figure the suggested retail price of their product, it is very common for a copyright holder to multiply the manufacturer's unit cost five times. For instance, a product that costs the copyright holder $4.00 will be resold to a distributor for $6.00, and will end up selling in a store for $20.00.

If you are having holograms made to your specifications, choose a company that produces holograms appropriate to your final application. Be aware that manufacturers have not yet standardized their pricing-some itemize production processes, others quote a finished price. Some quote by the square inch, others according to a sliding scale based on quantities ordered.

The Distributor

The distributor is a business that specializes in buying large quantities of a product from a copyright holder and distributing it to other businesses that cannot afford to, or are not interested in, stocking inventory. Distributors are often contractually obligated to order large amounts of merchandise, carry an entire line of their supplier's products, and not sell competing products. This alleviates many problems for the copyright holder and the manufacturer who are not usually set up to market their own products to numerous customers.

In return, the distributor commonly receives the sole rights to sell the product in a particular geographic region or to a particular group of customers and pays less than any other customer down the line. Distributors commonly pay 70% below the suggested retail price, resell these products for 60% below suggested retail, and depend upon a large volume of sales to make their profit. For example, if they pay $6.00 for an item from a copyright holder, they would resell it to a wholesaler for $8.00.

Many distributors also repackage goods under their own names, deal with import/export procedures and constantly work to expand the marketplace. A popular product can make a distributor a lot of money, due to the fact that potential customers have no alternative supplier. The distributor sells mostly to wholesalers.

The Wholesaler

The wholesaler connects distributors to retailers. The essential function of a wholesaler is to get the product into shops. Most wholesalers use a combination of in-house salespersons or independent sales reps working on commission to persuade buyers to try a product. Many rely on telemarketing departments, catalog mailings, and trade shows to establish new accounts and service existing ones. A good wholesaler will teach a shop owner how to best merchandise a product, provide point-of-purchase materials, restock displays, update product selection, and generally keep the customer happy.

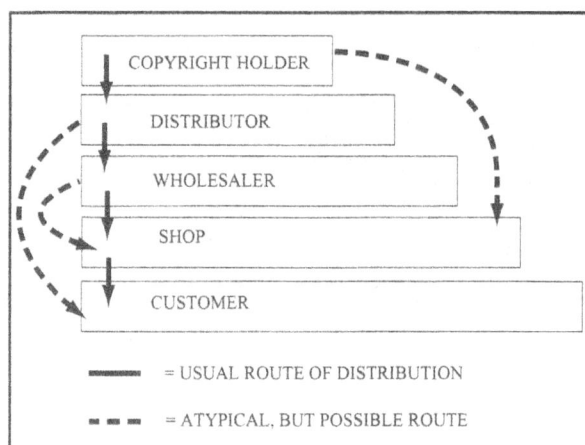

Chain of holography distribution

A wholesaler normally carries many different product lines, which is a convenience to retailers who want to consolidate the number of their suppliers. Also, wholesalers stock far less merchandise than a distributor, which allows them to react quickly to changes in the marketplace.

Wholesalers do not generally have exclusive rights to a product. They often extend payment terms to their customers (net 30 days is common) after a probationary period or credit check. For their efforts, wholesalers receive special pricing that allows them to make a profit when they resell the goods to retailers. Wholesalers typically work off 25% profit margins; if they buy an item for $8.00, they will resell it for $10.00. Some distributors offer additional discounts to wholesalers for larger orders.

The Retailer

Retailers are the point of contact with the public, the place where merchandise is displayed and purchased by the customer. Holograms and related products have been sold in a variety of settings, ranging from temporary tabletop setups (flea markets, trade show booths) to art galleries and department stores. Many entrepreneurial businessmen start with a small cart or kiosk in a busy mall during holiday season and graduate to a bigger store that runs year round. Several single-store operations have expanded to multi-store chains.

Wholesalers have placed merchandise in obvious locales like museum gift shops, technology stores, and poster shops; less obvious locations include airport shops, nature stores and stationery stores. Other dealers have targeted specific interest groups such as hobbyists and collectors, and sell licensed products to comic book shops, trading card stores, and the like. Larger suppliers have cut deals with amusement parks, national chains, and event promotion agencies. Holograms have even been sold by mail order using catalogs and classified ads, even though a written description or photograph does not adequately capture the wonder of a three-dimensional image.

All successful hologram retail businesses have several things in common: they are located in high-traffic (pedestrian) areas in places where people go to buy interesting items (often tourist destinations); they display the merchandise correctly and with flair; and they offer a high

level offriendly customer service. Although the retail price suggested by the copyright holder is only a guideline, most retailers double their costs to establish the final price a customer sees in the store. Therefore, an item that costs them $10.00 will end up on the shelf for $20.00.

CATEGORIES OF MERCHANDISE

Most successful hologram stores sell a selection of holographic merchandise, including pictures (stock wall decor images and limited-edition fine art), jewelry (and related fashion accessories), executive gifts (desktop accessories), toys, and optical novelties. There are several ways to categorize this merchandise-by price, by size, by manufacturer, and so on. For now, we'll discuss selection by "format" (which refers to the type of material the hologram is produced on).

Silver-Halide Glass Plates

Traditionally, most holographers have produced their holograms on sheets o f glass (glass plates) coated with a high-resolution light-sensitive emulsion called silver halide. This is similar but not identical to the emulsions used in conventional photographic films.

These glass plates are rather costly, but can be used to make the highest quality holograms-mostly because the glass plates are quite rigid and will not move during the exposure period (movement will ruin the hologram). Working with these glass plates is quite time-consuming, as each plate has to be handled with great care during each production step.

Due to the time and the cost involved in making holograms on silver-halide glass plates, they are mainly used for limited-editioll holographic wall art, archival images, or custom work. A finished 8" x 10" glass plate hologram typically retails at $300.00 or more for an open-edition work.

It is possible to produce striking "deep image" holograms on silver-halide glass plates - that is, images which display considerable projection in front of the hologram's surface, and/or appear to be a considerable distance behind the hologram's surface. To achieve such dramatic effects, glass plate holograms require the best possible illumination, which usually takes some foresight.

Glass plate holograms are extremely fragile and should always be securely wrapped and handled with care. When the hologram is framed, the actual emulsion is facing toward the back and is protected. The surface which we can see (and touch) is ordinary glass and can be cleaned by gentle polishing with a soft cloth.

Silver-Halide Film

Since glass plates were too expensive and impractical to mass-produce, holographers developed methods to produce holograms on silver-halide holographic film, which is cheaper and easier to handle. It is thin and flexible, similar to the film used in an ordinary camera. After exposure and processing, each piece of film is usually sandwiched in a cardboard matte, ready to package and/or frame as wall decor.

An 8" x 10" silver-halide film hologram typically retails for $150.00 (less than half the cost of a comparable hologram produced on glass), making this format much more suitable for the giftware market. The bestselling size has traditionally been a 4" x 5" film mounted in an 8" x 10" matte due to the retail price (which has hovered around $35.00). Other standard sizes include a 2" x 2.5" film mounted in a 5" x 7" matte (which typically sells for $15.00-$20.00), and a 5" x 8" film mounted in an 8" x 10" matte (which typically sells for $70.00).

Silver-halide film holograms do not display quite the resolution, projection, and depth of their glass plate counterparts. However, under proper illumination they are quite striking.

The emulsion of a film hologram is usually covered by a cardboard matte; however, the front surface can scratch easily. Fingerprints and dust can be cleaned by gently polishing with a soft cloth. Framing these pieces behind clear glass is recommended. Like all photographic films, they should be kept away from excessive heat and moisture.

Photopolymer Film

Several companies (notably Polaroid and DuPont) have developed a plastic film material that is especially suited for reproducing holograms. These photopolymers are extremely bright and durable, which makes them appropriate for a wide variety of commercial applications. The automated production process used to reproduce these holograms cuts costs considerably. An 8" x 10" matted photopolymer film hologram can retail for well under $100.00.

Image depth is a bit shallower and projection distance is a bit less than in silver-halide films. However, photopolymer holograms are a magnitude brighter and generally have a much wider viewing angle. Although this allows for more latitude when using a less than ideal light source, a single overhead light is still recommended for illuminating wall decor. The material works exceedingly well for items that will be illuminated by normal outdoor lighting, such as jewelry, bookmarks, and postcards. The plastic surface of the photopolymer films can be cleaned with a soft cloth.

Dichromated Gelatin (DeG)

Another photosensitive material used to make holograms is "dichromated gelatin." Unlike silver-halide glass plates or rolls o f film, this material is actually a chemical/gelatin mix that is coated onto a piece of glass. After exposure, a cover glass is securely attached to the base plate, creating a permanent seal - since moisture or excessive heat can dissolve the dichromate gel and the hologram will eventually disappear. Sometimes plastic is used instead of glass, lowering production costs.

Dichromate holograms are very bright and can be viewed under less-than-ideal lighting conditions, e.g., outdoors. Therefore, they are most frequently used as jewelry items and fashion accessories. The glass can be cut into any shape, but it is often cut into small discs. These hologram discs have traditionally been used in watch faces, broaches, key rings and belt buckles. For years, the $20.00 dichromate glass pendant has been a staple for retailers. An 8" x 10" framed dichromate plate usually retails for over $100.00, but they are seldom made anymore.

Embossed Foils

These holograms are reproduced by a process that is not optical and does not use costly photosensitive materials. In this case, the holographic image is transferred onto a mechanical stamping die, which is then used to emboss the image on rolls of very thin plastic films or metal foils. These holograms can be reproduced for cents per square inch, making it the least expensive way to copy an image. These foils are commonly hot stamped on paper or plastic, or backed with an adhesive layer. The material is very durable .

The embossing process creates very shallow images that can be viewed under poor illumination. They are usually silver-backed and reflect a rainbow of colors. Some of the bestselling holographic products are animated 3D embossed images which display fluid motion and realistic colors. These commonly retail for $15.00-$20.00, matted and packaged, ready to frame. Most of the embossed holograms on the market are called 2D/3D. They display multi-level graphics and are usually used as inexpensive stickers, or part of key chains, pins, and magnets.

Often a repeating prismatic pattern is embossed onto the plastic or foil, which creates a ever-changing rainbow effect. This colorful material is used in toys and optical novelties.

OPENING A SHOP-BASIC GUIDELINES

1. Choose the best location you can afford. Tourist destinations in "festival" type shopping districts located near dining and drinking establishments traditionally do very well. High-traffic locations ensure a steady stream of curious shoppers. Malls do very well around the holidays, but can be slow during the off-season. Do extensive market research.

2. Shop for a reliable supplier that can stock you in a timely manner. A good supplier will be able to replace the merchandise you have sold quickly, thereby reducing the number of units you need to backstock. A good supplier will also assist you with displays, inventory selection, and point-of-purchase materials.

3. Stock a wide selection of merchandise in various price ranges. Many hologram stores cover their overhead with the sales of inexpensive items in the $1.00-$10.00 range. Most retail sales average $5.00-$25.00. Wall art typically sells well at the $35.00 price point, as do $65.00 watches. Sales of more expensive pieces can be icing on the cake.

4. Plan your displays carefully. Due to lighting considerations and limited viewing angles, holograms need to be merchandised more carefully than many other products. Installing adjustable track lighting is more practical than using stationary fixtures. Halogen spot lights (50- 100 watts) work very well. Avoid fluorescent and recessed incandescent lights.

5. Create an entertaining atmosphere. No one needs to buy a hologram. A friendly, informative staff will boost sales dramatically.

Start-up Costs

The amount of capital you need depends on how large an operation you want to have. On the low end, you can stock a "cart" operation, as opposed to an actual shop, for as little as $5,000. Buying starting stock for a small shop (several hundred square feet) will cost several times that. A larger space of 800-1,000 square feet can easily hold $50,000 worth of goods. Plan on spending $10,000-$15,000 for high-quality lighting and displays for a store of this size.

Based on a double markup for inventory, $200,000 worth of annual gross sales should support an owner/manager making $30,000, a small staff (dividing another $30,000), rent/overhead in a high-traffic location ($25 ,000), and first year build-out ($15,000).

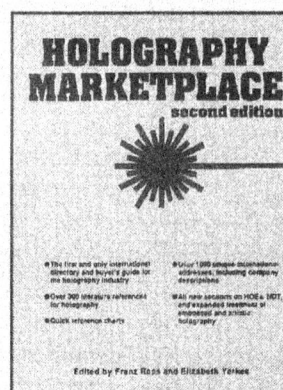

SECTION 2
New Technologies

Chapter 6 - Color Holography

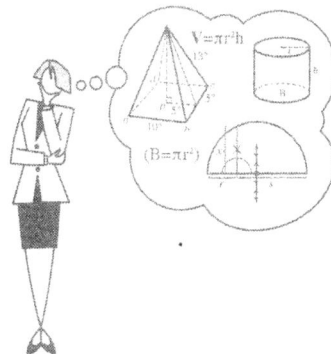

Chapter 7 - Emerging Technologies

DIGITAL FULL COLOR HOLOPRINTING

Update your mastering capabilities

Computergraphics and digitally manipulated images are common nowadays. In order to best realize your clients' ideas, you need a holographic setup which can handle this input.

Dutch Holographic Laboratory B.V. (DHL), known world wide as a supplier of high quality holograms and innovative holographic products, can supply you with a powerful holographic production tool:
the Digital Full Color Holoprinter®.

The DHL Digital Full Color Holoprinter® will greatly expand and update your mastering capabilities. This fully digital system combines several holographic techniques to produce the highest quality photoresist masters - ready for embossing. Using this system, holograms can be created from a variety of sources:
A sequential series of video pictures (MVGH), video footage (VAH), computer generated graphics (CGH), graphic art for 2D/3D - or even from combinations of these inputs. The new generation Pentium processors makes it possible to use the DHoX holoprinter programs formerly restricted to expensive SGI workstations, with more affordable, user-friendly personal computers.

This substantially decreases system cost; Estimated break even point is 15 - 20 mastering jobs. Digital images can be fabricated using your own 3D CAD software or DHL Traces®.
Imagine the possibilities!

Advantages
● Mastering: CGH / MVGH / VAH / 2D/3D
● Full Color rainbow photoresist masters
● Quality undistinctable from slides / film
● One table for both mastering / transferring
● Flexibility because of the modular design
● Customer installable

PC-based DHoX Holostudio suite consists of:
● Powerful holograb software
● Framesequencer
● Digital image processing
● Holo-editing
● Holoprint software
● Holotransfer

Interested?
Fax the coupon for more information or visit us at HTTP://WWW.iae.nl/users/walter/dh/tabs

DUTCH HOLOGRAPHIC LABORATORY B.V.

Dutch Holographic Laboratory B.V.
Kanaaldijk Noord 61
5642 JA Eindhoven
The Netherlands
e-mail: walter@iae.nl
Tel. +31-40-2817250
Fax +31-40-2814865
HTTP://WWW.iae.nl/users/walter/dh/tabs

Send me information on state of the art Holoprinters:

☐ Turn key system.
☐ Separate components, customer installable.
☐ samples DFCH

Company: ....
Name : ....
Address : ....
City : ....Zip...............
Country : ....
Telephone: ....
Fax: ....

6

Color Holography

This chapter introduces a major achievement in display technology-the high-volume production of full-color, fully dimensional reflection holograms reproduced on photopolymer film-and provides an overview of this emerging segment of the holography industry. Multi-color photopolymer hologram production is also discussed. The chapter concludes with an article describing the latest developments in using holography to color fabrics.

COLOR PHOTOPOLYMER HOLOGRAMS ARE HERE!

Holography MarketPlace

It's been approximately 50 years since Gabor first published the basic principles of holography. Since that time, holography has become a viable technology that has been successfully adapted to many commercial imaging applications, most notably in the packaging, security, decorative, and display-related industries. However, due to a lack of suitable recording materials and production equipment, high-volume commercial users have been unable to exploit what is probably the most visually impressive property of holographs-the ability to record and recreate images that appear truly lifelike. Images with true dimension and true color.

Over the years, analysts familiar with the evolution of commercial holographic technologies have predicted that if easily viewable, fully dimensional, realistically colored images could ever be mass-produced in a cost-effective manner, holography could revolutionize the imaging business-especially the multibillion-dollar graphics and display industries. Judging by the increasing use of current holographic technologies (in spite of their existing limitations), the commercial potential for even better holographic images is certainly enormous. People seem eager for holographic technology to catch up with what science-fiction has portrayed-holographic images that appear lifelike!

Imagine all the places a realistic-looking, full-color, 3D image could be used in place of a flat-color picture or photograph. ID cards, point-of-purchase displays, illustrations in textbooks, and portraiture are just some of the more obvious applications that come to mind. The demand for accurately colored 3D hard-copy from desktop computers would probably be very significant. A new palette of available colors might even reinvigorate the ailing holographic art scene.

And since not all holographic images have to be representational, new "holographically-colored" materials could have a significant impact on the packaging, decorative, and security industries. These market segments are already using substantial amounts of holographic materials; improved versions will surely be integrated in their products. Less obvious coloring applications with enormous commercial potential, such as holographic automotive finishes, are currently being explored, too.

PREVIOUS LIMITATIONS

Although many manufacturers of embossed holograms have already succeeded in mass-producing affordable, full- color images, their product inherently suffers from unnatural color shifting and lack of image depth. Other manufacturers have succeeded in producing large quantities of easily viewable film holograms with reasonable image depth, but the resulting images are only single color. Still other manufacturers have produced very limited quantities of fully dimensional, accurately colored holographic images on dichromate and silver-halide emulsions, but these fragile, hand-processed materials are relatively expensive and very impractical to produce in large runs. Most industry observers have long awaited the introduction of a recording material specially formulated for deep-image color holography that also lends itself to high-volume manufacturing techniques.

A Solution From DuPont

DuPont has developed a specially formulated photopolymer material that seems to offer hologram manufacturers a solution. It is a panchromatic film, well-suited for automated, high-speed exposure and processing. Preliminary tests conducted by DuPont in 1997 proved that a wide range of colors can be recorded and replicated accurately on this material (see the color picture "Earth/Spiral Rainbow" in Holography MarketPlace 7th Ed.).

Versions of this material have been tested for some time' however, due to market concerns, engineering problems: and the expense involved in setting up a full-color hologram production facility, DuPont's panchromatic recording material has hardly left its R&D laboratory. (Polaroid, formerly a major manufacturer of photopolymer holograms, I had reportedly also developed its own color hologram production techniques, but for some reason, never offered these services to its clients.)

In an attempt to offer customers a wider variety of colors to incorporate in their display holograms while the panchromatic distribution channels are being established by DuPont, alternative methods have been developed to produce color holograms using DuPont's monochromatic materials along with their color tuning films. In 1992, two researchers at MIT, Paul Hubel and Michael KIug, demonstrated a technique by which true-color holograms were produced in their lab using swelling agents and multiple layers of photopolymer. Although this approach has not been adopted for commercial production, at least one photopolymer production facility has used multiple layers of DuPont's material to create multicolor images.

Dai Nippon Printing Co. Succeeds

To date, only one company has demonstrated the willingness and the ability to combine all the components (panchromatic film, multiple lasers, specially designed replication equipment) needed to mass-produce bright, high quality display holograms that are fully dimensional and realistically colored-Dai Nippon Printing Co., Ltd. (DNP) of Japan. One or two companies have tried, one or two have even developed prototype systems, and one or two more are planning to start production in the near future, but only DNP is currently taking orders and delivering products to customers.

DNP has been developing holographic technologies for many years. In the 1970s and 1980s, the company worked With embossed holograms and volume type holograms, including silver-halide and dichromated gelatin materials. In the 1990s, research and development mostly shifted to volume type holograms using DuPont photopolymer materials, including its panchromatic ones.

It is a great honor for Ross Books to feature one of these magnificent new DNP holograms on the cover of the Holography MarketPlace 8th Ed. It is a first for the holography industry and the publishing industry as well. We are proud to commemorate the first 50 years of holography With such a hologram and sincerely believe that it represents the future of the field. Following is an article from DNP that describes its newly developed process.

PRODUCTS & SERVICES

Security Hot Stamp Foils

Security hot stamp foils are designed to protect credit cards, transaction cards, and paper-based documents. Security foils manufactured by ABH can be found on Visa, MasterCard, Europay, Discover, and Diner's Club credit cards, as well as a wide variety of paper-based products such as currency, passports, traveler's cheques, and gift certificates.

Security Labels

Tamper-Apparent security labels are an excellent anti-counterfeiting measure. These labels are pressure sensitive and adhere to all forms of product packaging. ABH offers a full range of security features in its labels which make them extremely difficult to duplicate, and also allow the consumer to easily verify a product's authenticity. Current users include pharmaceutical companies, audio/video companies, software publishers, and merchandisers.

Holographic Cards & Hang Tags

Full-face holograms are used on prepaid telephone cards, authentication hang-tags, and trading cards. Our holographic foil can be applied to both PVC and paperboard. ABH is one of the only companies in the world to offer this product, which provides not only a high level of security, but also offers an incredible visual impact.

ID Laminates

ABH produces high security laminates designed to protect passports, ID cards, and driver licenses against fraud. Our patented demetallized holography can be incorporated in either pressure-sensitive or heat sealed laminates, providing the most advanced protection against fraud available today.

Hologard™

A complete security program combining proprietary technology with diligent corporate and compliance investigation. Customized for the protection of your products and corporate standards in the global market.

Machine Readable Features

ABH incorporates proprietary technology to easily track and verify optically variable devices, combining both invisible and visible features for field authentication of customer's products.

The world leader in secure imaging systems

ABH

AMERICAN BANK NOTE HOLOGRAPHICS, INC.

399 Executive Blvd. Elmsford, NY 10523 • 1-800-966-2264
www.abnh.com

Visual and Collection

Full color and Full 3D

- Premium
- Fancy Goods
- Sports Equipment
- Promotional Display
- Book Decoration

ASTRO BOY

Brand and Security

High security

- Brand Identification
- Security Label
- Membership Card/Badge

DNP

Contact to :

Full Color Hologram Label

DNP offers Full Color Hologram Label, the new type of label product, which contains three-dimensional holographic images in full color. As well as its attractive appearance, Full Color Hologram Label provides high security performance owing to high degree of technical skill required for the manufacturing.

Visual

The holographic image reproduced on the label appears true to life, just like the original object, with its vivid color and 3D effect.

+

Security

DNP has developed the original method to reproduce colorful three-dimensional images on the new base material. Counterfeits of security products are prevented effectively by applying them .

See the DNP hologram on the cover !!

World's first commercially viable full color 3D hologram using photopolymer material.

The article inside of this book explains more about it.

DNP Produces "True Image" Holograms Using DuPont's Panchromatic Material

by Tsuyoshi Houa and Masachika Watanabe (DNP)

In 1998, Dai Nippon Printing Co., Ltd. (DNP) put the world's first mass-produced, full-color, photopolymer holograms on the Japanese market under the name "True Image" holograms. Our company intends to promote these state-of-the-art holographic products and manufacturing services worldwide for a variety of applications including: product enhancement; packaging; promotional displays; book decoration; and premiums. We are receiving many inquiries about our new product every day.

As of this time (October 1998) we have already produced 3,000 "Pocket Monster" True Image holograms for premium card cases; 12,000 True Image holograms of the "Astro Boy" character for commemorative cards and sales promotions; and 5,000 True Image holograms for trading cards. Plus, we produced the "Parrots in the Jungle" color hologram that appears on the cover of this book. (Editor s note-See the color pages in this book for photographs of full-color photopolymer holograms.)

In addition, since these unique True Image color holograms are produced only on our premises using proprietary technology, they are also effective as authentication devices for use as brand identification, security labels, and membership cards.

DNP plans to implement a comprehensive sales and marketing campaign and continue its R&D efforts in order to keep its dominant position in the world as a manufacturer of these unique holograms. In order to accomplish this feat, our laboratory conducted research in three main areas:

1. The design and production of appropriate "master" holograms;

2. The design and manufacturing of automated high-speed replication machinery;

3. The development of an environmentally stable finished hologram product that incorporated a pressure-sensitive, adhesive backing.

After many trials we succeeded in producing a high quality product that could be integrated easily into a wide variety of commercial applications. We developed reliable mastering techniques, high precision replicating equipment, and an affordable, easily utilized finished product. We may or may not license or franchise the technology in the future.

Here is some additional information related to our work:

Holographic Recording Materials

Compared to other recording materials, DuPont photopolymer film is the logical choice for automated high-speed production of reflection holograms. This is partly due to the fact that the DuPont photopolymer films are processed using only UV light and heat, which makes them much easier to work with than materials that require "wet" processing. Large quantities of holograms with stable characteristics can be produced on a continuous roll using "in-line" opto-mechanical systems.

DuPont has been researching and developing panchromatic photopolymer holographic recording materials for some time. We would not have been able to produce our True Image holograms if DuPont had not developed these particular formulations. Over the past decade DNP has continually evaluated DuPont's holographic materials and provided their scientists with valuable feedback. We expect to see further refinements as our work progresses.

Recently, DuPont produced a panchromatic photopolymer (HRF800X001-15) with a high sensitivity and high "delta n"* value. A high "delta n" results in greater brightness and a wider viewing angle, making this particular film especially attractive for commercial imaging applications. We are hoping to improve the delta n value even more.

This gives our customer two options when using photopolymer. One is using the full-color film when accurate color reproduction is needed. The other is using the monochromatic OmniDex 706 (along with the Color Tuning Film) when maximum brightness is required (as the single color holograms can be "tuned" for an eye-catching gold or orange playback).

delta n is a measurement which expresses the difference between the lower refractive index and higher refractive index part of interference fringes. We divide that value by I two to get delta n. For example, if the higher refractive\l index = 1.50 and the lower refractive index = 1.40, then delta n = 0.05. In general, silver-halide emulsions have high sensitivities and a lower delta n. Dichromated gelatin have low sensitivities and a higher delta n. The newest photopolymers have a reasonably high sensitivity and a reasonably high delta n value.

Lasers and Optics

Both Coherent and Spectra-Physics provide lasers that we have used in our mastering and replication systems. We have about ten large frame lasers and combine them for mastering and replicating. We use argon-ion for blue; argon-ion, Diode-Pumped Solid State (DPSS) and Dye for green; and krypton for red.

The newest generations of lasers are quite stable-which considerably reduces the time we spend adjusting them. The availability of reliable DPSS 532 nm lasers was especially encouraging. Improvements of optics such as precision mirrors and lens also contributed to color hologram development.

THE MANUFACTURING PROCESS

Currently, all hologram production is done "in-house" on custom-built equipment. Here is a brief description of the DNP manufacturing process:

1. Design: We meet with the customers to ascertain their needs and determine how we can best meet their criteria;

2. Model making: If three-dimensional models are required, we make them and color them correctly. If two-dimensional (plane designs) are required, we generate the appropriate masking films from the artwork provided;

3. Mastering: Mastering techniques for recording color images have been refined for our process;

4. Replication: Photopolymer materials are exposed on our specially engineered machinery;

5. Processing: The photopolymer holograms are UV-cured and heat processed to produce bright, stable images. We have a DuPont laminator and oven;

6. Post-processing: Laminating, cutting, and other converting processes are performed in accordance with the client's specifications. We can provide transparent or black type backing and many suitable adhesives, depending on the requirement.

PRODUCTION CAPACITY

Our production capacity is increasing remarkably. We are currently able to produce several rolls of holograms every day. A roll is 1 foot wide and 500 feet long.

We have produced thousands of identical high quality holograms up to 5.5 inches by 8.6 inches. (The "Parrots in the Jungle" hologram on the cover of this book was originally 14 cm in length and 22 cm in width, before it was cut in half.) Our waste is very low and quite acceptable by commercial standards.

PRICE

The size and quantity of holograms ordered determine the price. We anticipate that we will be able to sell a hologram measuring 2 inches by 2 inches for about 70 yen (which is a half dollar at the recent currency rate.)

CONCLUSION

We strongly believe that potential customers all over the world are awaiting a commercially viable method to mass-produce holograms with deep, true-color, full-parallax images. Because we aimed at a goal that nobody had previously accomplished, we often encountered skeptical questions such as, "Do you really think anyone will buy them?" and, "What will people buy them for?"

The fact that other photopolymer hologram manufacturers (Krystal Holographics International, Inc., Polaroid Corporation) have already sold millions of single-color photo-

polymer holograms for display applications has certainly convinced our research group that a market exists and has encouraged us to proceed with our work. We have always been confident that if we could produce full-color holograms, we could sell them. Our coworkers within the company have had similar faith and have worked very hard to continually improve our holographic products. This positive attitude among all the members of our team resulted in the success of our project.

ACKNOWLEDGEMENTS

We would like to thank the management of DNP for supporting our research and development efforts in spite of the many obstacles that we faced, and thank all of our co-workers who helped us overcome these obstacles. We would like to thank DuPont, especially Dr. Weber and Dr. Stevenson, for working with us to develop photopolymer materials for our use. We would also like to thank Mr. Alan Rhody at Ross Books for giving us the opportunity to feature DNP's color holograms on the cover of Holography Marketplace's 8th Ed. to commemorate "the first 50 years of holography".

For further information, contact:
Tsuyoshi Hotta Central Research Institute
Dai Nippon Printing Co., Ltd.
Phone: 81-471-34-1213
Fax: 81-471-33-2540
Email: Hotta-t@mail.dnp.co.jp

Important Photopolymer Hologram Mass-Replication Technologies Patented in USA by Dai Nippon Printing

DNP has been researching various methods of mass-producing photopolymer holograms for some time. High-volume commercial reproduction necessitates fast machinery that is capable of making quality holograms consistently. In order to precisely replicate bright, stable holograms, we developed several unique production methods, such as incorporating a cushioning layer during the contact copying process and direct lamination of the material.

Here are some further details of the replication technologies we have patented in the United States that address these issues:

Patent USP5504593
Continuous and uniform hologram recording method and a uniformly recorded hologram.

Major Benefits: No master holograms are required to produce mirror holograms. Very suitable for producing long wavelength reconstruction holograms such as infra-red reflecting holograms.

A hologram recording method wherein a transparent member (for introducing a light beam into a recording material) is designed so that it and the recording film can be uniformly brought into close contact with each other through an index matching liquid without damaging the recording film, *and it is possible to carry out uniform and continuous exposure in a stable fashion without failure of recording or occurrence of unnecessary interference fringes.*

Also disclosed is a recorded hologram. A transparent member is placed at one surface of a recording film . The surface of the transparent member brought into close contact with the recording film is convexly curved only in the direction of the feed of the recording film.

In addition, the space between the contact surface and the recording film is filled with an index matching liquid, so that the recording film is brought into close contact with the transparent member through the index matching liquid.

In this state, a light beam is made incident on a surface of the transparent member other than the contact surface, so that the incident light beam reaching the recording film (through the contact surface) and the light beam reflected from the interfacial boundary (between the reverse surface of the recording film and the air) interfere with each other in the recording film, thereby forming and recording interference fringes in the recording film.

Patent USP5453338
Hologram and method of duplicating; and apparatus for producing the same.

Major Benefits: By using cushioning layer, defects such as dust and particles are reduced markedly.

A duplicating photosensitive material film is placed in close contact with a thin flexible ND glass or an ND glass coated with a cushioning layer through an optical contacting liquid containing a surface active agent. A spacer is interposed between a hologram original plate and the duplication photosensitive material, and the space de- fined by the spacer is filled with an optical contacting liquid (the spacer thereby regulating the thickness of the optical contacting liquid layer).

In addition, one cushioning layer is provided on the inner side of an AR-coated ND glass (or on the upper side of the photosensitive material film) and another cushioning layer is provided on the side of an original plate-protecting the glass surface of the original plate, which is closer to the optical contacting liquid (or on the lower side of the photosensitive-material film). With this arrangement, even if dust enters, it can be held effectively inside the cushioning layers. *Thus, it is possible to prevent undesirable flow of the optical contacting liquid and lifting of the film due to dust and hence perform duplication effectively without any hindrance.*

Patent USP5798850
Method of, and apparatus for, duplicating a hologram; and a duplicate hologram.

Major Benefits: This is a high precision replication technique without the use of index matching fluid. (Index matching fluid tends to move, leave residue, and cause lower diffraction efficiency.) Also, by removing the polyester layer (with higher bi-refringence) before laminating onto the masters, undesired fringe patterns are not recorded.

This hologram duplicating method and apparatus brings a photosensitive-material film continuously and smoothly in close contact with the surface of a hologram original plate, and the film is effectively delaminated after a duplicating process.

The apparatus includes a film supply part for supplying a photosensitive-material film, a film laminating part for successively laminating the supplied film on a hologram original plate with the film being squeezed by a roller from the upper side, a film delaminating part for successively delaminating the film from the original plate from one end thereof with the film being pressed by the roller, and a film take-up part for taking up the delaminated film.

The film can be laminated on the original plate without trapping air bubbles, and the film can be delaminated from the original plate without causing peel unevenness and thus undesired lines. Also disclosed is a hologram produced by the aforementioned duplicating method.

Information provided by:
Dai Nippon Printing Co., Ltd.

GENERAL INFORMATION

RECORDING COLOR HOLOGRAMS ON DUPONT'S PHOTOPOLYMER

by Hans Bjelkhagen

The color holography photopolymer materials from E.I. du Pont de Nemours & Co. are very interesting and easy to use for color holography. In particular, this material is the main candidate for mass production of color holograms. Although less sensitive than the ultrahigh-resolution silver-halide emulsion, it has the special advantages of easy handling and dry processing (only UV-curing and baking). It is a very suitable recording material for mass-replication by contact-copying color holograms and color HOEs recorded on silver-halide masters.

The DuPont color photopolymer material has a coated film layer thickness of about 20 Ilm. The photopolymer film is generally coated in a 12.5" width on a 14" wide Mylar®polyester base which is .002" thick. The film is protected with a .00092" thick Mylar®polyester cover sheet. Such experimental materials have been manufactured by the company, e.g., HRF-700X07l-20. However, these materials am not yet introduced on the market. The latest experimental version, HRF-800XOOI-15, has a film thickness of 15 μm and is improved over the earlier film.

The recording of a DuPont color hologram is rather simple. The film has to be laminated on a piece of clean glass or attached to a glass plate using an index-matching liquid. (Some practice with laminating techniques is necessary in order to not get any air bubbles or dust particles trapped in between the film and the glass plate.) To obtain the right color balance, the RGB ratio depends on the particular material, but typically it is about 4:1:2. It is difficult to obtain high red-sensitivity of the photopolymer film. Simultaneous exposure is the best recording technique for photopolymer materials.

Holograms can be recorded manually, but in order to produce large quantities of holograms, a special machine is required. For hologram replication the scanning technique can provide the highest production rate. In this case, three scanning laser lines are needed, which can be adjusted in such a way that all three simultaneously can scan the film. The color photopolymer material needs an overall exposure of about 10 mJ/cm2.

After the exposure is finished, the film has to be exposed to strong white or UV light. DuPont recommends about 100 mJ/cm2 exposure at 350-380 nm. After that, the hologram is put in an oven at a temperature of 120 C for two hours in order to increase the brightness of the image. The process is simple and very suitable for machine processing using, e.g., a baking scroll oven.

RGB HOLOGRAPHICS

A NEW PRODUCTION FACILITY

Paul Kruzel is the founder of RGB Holographics, a new U.S. hologram production facility based in New Hampshire. The company intends to specialize in making full-color display holograms on silver-halide glass plates and on DuPont's panchromatic photopolymer. Kruzel, a relative newcomer to the industry, purchased a complete mastering lab and production line from the now defunct Holos Corporation. The acquisition provides him with all the necessary lasers, mastering materials, replicators, and finishing equipment to produce full-color holograms and to mass-produce them on DuPont's photopolymer materials. Kruzel recently hired holographer Jeffrey Murray.

For further information, contact:
RGB Holographies
10 Shell Meetinghouse Rd.,
Canterbury, NH 03224
Phone: 603-783-9238
Fax: 603-783-9255
Email: laserhene@aol.com

H SPACE ™

EXPERIENCED TEAM ASSEMBLED

Andrew Laczynski and his partners Gus Angus and Larry Lieberman have recently completed a year-long in- depth study of the commercial marketplace for display holograms and plan to set up a full-color production facility based around DuPont's panchromatic photopolymer material during the 1999 year. The company is named "H Space". Laczynski has been involved in the art and science of display holography since 1979 and has instigated many notable projects that often incorporated the newest technology available. Lieberman has been a professional holographer for over 20 years and is one of the few people who has worked in DuPont's lab to produce masters for full color replication on their photopolymer. The team plans to develop new and exciting products that feature full-color, fully dimensional hologram images.

For further information, see: Web site: www.hspace.com

ZEBRA IMAGING, INC.

NEW TECHNIQUE CAN PRODUCE FULL-COLOR, LARGE FORMAT HOLOGRAMS ON PHOTOPOLYMER

Zebra Imaging, Inc., is currently offering a hologram production service for full-color, full-parallax images which are unlimited in size and have an extremely wide angle of view (approximately ninety degrees). Large format displays and a desktop device that outputs 3D hardcopy are being developed for medical, engineering, advertising, product design, and security applications.

The company was formed in 1996 by three graduates from the Media Lab at the Massachusetts Institute of Technology: Michael Klug, Mark Holzbach, and Alejandro Ferdman. While at MIT, the founders played important roles in the digital holography research program directed by professor Stephen Benton, which has formed the basis of Zebra's own research and development efforts. Klug's experience with color holography has also proven valuable.

The first public showing of a Zebra Imaging hologram was at an automotive trade show (SEMA) in November, 1998. The company supplied a hologram measuring 4 feet by 2 feet that switched between interior and exterior views of a NASCAR racer as a viewer walked past it.

LARGEST HOLOGRAM TO DATE
In 1998, the company created what it claims is the largest hologram in the world. A 6-foot-by-18-foot, full-color, full-parallax prototype hologram was produced for a Fortune 10 corporation. Intended for an indoor display where lighting and viewing angle can be controlled, the wall-sized holographic image is quite impressive according to published reports from visitors to the company's production facility near Austin, Texas.

"The largest holograms ever made before this measured approximately 1 by 2 meters," says Ferdman, "and holographers have always been limited by the size of film available to them. We're not limited in that sense because we've found a way of 'tiling' image components together so you can make a holographic image of any size."

FULL-PARALLAX, SCALABLE IMAGES
The company's patent-pending system utilizes a computer-modeled image or digitized scene as input. It employs the simultaneous output of three lasers (red, green, and blue) within a proprietary computer peripheral device to expose an array of variably-sized holographic elements, called "hogels," which together comprise a single, larger hologram. These hogels are analogous to pixels, and can range in size from hundreds of microns to millimeters, depending on the size of the hologram. Holograms are typically recorded onto a sheet of DuPont panchromatic photopolymer film measuring 60 by 60 cm. After dry processing, the film is usually laminated to a more rigid substrate for ease of handling.

According to the company, these approximately 2-foot-square pieces of film (tiles) can be used "as is" or they can be continuously pieced together to create larger displays. The tiles can also be used as "masters" for mass replication.

Unlike other automated hologram production devices that sequentially expose a series of narrow "slit" holograms to produce conventional holographic stereograms-which only display horizontal parallax-Zebra's system creates images that are fully dimensional. Their system is also unique in that holograms are recorded in a single step; separate mastering and transferring procedures are not employed. "Our method is much more practical for large format work and results in an image with an exceptionally wide viewing angle-a highly desirable property for display applications such as signage," explains Holzbach.

The company's expertise lies in appropriately digitizing a client' artwork (graphics or video) and precisely control- ling the opto-mechanical systems to produce a "seamless- looking" finished 3D image. The group is now focusing on increasing the speed of its output device.

Central to the creation of Zebra's holograms is the panchromatic photopolymer film from DuPont Holographic Materials. Holzbach enthuses, "It's a real revolutionary material, capable of accurate color renditions (including human skin tones) and very high contrast,"

COMPUTER PERIPHIAL FOR 3D OUTPUT
The company also intends to incorporate DuPont's panchromatic films and its newly developed full-parallax technology in its desktop "holographic-hard-copy computer peripheral device" project. The company believes that its system offers a new visual medium for engineers, artists and designers. Potential commercial applications include advertising, marketing, industrial design and analysis, architecture, medical imaging, security applications, and portraiture.

Ferdman explains: "Our goal is to get it to be a peripheral (device off your computer, so somebody who's designing something in 3D just pushes a button and gets a hologram. That should be feasible in about two years. We want to put this in everybody's hands."

For further information, contact:
Zebra Imaging, Inc.
PO. Box 81247, Austin, Texas 78708
Phone: 512-251-5100
Fax: 512-251-5123
Email: ferdi@zebraimaging.com

KRYSTAL HOLOGRAPHICS, INC.

MULTICOLOR HOLOGRAMS AVAILABLE IN HIGH-VOLUME PRODUCTION RUNS

Krystal Holographics, Inc. is the primary high volume manufacturer of photopolymer holograms in the world. The company draws heavily on the marketing, design, and production expertise of company vice-presidents Doug Miller and Gerald Heidt who work at the company's U.S. production facility located in Logan, Utah. Both are highly experienced in the holography industry and were instrumental in the development and mass production of full color dichromate holograms before photopolymer film was available to the market.

Today, Krystal Holographics specializes in producing large numbers of reflection holograms on DuPont's photopolymer film for top corporate clients such as Nike and Speedo, primarily for product enhancement, security, and brand identity applications . Although most of the company's production to date has been display holograms, Krystal has begun manufacturing a greater number of Holographic Optical Elements (HOEs) and recently installed a "certified clean" production line to meet an increased demand for this product: Krystal reports that it is currently producing holograms "around the clock" to meet the needs of all its U.S. and international customers.

The company has now begun to offer multicolor (as opposed to natural-color) production services for photopolymer. Holography MarketPlace (HMP) interviewed international sales manager Dave Rayfield (DR) and Miller (DM) about this exciting new development. Rayfield indicated that the first major consumer product that incorporates this new technology will hit the shelves in 1999.

HMP: Can you briefly describe your multicolor production process?

DR/DM: We use a proprietary mastering process (which draws on our existing expertise with DCGs and photopolymer), to make high quality multi-color holograms using DuPont's Omnidex photopolymer film products. The holograms are reproduced on our own high speed replication machinery, thereby making high volume production of this new product both feasible and economical.

HMP: What colors and sizes are available?

DR/DM: We currently can produce finished two-color holograms measuring up to 4 inches by 4 inches in either green and blue or green and red, depending on whether DuPont's color tuning film is employed. We are also experimenting with other ways to expand the choice of colors and produce interesting effects.

HMP: What should the client bring to you for multicolor production? Computer files? 3D models? How can the client be best prepared?

DR/DM: We consult with each client individually to determine the feasibility of producing a multicolor hologram. We have an in-house design and art department which can produce the appropriate 2D or 3D artwork.

HMP: What are the associated costs? How do prices compare to monochromatic production?

DR/DM: Costs are approximately 30 percent higher than for single color production, though we strongly encourage potential clients to meet with us, as each job is quoted individually. Unit costs are based on volume and size, just as in single-color work.

HMP: What is the typical turn around time for a multi- color production run?

DR/DM: With the added design, art work, mastering, and proofing time that is required for multicolor production, figure six to eight weeks for a typical production run. We quote the same time frame for single color work, though rush jobs have been successfully completed.

HMP: Are you planning to offer full-color hologram production anytime soon?

DR/DM: Yes, as soon as the marketplace requests it. Our high-speed replication machines can be adapted for panchromatic photopolymer reproduction.

For further information, contact:
Krystal Holographies, Inc.
555 West 57th Street
New York, N.Y. 10019
Phone: 212-261-0400
Fax: 2 12-262 -0414
Email: hologram@krystaiteeh.com

COLORING FABRICS WITH HOLOGRAPHY

COLORING FABRICS BY USING HOLOGRAPHIC TECHNOLOGIES

Fabrics are traditionally colored by employing one of three methods: immersing yams or fabrics in a liquid bath using chemical dyes; printing the fabrics with chemical dyes or pigments; or adding pigment particles to thermoplastic-polymer-melt before spinning the mix into fibers. These technologies are relatively simple and broadly applicable. They can be used on fabrics made from synthetic and natural fibers. The resulting colors can be seen easily under all types of illumination. Most importantly, these coloring techniques do not affect the basic qualities of the fabric, such as its durability, its permeability, and its flexibility. However, these three traditional technologies can only produce static colors and two-dimensional images. In addition, they release a large amount of toxic chemicals into the environment during the manufacturing process.

Another approach to fabric coloring is to exploit the physical phenomenon of light diffraction from microscopic grooves that are embossed into an appropriate substrate. This opto-mechanical technology is already well-established for imparting prismatic colors and dimensional images onto plastic films and metal foils, but it has not yet been fully adapted for direct application to fabrics. If fabric could be colored using holographic technology, many existing products could be enhanced and many new commercial applications would certainly arise. A whole new palette of colors could be introduced to the fashion and decorative industries and new light-reflective materials could be produced. As important, the use of hazardous materials by fabric manufacturers would be significantly reduced.

To those familiar with embossed holography, it might seem illogical to even attempt to press complex holographic interference patterns, or even simple diffraction gratings, into a flexible material that is intended to be made into a garment. Wouldn't stretching, folding, washing, and just wearing such a material alter or destroy the embossed microscopic interference pattern, and thus the desired effect?

The results of recent experiments by a former DuPont researcher, Munzer Makansi, suggest otherwise. Makansi recently demonstrated the feasibility of embossing "holographic colors" and/or dimensional images directly on fabrics, while maintaining the fabric's original attributes. He has formed a company, Fiber Engineering, Inc., and is currently exploring ways to mass-produce these newly processed materials.

CREATING GLITTER AND SOLID COLORS

Makansi's preliminary research with embossing diffraction gratings directly into a piece of fabric revealed that if the embossed grooves are aligned only within each filament, or within filament segments but without the additional filament-to-filament alignment on the fabric's surface, it is possible to produce colors, but they are limited to randomly distributed glitter points and/or short line segments. This results from the destructive cancellation of the diffraction waves in a random manner corresponding to the random distribution of distances between filaments in fabric.

In order to produce a rainbow of "solid" holographic colors and/or dimensional images on fabric, Makansi found that the diffraction grooves must be aligned, not only within each filament, but also filament-to-filament. The patterns of fine grooves on individual filaments of the fabric must continue to neighboring filaments, in phase, such that the distance between the last groove on one filament and the first groove on the neighboring element is an integer multiple of the groove width. This alignment is necessary in order for the diffraction waves to constructively reinforce one another and produce coherent and full images. Makansi's experience with synthetic fabrics at DuPont led him to believe that certain fabrics could be successfully embossed to achieve the desired outcome.

THREE MANUFACTURING METHODS

Currently there are three different methodologies for imparting diffraction colors to fabrics. They are:

1) attaching holographic materials onto fabric,

2) preparing fabric from diffraction filaments, and

3) directly embossing holograms on fabrics.

Film/Fabric Laminates

(This relatively popular methodology includes three sub-groups: (a) the continuous lamination of a pre-embossed, holographic film on a fabric's entire surface, (b) the spot lamination of pre-embossed holographic film or foil labels on to a fabric surface, and (c) coating a fabric with film, then embossing it with holographic interference patterns. (In the aforementioned cases, lamination refers to the process of joining two materials, which frequently involves the use of an adhesive, pressure, and/or heat.)

The continuously laminated fabric came onto the market about eight years ago. It consists of nonwoven, felt-like fabric, coated with a very thin layer of plastic film, pre-embossed with holographic images.

The spot laminated fabric is produced by attaching a piece of film or foil (pre-embossed with a hologram image) to a larger area of fabric, usually a garment, by one of several available means. One example is U.S. Patent 4,956,040 (issued in 1990), which discloses a method in which a pre-cut foil (which has a hologram pattern embossed on its surface) is laminated between a clear polyester coating and an adhesive scrim backing to form a unit which is then adhered to a woven fabric.

Textile Graphics

One U.S. company, Textile Graphics, Inc., specializes in this spot lamination technique and even offers customers a selection of stock image transfers, called Holography Presses On. The company has developed a way to cut holographic foils into intricate shapes (often logos) which can be easily ironed on to a T-shirt or other garments. The company's president, Jan Bussard, has gradually perfected a method that prevents the transfers from peeling apart after they are attached to a garment-the holographic "labels" (each a 5 mm multi layered sandwich made up of a polyester carrier, vacuum-deposited aluminum bits and an appropriate adhesive) are simultaneously cut out and sealed along the edges by a laser.

In a series of previous attempts, Bussard had tried sealing the edges of the holographic material by imbedding pre-cut holograms in plastisol printing inks during the transfer process or by employing a hot cutting tool when carving customized holographic designs. Another successful transfer method, "screen printing," involves chopping the holographic material into tiny particles that are then mixed with a clear plastisol adhesive base and squeezed through a coarse mesh on to a fabric before the mix cures. A rainbow of iridescent colors results.

Bussard claiming that her patented sealing technologies prevent delamination and degradation after exposure to normal wear and tear, outdoor weather conditions, laundry, and dry cleaning. In a paper presented at an industry conference in 1995, Bussard writes, "The ability to cut diffraction grating materials in simple and intricate patterns, produce the product in large volume at a lower cost than ever before, and apply the product in a variety of forms without difficulty...has greatly increased the potential use of holographic materials for promotional purposes on fabrics." In addition to decorative and promotional applications, the technologies she developed are currently being used to securely attach labels onto fabrics for authentication purposes.

Holotex

An example of the third technology is the "Holotex" fabric (produced by the German mill Scheibler and sold in the U.S. by European Textile Trading Co.), which came onto the market about three years ago. In the Holotex process, urethane polymer (latex) is first coated on poplin fabric backing. The coated surface is then embossed with holographic images. In 1995, the retail price of"Holotex" fabric was $10.00/sq. yd.

Advantages

One main advantage of these film lamination processes is that they can be used on any kind of fabric, including cotton and wool. Another is availability- several foil and film manufacturers can supply pre-embossed holographic

materials with suitable topcoats and backings. A third is ease of use-low-volume production may require nothing more than an iron and ironing board, making household decorative applications possible.

Limitations

However, the holographic film/fabric laminates do have several qualities that make them undesirable for some fashion and decorative applications: the hologram images are rather bright and metallic looking (an exception is the growing use of photopolymer materials); the fabric has excessive stiffness; it rustles upon handling; and is impermeable to air or moisture. Due to these features, the use of this fabric is mainly limited to specialty garments, ID tags, and emblems. In addition, the processes used to make these laminated fabrics employ chemical solvents and adhesive compounds. These can pose public health hazards and pollute the environment.

Fabrics Prepared From Diffraction Filaments

This category is represented by two different methods. The first is described in international patent application W089101063 and Japanese patent application SH063-309639. In this first method, a hologram pattern is embossed on a thin plastic film. The film is then slit into long ribbons of 1-20 mm in width. The ribbons are then formed into yams from which fabrics are made. Unfortunately, the dimensions of these slit-ribbon yams are usually too large for them to be used for apparel fabrics, though some suppliers have developed holographic threads and decorative gimps that fit in household sewing machines.

The second method is described in the Japanese patent SH062-170510. In this method, two molten polymers are co-spun simultaneously into bicomponent filaments. One of the polymer components is then dissolved out from the filament surfaces, leaving fine grooves on the surface of the remaining filaments. These grooves are parallel to the filament axes. The filaments are then converted to fabrics.

Unfortunately, both methods place diffraction grooves on the fibers before their conversion to fabric, and thus cannot produce full and coherent rainbow and/or holographic images. As a result, the diffraction colors produced in these fabrics are limited to random distribution of glitter points and/or glitter line segments.

Direct Embossing on Fabric

Two very recent patent applications by Makansi (U.S. Serial number 08/777, 821 and its international counterpart pctlUS97/23961) disclose methods for the direct embossing of rainbow and hologram images on fabrics. Although the technology of embossing diffraction patterns directly on fabrics does not yet exist for mass production, a joint research and development program is underway between Makansi and the Georgia Institute of Technology's School of Textile and Fiber Engineering. The team hopes to move this development from the demonstration stage towards full-scale commercialization by developing fabric-embossing machines capable of handling factory production.

The initial demonstrations were carried out using thermal embossing at high pressures, with a (narrow web) metal stamper in a platen press heated to a temperature between the softening point and the melting point of the polymer from which the fabric was made. The temperature and pressure were optimized to produce images with good colors, while retaining the desirable fabric attributes of breathability, flexibility and tactile aesthetics.

The thermal embossing of diffraction groove patterns on fabric at pressures sufficiently high to transfer the fringe patterns from the stamper to the fabric surface insures that the conditions of alignment required for full image transfer are met. In fact, according to Makansi, embossing is the only method now available which can insure that these alignment conditions are satisfied on fabric surface filaments.

Unique Attributes

Makansi reports that this direct embossing process was successfully used to produce rainbow and hologram images on fabrics of different constructions including woven, knit, or spun-bonded materials made from different organic, synthetic polymers including polyester, nylon, polypropylene, and polyethylene of different colors. The images produced become visible on exposure to direct light. They resist handling, washing, and dyeing and remain visible even when the fabric is wet, as in swimwear. The direct embossing technology can be used by itself, or in combination with traditional printing or dyeing.

These holographic fabrics have unique and attractive visual aesthetics, and are especially suited for fashion and decorative applications. They have relatively subdued colors with brightness equal to, or slightly higher than, dyes or prints. They are dynamic and create subtle dimensional effects. (Editor-Early production samples which were examined by the staff of Holography MarketPlace felt "normal" and displayed prismatic colors which shimmered under normal lighting.) When the fabrics are made into a garment, the width of the rainbow lines of color and the spacing between them change in accordance with body contours. The spectral patterns which are produced move graciously in response to the motion of the human body (relative to the incident light) in a manner unique to the person wearing the garment, thus creating his or her own personalized appearance.

Existing Limitations

The limitations of diffraction colored fabrics relative to color-dyed and color-printed fabrics are:

(1) the colors are visible at certain specific angles relative to the incident light which should be incandescent light or sunlight; and

2) the technology is (so far) limited to synthetic fabrics made from melt spun fibers like polyester, nylon, polypropylene, and polyethylene.

For further information, contact:

Fiber Engineering, Inc.	Textile Graphics/HPO
106 Stratford Way 201	Fruitport Road
Signal Mountain, TN 37377-2521	P.O. Box 193
Phone: 423-886-3783	Spring Lake, MI 49456-0193
Fax: 423-886-7865	Phone: 616-842-5626
Email: rnrnakansi@aol.com	Fax: 616-842-5653
and	

THE POTENTIAL MARKET FOR HOLOGRAPHIC FABRICS

by Munzer Makansi (Fiber Engineering, Inc.)

The direct-embossing, coloring technology is expected to find applications in a variety of end uses, including: swimwear, skiwear, sportswear, patio and beach furniture, sun umbrellas, fashion accessories (scarves, handbags), tents, clothing, luggage, upholstery, draperies, and decorative articles.

EXISTING MARKETS

The world market volume of embossed hologram sheets is estimated at $300 million, of which $55 million is laminated on fabrics. The cost of sheet embossing is estimated at $1.60/sq. yd. for a 300,000 sq. yd. volume, decreasing to - $0.50/sq.yd. for much larger volumes. This cost is comparable to fabric printing and dyeing costs. The estimated selling prices of laminated fabrics, with holograms embossed on them, is in the range of $5-$15/sq. yd. depending on the base fabric material, the artistic value of the hologram images, and the stage of novelty for this product.

The dyed and printed fabrics are discussed here because they represent the market penetration opportunity for these newly colored, holographic fabrics. The world production of nylon and polyester apparel and industrial textile fabrics is estimated at 16 billion pounds, equivalent to - 85 billion sq. yd. Assuming that 50% of this yardage is color dyed or color-printed, then the total volume of color dyed and color printed fabrics of nylon and polyester calculates to - 43 billion sq. yd. At an average coloring cost of - $OAO/sq. yd., excluding scouring and heat-setting costs, the total dyeing and print coloring business becomes - $17 billion. The estimated retail price for these colored fabrics is in the range of $1.20-$6.00/sq. yd., depending on fiber type, fabric construction, the complexities of the dyeing and printing processes used, and the specific end uses . For example, the novelty, the specialty, the high fashion and the sport clothing industry markets command relatively high selling prices.

HOLOGRAPHICALLY COLORED FABRICS

Successful commercialization of direct hologram embossing technology on fabrics would result in a manufacturing cost which is comparable to the cost of state-of-the-art hologram embossing on films. The higher initial processing cost of diffraction colored fabrics (due to new investment and start up expenses) will be offset by savings in lamination material, labor costs, and by the anticipated larger production volume of the diffraction-colored fabric market. At a 0.5% penetration of the state-of-the-art markets of $17 billion, and a price mark up of 200% from the estimated cost of $0.60/sq. yd., the fifth-year market of the product produced by this newly developed coloring technology calculates to $387 million/yr. (The mark up of 200% can be justified on the basis of these competitive advantages: novelty, styling, image, aesthetics, patent protection, and environmentally clean technology. These advantages should also speed up market penetration.)

VIRTUAL IMAGE®

a Printpack company

Virtual Image's revolutionary embossing technology has greatly impacted the holographic industry. Using our proprietary process, Virtual Image has succeeded in substantially reducing the high cost associated with holography, making holograms affordable for everyday packaging applications. Our many capabilities and advantages include:

- ◆ High quality, wide web embossing up to 48"

- ◆ Flexibility and decreased cost of an oriented polypropylene base sheet

- ◆ Heat sealable, PVdC, or acrylic coated films

- ◆ Expertise in both flexible and rigid applications

- ◆ Expanded in-house metallizing capacity

- ◆ Financial stability of a $1 billion flexible packaging parent company

- ◆ Process approved for FDA direct food contact

Virtual Image's products have won several packaging awards, including two WorldStar Awards, three AIMCAL awards, and a "Food and Beverage Marketing Magazine" Award.

Contact us today to insure that your next package design is a winner!

Sales Department
1050 Northfield Court
Suite 300
Roswell, Georgia 30076

(770) 751-0704
(770) 751-0806 fax

Holographic Package Enhancement

Visit our web site at www.virtimage.com

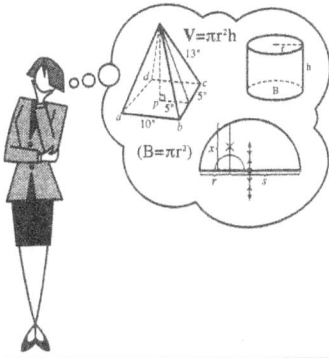

7

Emerging Technologies

This chapter describes two emerging holographic technologies that will probably have significant impact in the field of dimensional imaging display equipment. One has seemingly obvious and tremendous commercial potential as a consumer product and could be available in the near future. The other points the way toward imaging systems that may be common in 21st century image display devices.

EMERGING TECHNOLOGIES

Holography MarketPlace

There are a number of emerging technologies in the field of holography that have not yet made it to market but they have developed to a point that they can be discussed. Generally, these technologies seek to improve existing recording and production methods. Occasionally, a novel idea is introduced and pursued.

Since many of the research and development programs that produce working models are quite costly, the groundbreaking work is typically performed by inspired and dedicated holographers employed by large corporations and government institutions. Unfortunately, much of this innovative and fascinating research remains unseen and unappreciated by the general public until the marketing and administrative departments do their work. Often the finished product does not reveal the development process that led to it or its inner workings.

However, in order to best educate our readers about future trends in our industry, we have selected two emerging technologies that we believe will have significant commercial potential in the near and distant future. Both seek to record and display 3D holographic images faster and better than ever before. Both could only have been described in science fiction stories just a decade or so ago. Today, they exist as working models in the laboratory. We eagerly await their commercial introduction in some modified form.

SONY'S RESEARCH PROGRAM

Sony research and development program used a combination of new and existing ingredients to produce a product that is still in development, but seems to have tremendous commercial potential-a relatively small, self-contained holographic printer that outputs an easily viewable, 3D image in just a few minutes. It is notable that Sony engineers adapted and improved existing automated recording devices by employing transport systems similar to those used in portable cassette players and VCRs. They also integrated the newest recording materials and lasers in their machine.

The device can be hooked up to a digital camera in order to produce holographic portraits of people or it can conceivably be attached to a desktop computer to produce 3D hardcopy of digitized information. It can be used indoors or outside. It is intended to be easy to operate: film cartridges can be loaded in and after a few buttons are pushed, a finished hologram comes rolling out.

THE U.S. ARMY'S RESEARCH PROGRAM

The U.S. Army has demonstrated a method for "instantaneously" recording and projecting holographic images using a strontium barium niobate (SBN) crystal. The 3D image that results is full-color, realistic, and can be viewed over a large field-of-view. Multiple 3D images have been stored and read out of the crystal via wavelength multiplexing. The holograms were also noted to persist without any external processing or fixing mechanisms.

SONY CORPORATION

INSTANT HOLOGRAPHIC PORTRAIT PRINTING SYSTEM

information contributed by Sony Corporation

Watch out world consumer electronics giant Sony is into holography! Last year, company engineers previewed an instant holographic portrait printing system at the annual "Photonics West" conference in San Jose, California. It makes affordable, 3D pictures of people in less than three minutes at the push of a button. The final version of the system is intended to be relatively portable (desktop-sized) and as easy to operate as an office copier.

At the 1999 conference, Sony researchers announced an important refinement to the prototype system: its internal optical geometry had been altered to allow "edge-lit" holograms to be recorded. This development is very significant as it would allow the finished holograms to be inserted into a relatively slender, self-illuminated viewing device. Creating an instant hologram machine that makes easily viewable holograms is a long-awaited achievement.

Holograms produced by the Sony system were exhibited and the results were quite impressive. The sample hologram portraits were bright and realistic looking. The images had depth when viewed directly, and when viewed off-center you could see either side of the person's face, just as you would in real life (i.e., the images displayed horizontal parallax).

Due to their size and color (they were small and green-tone), the hologram portraits were reminiscent of the wallet-size black-and-white pictures teenage couples take in those instant-photo booths--except these were 3D! One hologram even showed a person sneezing as the viewer walked past it, as the system is capable of recording a few seconds of subject movement, too. Smaller, faster machines that will produce larger, true-color holograms are in development.

HOW THE SYSTEM OPERATES

The system works like this: a person is positioned several feet in front of a stationary nine-foot-long rail. A specially engineered digital camera is mounted at one end of the track. The camera moves quickly, but incrementally, along the rail from right to left. Over two hundred different views of the subject are recorded during a single pass. This sequence of digital images is fed into Sony's holo-printer, which is about the size of a desktop copy machine. The holo-printer contains digital processing boards, a laser, and various optical and mechanical components. Minutes later, out pops the hologram on a small sheet of plastic. The plastic is actually a piece of holographic recording material made by DuPont called "photopolymer film" that was automatically exposed inside the machine. Unlike conventional photographic film, the DuPont material does not require chemical developing. Instead, a simple processing procedure (exposure to UV light and baking) is all that is needed to finish the hologram. When properly illuminated, the hologram displays a 3D image of the subject.

Other holographic portrait systems exist; however, most are built around powerful pulsed lasers that require highly trained laser technicians to maintain and professional holographers to operate them. Furthermore, each hologram must be exposed, developed, and copied by hand, which makes the process relatively expensive and time-consuming. In addition, these fragile pulsed lasers are not very portable; subjects usually have to travel to a custom-built holography studio in order to have their portrait shot.

THE HOLOGRAM PRINTER'S EVOLUTION

Sony's new system might seem revolutionary. It is in fact, based on thirty-year-old ideas developed by holographers who could only imagine digital cameras, miniature solid state lasers, LCD screens, and instant hologram films. These pioneers did however, realize that it was possible to make easily viewable, moving 3D images by combining cinematography with holographic recording technology using stereographic techniques. The holograms that they produced were called "holographic stereograms" or "integral holograms".

The term "stereographic" refers to the method by which 3D images are created when our eyes are presented with a pair of related images. The term "integral" refers to the fact that multiple exposures are used to construct one composite hologram. Early versions of these "holographic stereogram printers" or "holographic optical printers" employed specially shot film footage, finicky Helium-Neon lasers, modified movie projectors, homemade liquid-filled lenses and expensive silver-halide emulsions. Special displays were needed to view the results. Subjects were occasionally filmed outside the studio, but hologram production was restricted to the laboratory. The manufacturing process was tedious and time-consuming.

Over the years, holographers (most notably those at the Massachusetts Institute of Technology) developed variations of the early systems. Video cameras and digitized graphics replaced movie cameras and photographic film. Ever smaller, more efficient lasers were utilized. LCD devices replaced movie projectors. Holographic optical elements were substituted for bulky lenses. New recording materials were developed. Displays were improved.

In a paper prepared for the 1998 SPIE conference, Sony's research team reviews some technological developments that led to their device:

Camera Controller

Digital I/O

DOS/V PC

Digital I/O

Printer Mechanism

Parallax image shooting

CCD Camera

NTSC parallax images

Data Converter

RAM

Frame Buffer

VGA reconstructed images

LCD Driver

LCD

Image reconstruction

Image Capturing Camera

Digital I/O

Memory Controller

LUT

Hologram Printer

Hologram exposure

Image Processing Board

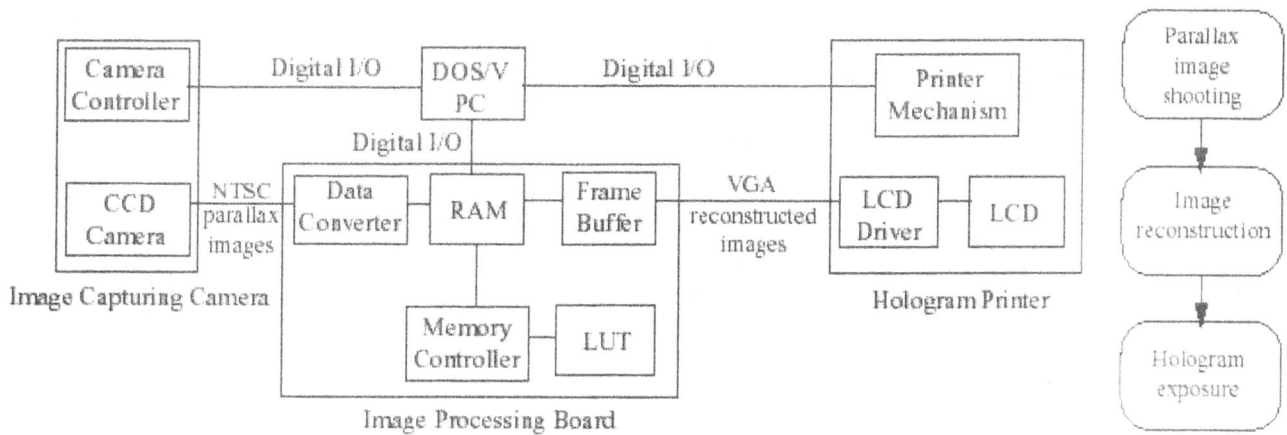

Fig.1. Sony's holographic portrait printing system

In 1969, DeBierto proposed an angular multiplex holographic stereogram recording method. In this case, cinematic photographs [movie footage-ed.] captured from equally spaced positions along a horizontal line are projected onto a ground-glass screen and a holographic plate is masked to record each elemental hologram image in a narrow strip. In 1969, King, Noll, and Berry reported a two-step* white-light viewable holographic stereogram as a transfer image from a spatially multiplexed master hologram of sequential exposures. As far as one-step** recording is concerned, Cross developed the Multiplex™ Hologram using a one-step recording method for cylindrical holographic stereograms in 1976. To get rid of various distortions of the Multiplex™ Hologram in reconstruction, several significant studies were made in which the perspective images for recording were remapped prior to exposures. This "predistortion" image processing was applied to alcove holograms and flat holograms and finally led to "Ultragram," which is not heavily dependent on the recording condition. As far as the practical issues are concerned, E. van Nulandreported about an "Office Holoprinter," and Klug developed a compact prototype Ultragram printer.

The new printer we have developed is basically a modification of the Ultragram printer toward improving the image quality and prototyping for on-site desktop printing operation under non-laboratory ambient conditions. Recently, holographic stereogram research seems to be moving toward a full-parallax approach instead of horizontal parallax only (HPO) such as Klug's work and Yamaguchi's work. However, there may be still more difficulties in prototyping the full-parallax printer than HPO as far as the data size and printing speed are concerned. We have been concentrating on improving the image quality and prototyping a practical, fully automatic hologram printer using one-step HPO technology.

There are two different methods of making white-light viewable holographic stereograms. One is a two-step process, in which the holographic stereogram is optically duplicated from an original "master" hologram. All the holographic stereograms to be mass-produced have so far been made using the multistep process.

*** The other is a one-step process, in which the holographic stereogram is directly recorded on the photosensitive material. The one-step process has the benefit of providing economical, small-print runs.*

Holographic Printers/or Embossed Holograms

Several holography companies have integrated a few of these improvements into machines that are specifically geared for making embossed holograms. However, holoprinters designed for embossed holography output a photoresist plate that needs to be factory processed before a finished hologram is produced. This additional manufacturing step increases turnaround time and expense. Also, embossed holograms typically display very limited image depth, they shift color and they require a mirrored backing. People often associate them with kids' stickers or product labels.

SONY INITIALLY MAKES REFLECTION HOLOGRAMS

Sony is the first company to refine each component used in the holographic stereogram production process and integrate them all into a fully automated, commercially viable system that produces high quality "reflection" holograms. (The term "reflection" means that an image appears when light bounces off the front of the hologram.) These reflection holograms reproduce highly detailed, deep images that appear more "solid" and are easier to see than their embossed counterparts. They are often framed and hung on the wall for artistic or commercial applications.

Why would Sony even want to produce hologram portraits? The system is actually a working model of a new type of high-speed printer capable of producing 3D hardcopy from digital data. Apparently, the company believes holographic stereograms are the best way to display digitized 3D images due to the amount of visual information that can be presented and the speed at which they can be produced. Since the hard-copy output is a hologram, no special glasses or viewing apparatus (aside from a light bulb) is needed for viewing 3D imagery. And since the machine makes reflection holograms using photopolymer film (which allows for quick, automated processing), bright, easily viewable holographic images can be produced in a matter of minutes.

According to Sony's literature, a fully automated one-step holographic printer system that can operate in ambient conditions would be valuable "for use in such areas as computer-aided design, medical imaging, and information visualization." In fact, for many of these applications, the printer would not have to be hooked to a camera as long as the digital input is formatted correctly. That hologram portraits are being produced is somewhat incidental, although many security and packaging applications for the existing system come immediately to mind.

DETAILS ABOUT THE SONY SYSTEM

As stated, Sony started with an existing technology, updated it, and solved some major optical and mechanical problems along the way. Here are some details about the prototype system:

Digital Input The input device is a digital video camera that moves along a stationary rail (2700 mm in length). In order to minimize optical distortions that occur using this recording geometry, the 2/3" CCD unit inside the camera changes position in relation to the lens (f= 20 mm) as the camera travels along the track. During one scan (which takes 7.5 seconds) the camera sends 225 frames to an image processor.

Processing A high-speed image processing device (with a 256 Mbyte memory) receives the 225 frames, remaps them, and sends 295 reconstructed images to a liquid crystal light valve located inside the printer. The processed images, condensed into vertical strips by a cylindrical lens and displayed on the LCD, is the "object" recorded onto the hologram.

Exposure A 400 mW doubled Yag laser (532 om green) sequentially exposes each elemental image onto a piece of DuPont HRF700X071-20 photopolymer recording material. Since conventional step-and-repeat film transport

systems (see figure 2a) induce vibration and are relatively slow, Sony engineers developed a high-speed film transport device based around pinch rollers and a torsion spring coil that holds the film stationary during each exposure (see figure 2b). Each exposure measures 114 second. Each complete exposure cycle, including film transport and vibration damping, lasts 112 second. Recording the 295 elemental holograms takes 147 seconds.

Printer In order to make daylight or portable use practical, the printer assembly is light tight and the optical setup is mounted on an anti-vibration breadboard table. The laser operates off household current and needs no external cooling. For ease of operation, a roll of film can be preloaded into a cartridge.

After exposure, the film is cured by UV light to stop the photo reaction and baked to make the refractive index modulation greater (i.e., make the hologram brighter). Due to the fact that ventilation is required during the baking process and fans would create unwanted vibration, the Sony engineers physically separated the exposure unit from the processing unit in their prototype machine. They are hopeful that the next generation of recording materials will not require ventilation. In addition, the size of the unit is being reduced, though it is still intended to be a desktop unit.

Holograms The machine outputs a horizontal parallax image measuring 78 mm by 59 mm (approximately 3 inches by 2 112 inches). Based on existing price structures, the material costs would be less than a dollar per hologram.

SYSTEM SPECS

Reconstruction Color–532 om Green
Printing Area–59 mm (horizontal) x 78 mm (vertical)
Viewing Angle–57 degrees (h) x 40 degrees (v)
Number of Elemental Holograms–295
Size, Elemental Holograms-O.2 mm (Pitch) x 78 mm (v)
Resolution, Elemental Holograms-480 x 640 pixels (v)
Printing Speed-3 minutes
Printer Size (version 1)--1100 mm x 700 mm x 300 mm

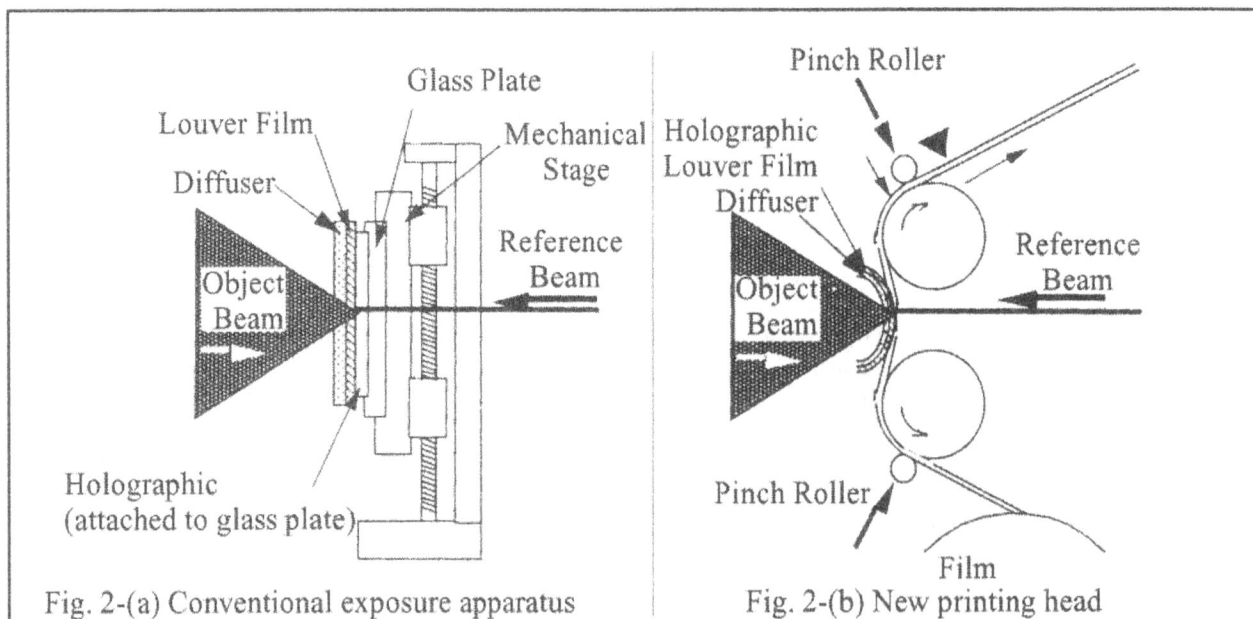

Fig. 2-(a) Conventional exposure apparatus   Fig. 2-(b) New printing head

Sony's Newest Development: an Instant Printer That Produces Edge-Lit Transmission Holograms

In 1998, Sony engineers introduced a one-step holographic stereogram printer that produced reflection holograms. This was a very admirable achievement; however, as with most reflection holograms, it was difficult to appreciate the quality of the holographic images output from their newly developed machine unless the holograms were illuminated correctly. Since typical lighting conditions outside the laboratory are not optimal, viewers frequently only saw a blurry image. This was not only frustrating to the engineers, it hindered the potential wide-spread commercialization of the technology.

In order to overcome this problem, the Sony team has developed a version of their printer that makes edge-lit* transmission holograms that are designed around the use of a "built-in" illumination source. (Transmission holograms are lit from behind rather than from the front; an optical configuration that works better if the light source and hologram are to be self-contained.) This new method eliminates the necessity of employing the unwieldy lighting arms or unaesthetic front-mounted lighting fixtures that are used to illuminate reflection holograms.

In addition, a transmission edge-lit hologram has an advantage compared to a reflection edge-lit hologram because the resulting holographic image is reconstructed around the front surface of the glass block upon which the finished hologram would be mounted, rather than behind the glass. In other words, if a viewer were looking at a finished edge-lit transmission hologram mounted in a lighting device, the three-dimensional image would appear on the surface of the "screen," rather than under it-a much more impressive visual effect.

According to the Sony researchers, it is usually very difficult to make a transmission edge-lit hologram using a one-step holographic stereogram printer because of the interference between the reference beam and the object beam's converging lens when using the existing optical geometry. They solved this problem by employing a conjugate beam reconstruction. The desired beam path was obtained by making a small change to the reflection type hologram printer head. By using this reconfigured optical set up, they succeeded in making an instant edge-lit hologram printer.

*Editor-The term "edge-lit" refers to the position of the light source which illuminates the finished hologram. An edge-lit hologram is mounted onto a sheet of glass that has a light source attached to its outer edge. The beam of light used to reconstruct the image travels "inside" the piece of glass, eliminating the need for an external light.

Presumably, the hologram which is output from the Sony hologram printer would be inserted into a slender, hand-held display device that would contain an inexpensive light bulb.

Future Developments

According to Sony's engineers, they would like to develop a one-step holographic printer that would produce full-color images with full-parallax. This would involve employing a three-laser array and panchromatic film for true-color reproduction. Making full-parallax holograms without effecting printing speed is another obstacle to overcome. Eventually, the size of the printer will also be reduced.

For further information, contact:
Akira Shirakura, Nobuhiro Kihara or Shigeyuki Baba
Research Center, Sony Corporation
6-7-35 Kitashinagawa, Shinagawa-ku, Tokyo 141, Japan
Phone: 81-3-5448-2945
Fax: 81-3-5448-7907
Email: sirakura@devd.crl.sony.co.jp

PHOTO-REFRACTIVE CRYSTALS

U.S. Army Researches Real-Time, Full-Color, Projected Holographic Displays Using New Recording Material

information contributed by Christy Heid

The United States Army Research Laboratory has been conducting research to study the feasibility of using the newest holographic techniques along with photo-refractive crystals to project remote-controlled, life-size 3D images in "mid air." Presumably, these full-color, 3D projections would be employed on the battlefield as decoys in order to confuse enemy troops. According to Dr. Christy Heid, a scientist at the Army Research Laboratory (ARL) in Maryland, the research team she worked with made important progress in the development of new holographic recording materials, playback techniques, color reproduction and real-time imaging systems. Though the program has been postponed, the Army's research has already yielded information with significant potential to those developing new imaging processing systems for industrial and commercial applications.

Heid explains her team's motivations in her 1998 post-doctoral report, "Nonlinear Optical Studies for the Army: Three-Dimensional Color Holographic Display":

Present displays used by the military (...for such purposes as training and education...) tend to be two-dimensional and monochromatic in nature, which can give an incomplete picture or supply inadequate information. Therefore, three-dimensional, color images are needed to produce a realistic, fully informative display for the Army. We have chosen to produce an advanced display using holography since it offers the capacity of projecting a real, full-color, three-dimensional image in free space. Conventional holographic displays, such as those generated by computers using emulsion films, are usually time-consuming and cumbersome, requiring intermediate processing steps and are often virtual images. However, the use of photo-refractive crystals as a holographic storage medium eliminates these and other limiting factors.

The ARL Holographic Set -Up

The researchers employed a standard holographic recording geometry as the basis for their system. In brief, a coherent beam of light was split into a reference beam-which traveled directly to the recording material, and an object beam-which was used to illuminate small 3D objects (such as dice). As the reference beam and the light reflected from the object converged at the recording material, a hologram was formed. When this hologram was later illuminated by the reference beam alone, a 3D (full-parallax) image of the dice was formed where the physical object once rested. If the physical object were removed, a viewer looking through the crystal would see a realistic-looking image (in this case of the dice) floating in "mid air." (See figure 1).

Heid elaborates:

We were using an Argon ion and/or a Krypton ion laser, and simple lenses and beam splitters. We started with 1 W of power and we expanded the beam to reflect enough light off the object. The actual power used to record the hologram was much less. The object beam power (reflected off the object) and the reference beam power were about 0.1 to 10 mW If we used a lower power, we increased the recording time. In the future we hope to incorporate diode lasers in order to make the system less cumbersome.

In order to meet the project's unique requirements, the holographers introduced several major variations to the conventional recording/playback set up, the first being their choice of recording material.

New Recording Material

The Army's experimental system uses a specially manufactured photo-refractive crystal as a recording medium, unlike conventional holographic recording systems that capture the interference pattern on a glass plate coated with either silver-halide emulsion, photopolymer, photoresist or DCG. These unique photosensitive crystals (grown by Dr. R. Neurgaonkar at the Rockwell International Science Center) have the ability to record an interference pattern, store the pattern for some time, and erase themselves.

"Due to availability, we used strontium barium niobate doped with cerium for our studies and found it to work quite well," states Heid. "The crystals we used measure 20 by 20 by 1.3 mm. We put two crystals together in a mosaic to get a larger surface area to record the hologram and thereby increase the field of view of the hologram. The crystals are cut to a particular size so I'm not sure what the size limit is. Some people have claimed to grow a 1-to-2- inch boule from which the crystals are cut." According to Heid, the crystals currently cost a few thousand dollars each, but the price would be reduced drastically when produced in bulk.

Focus on the Future

Krystal Holographics International, Inc.

555 West 57th Street New York, N.Y. 10019 USA 212-261-0400
365 North 600 West Logan, UT 84321 USA 435-753-5775

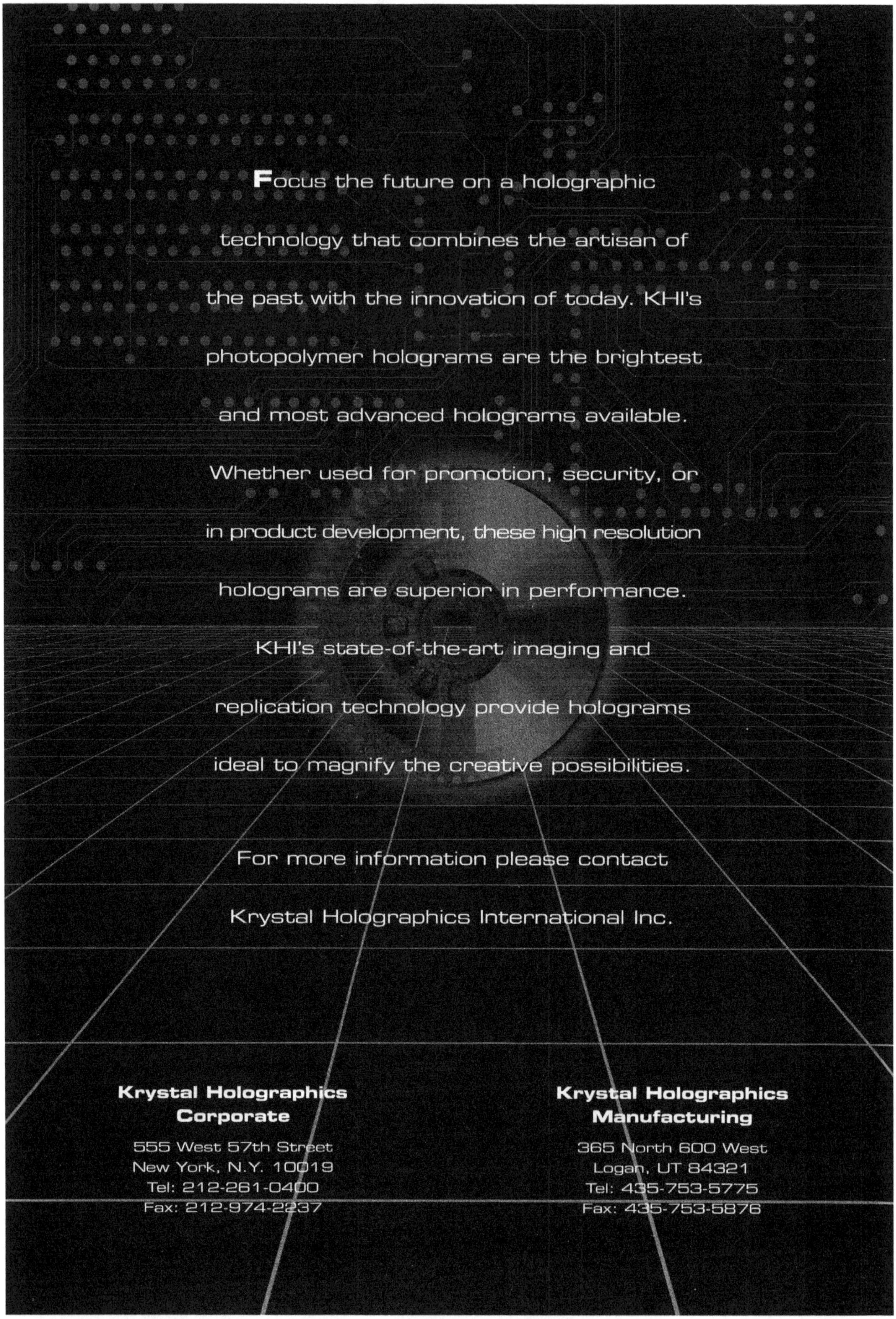

Focus the future on a holographic

technology that combines the artisan of

the past with the innovation of today. KHI's

photopolymer holograms are the brightest

and most advanced holograms available.

Whether used for promotion, security, or

in product development, these high resolution

holograms are superior in performance.

KHI's state-of-the-art imaging and

replication technology provide holograms

ideal to magnify the creative possibilities.

For more information please contact

Krystal Holographics International Inc.

**Krystal Holographics
Corporate**

555 West 57th Street
New York, N.Y. 10019
Tel: 212-261-0400
Fax: 212-974-2237

**Krystal Holographics
Manufacturing**

365 North 600 West
Logan, UT 84321
Tel: 435-753-5775
Fax: 435-753-5876

How Do the Crystals Work?

Row do the crystals work? Very simply, when the reference beam and object beam meet in the crystal, the incoming waves of light either amplify (constructive interference) or cancel each other out (destructive interference), creating a unique fringe pattern. Reid elaborates, "The laser light causes free charges [electrons] to move about, and they diffuse into darker regions and recombine. This upsets the local charge balance, creating a space charge field. This field in turn changes the crystal's refractive index." The sum of the changes creates a recording of the fringe patterns, i.e., a hologram.

The change takes several seconds to occur. When the light entering the crystal is removed (ending the "exposure"), the crystal gradually returns to its original state. Depending on the background illumination, temperature, and the crystal's composition, the fringe pattern can last from a fraction of a second to days. During its existence, the resulting diffraction grating acts like any transmission hologram. If it is immediately re-illuminated at an appropriate angle by the reference beam only, the object beam will be regenerated. The resulting wavefront creates a three-dimensional image.

In contrast, all of the other aforementioned recording materials require some sort of processing before the holographic image can be viewed. In practice, this entails removing them from the recording set-up, which takes time; and once processed, they cannot be exposed again.

Reid's thesis further explains that: "When using a photorefractive storage medium, holograms may be recorded and projected immediately without time-consuming processing and with greater storage capacity through various forms of multiplexing. In addition, the photo-refractive recording medium is sensitive to low level intensities and is reusable such that previously stored holograms may be erased and the crystal reused to store other unique holograms."

OPTIMIZING IMAGE PLAYBACK

The best way to playback any holographic recording is to illuminate it with the reference beam that made it. Ideally, the playback beam would have the exact characteristics of the original-the same polarity, phase, and angle. Since this is impractical in most instances, the Army team devised an elegant solution: build a "mirror" that would use the original beam to make an exact duplicate of itself for playback purposes.

This type of mirror is actually a special optical feedback system (called a "phase conjugate mirror"). By using such a device, the researchers could ensure they were getting the best possible output from the hologram. "Using a phase conjugate mirror saves us from making extremely precise and painstaking beam alignments and it generates a very uniform image," explains Reid.

COLOR REPRODUCTION

"To produce a realistic display, we want to reproduce the actual colors of the object such that the object and the hologram are indistinguishable," writes Reid.

In their quest to replicate colors found in the real world, the ARL team successfully recorded and played back the three primary imaging colors: red, green, and blue. A krypton laser was used to generate the red wavelength, an argon-ion laser was used for the blue and the green. Although the researchers have not attempted color mixing (white), they did record a red object and a blue object simultaneously. Reid explains that the" photo-refractive crystal we used has a broadband wavelength sensitivity. Color comparison of the laser illuminated object with the resulting holographic image has shown good agreement; therefore we have demonstrated 'true' color holography."

A "REAL TIME" 3D IMAGING SYSTEM

A very exciting aspect of the research was the ability of a single crystal to be both a recording and a playback device within a matter of seconds. A fully developed crystal-based holographic imaging system devised to generate 3D images in "real time" could have tremendous potential in both entertainment and education.

Currently, in order to show a sequence of images (i.e., to create movement) using conventional holographic techniques, one has to rapidly illuminate an entire sequence of prerecorded holograms. In theory, if each hologram depicts a "frame," a whole roll of material would have to be passed by a projector of some sort to create just a few moments of 3D animation. This probably would be costly and unwieldy.

However, in a system based on a photo-refractive crystal, the same crystal could be used repeatedly. If the record storage/decay cycle time was quick enough, a series of related holograms might be recorded and played back quite rapidly. For instance, if the object being recorded was placed on a rotating stage, the crystal would record one view (which could be quickly re-illuminated to create a holographic image), and then erase itself. The object would

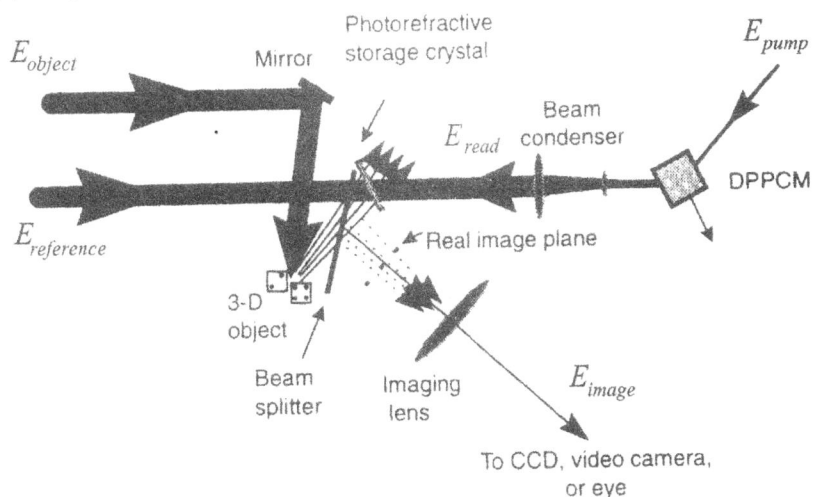

Fig. 1 An ARL Recording/Playback Set-Up

rotate and a new view would be recorded and played back. If the turntable could be synchronized with the crystal's response time, a whole sequence of related images could be recorded and "instantly" played back "in real time." A properly positioned viewer would see the object turning.

UNSOLVED PROBLEMS AND LIMITATIONS

Isn't this what the Army's after? A full-color, real-time 3D image? Yes, but the problem still remains: how do you get the image projected in "mid air" for everyone to see? Holograms by their very nature focus light in a very specific direction. If you are out of the viewing zone, you will see nothing. The hologram itself must be in the line of sight of the viewer. Additionally, the amount of image projection is limited by the coherence length of the laser that recorded the hologram and the size of the recording material.

The Army tried various ways to redirect the holographic image onto a scattering medium (the light has to bounce off something so we can see it!), but creating an appropriate volumetric "screen" proved difficult. Contained "boxes" of smoke or fog proved impractical to produce, as one might guess. A scattering liquid was tried but proved ineffective. Rotating helixes and photonic emission systems were proposed, but not tested. Clearly, there are important problems left to solve.

In Heid's opinion, "The major limitations of the present system are the accessibility of the photo-refractive crystals, laser power requirements, scaling to life size, and simplifying and miniaturizing the optical setup."

CONCLUSION

"Our method of recording and reading the hologram is consistent with conventional holography," Heid concludes, " but is simpler in that external processing is not required for immediate viewing of the holographic image. The unique aspect of our approach is that we are storing a 3D image in a photo-refractive crystal. Previous research has focused on storing 2D information. The resulting image is a 3D, realistic image, which is visible over a large field of view and limited only by the crystal size and object size."

How does Heid see this system being used in the future? "The 3D holography systems have many potential uses in the future, such as: quality control, comparative studies, pattern recognition, information storage, holographic assisted design (H.A.D.), and even holographic video recording is a possibility."

About the contributors:

Christy Heid received her Ph.D. from Lehigh University in physics. Her thesis work was in nonlinear optics. Her postdoc position at ARL working on the 3D holography project was arranged by the American Society for Engineering Education. Her fellow team members included Mr. Brian Ketchel, ARL; Dr. Gary Wood, ARL; Professor Greg Salamo, Univ. of Arkansas; and Dr. Richard Anderson, National Science Foundation.

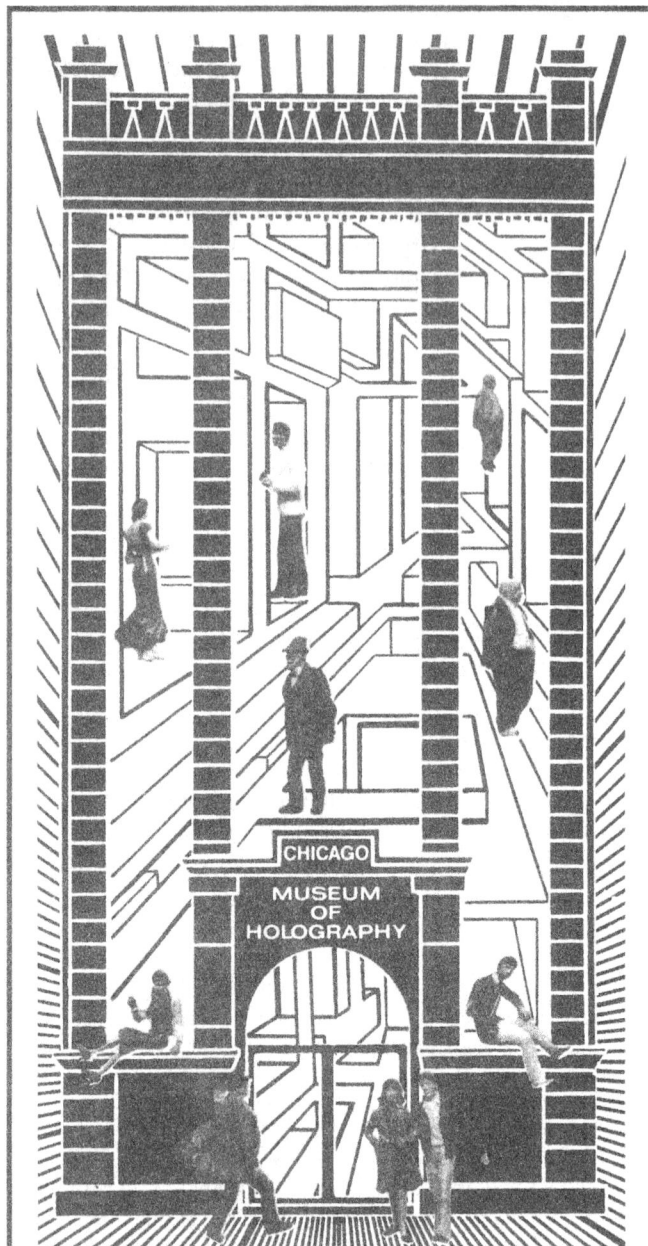

SECTION 3
Materials and Equipment

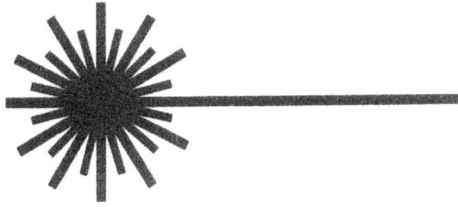

Chapter 8 - Laser Fundamentals

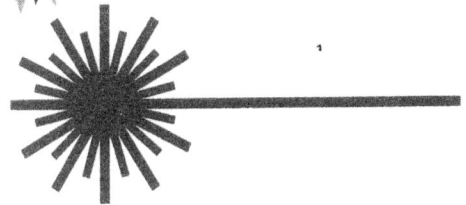

NEW

Chapter 9 - CW Lasers

NEW

Chapter 10 - Pulsed Lasers

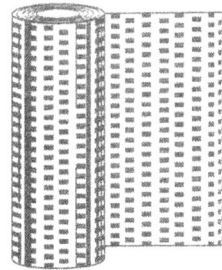

Chapter 11 - Production Equipment

Chapter 12 - Recording Materials

8

Laser Fundamentals

This chapter could be titled "from photons to etalons." It begins with a review of the fundamentals of electromagnetic radiation and then proceeds to a more detailed explanation of how the lasers that are most commonly used in holography (CW lasers) work.

LASER FUNDAMENTALS

To sufficiently understand the operation of lasers, their many advantages and their necessity in the production of holograms, one must first comprehend certain properties of our physical world.

The entire universe consists of only two things: **matter and energy.** Matter is all things that have physical substance; energy is the mover, or potential mover, of physical substance. Matter is the stuff we see, smell, and feel. It has mass and occupies space.

Energy, on the other hand, is more abstract. It is most often invisible, though sometimes not. Yet, it is everywhere. It lurks in the crevices of every molecule and sweeps the skies with its magnificence. A master of transformation, energy easily converts itself from one of its many forms to another, all without sacrifice.

Energy is the driving force behind all forms of motion: the motion of our car, the motion of planets, the motion of atoms. Nothing moves without it. Matter, without energy, is reduced to a dark, frozen lump of nothingness. In a dynamic universe, matter both possesses energy and is affected by it.

Energy not only changes form; it is also easily passed from one object to another. Interestingly enough, no matter how many times it transforms or transfers, the amount of energy involved in any given transaction never changes. The **law of conservation of energy**, one of the most important laws in the universe, dictates that energy is nev-er created or destroyed; it can only be transferred to another object or converted into a different form of energy.

Because the amount of energy in the universe remains fixed, phrases such as "energy shortage" and "depleted energy" are misnomers. You can not lose energy, nor can you be in short supply. The amount of energy in our environment is so great that it is beyond our comprehension. The discomforts in past decades from "energy shortages" were created only by our inability to either convert energy to a usable form or distribute usable energy to where it was needed.

Energy is measured in **joules**, in honor of the British scientist James Joule. One joule is roughly the amount of energy required to lift an apple from your kneecap over your head.

A glass of apple cider has 502,092 joules (equal to 120 calories-the **calorie** is another unit of energy often used when referring to the content of food) of food energy. A gallon of gasoline has over 200 million joules of energy.

The process of applying energy to matter is called **work** (also measured in joules). Work is the mechanism that transfers energy through a system. It is produced by applying a force on an object such that motion occurs over a distance. For example, when you pick up a book, you have performed work on the book. The heavier the book, the greater the force that is necessary to raise it-therefore, the more work done. The farther you raise the book, the greater the distance in which the force must be applied. Again, more work is accomplished.

Energy is formally defined as the ability to do work. It can be classified in two categories: stored energy and motion energy. Stored energy is more commonly called **potential energy**. When raising a book in the air, work must be performed on it. While suspended in air, however, the book has the ability to perform work in the opposite direction-courtesy of the earth's gravitational field. This "stored" energy may be released simply by dropping the book.

The potential energy an object possesses due to its position in a gravitational field is called, predictably enough, **gravitational potential energy** (or **GPE**). Any object raised above the earth's surface has gravitation potential energy. Water behind a dam has a significant amount of GPE that may be converted (as mandated by the law of conservation of energy) into electrical energy.

Other forms of potential energy are also commonplace. A coiled spring is the good example of **mechanical potential energy**. By performing work on the head of a Jack-in-the-Box, one can push it into the box. With the lid secured, the box has stored energy (mechanical potential energy): hence, the ability to do work. By unlatching the lid, the stored energy is released and work is performed in the reverse direction on our friend Jack.

The axiom that "opposites attract" is especially true for electric charges. Electric charges that are positive attract those that are negative, and vice versa. Equally, two electric charges of the same type (both positive or both negative) repel each other. The attractions and repulsion of electric charges are caused by invisible **electric fields** produced by each charge. An electric field permeates the territory around each charge, affecting all other charges that occupy its space. The larger the charge, the more influence its electric field exerts on its occupants. We encounter electric fields to varying degrees throughout a typical day. We witness them when we use our dryer, for it is electric fields that cause static cling.

To separate two opposite charges-or unite two like charges-requires work. Like the Jack-in-the-Box, when work is applied to bring like charges together (or to separate opposite charges), **electric potential energy** is created. Remove whatever constraint holds the stored energy (in the Jack-in-the-Box the constraint was the lid, in electric charges it is usually a nonconductive material like air or plastic) and work is performed in the opposite direction.

A familiar device utilizing electric potential energy is the battery. The stored energy in a battery can be released by placing a conductive path between the positive and negative terminals. The performance of battery is stated in terms of **voltage** (also called **electric potential**, abbreviated with a **V** and measured in **volts**). Voltage, an important element in laser operation, is the ratio of electric potential energy (EPE, not to be confused with electric potential) to the amount of charge (abbreviated in equations with a **q**, measured in **coulombs**). In equation form:

$$V = (EPE)/q$$

Another common source of voltage is the generator. Generators are machines that convert various types of energy into electric potential energy. Generators in dams convert gravitational potential energy from elevated water. The energy is transported to homes and businesses, readily available for those who wish to do a little work. Since generators produce higher voltages than batteries, they are used to supply power to all gas lasers and most others.

Matter is the greatest repository of energy. Atoms arranged together have binding electrical forces (called bonds) that act much like infinitesimal coiled springs. When bonds are broken, stored energy is released. This stored energy in molecules is called **chemical potential energy.** The gas we pump into our cars and the food nourishing our bodies are two common forms of chemical potential energy being utilized in our lives. Forces holding the nucleus of an atom together store an astounding amount of **nuclear potential energy,** as witnessed on July 16, 1945, when the Manhattan Project unveiled the atomic bomb.

Matter in motion possesses **kinetic energy**. An object will gain kinetic energy when work is done to it. An object will lose kinetic energy when work is done against it. The amount of kinetic energy an object gains or loses is exactly the same as the amount of work done on or against it.

For example, the engine of a train converts chemical potential energy into kinetic energy and performs work on the train. The train will move; it now has the amount of kinetic energy equal to the net gain of work done on it. If you turn off the engine, the train eventually stops, even if it is riding on a perfectly level set of tracks. This is because friction (between the wheels and the track; between the wheels and their axles; and between the air molecules and the front of the train) is performing work against the train.

When a train in motion hits a stationary object in its path, work is performed on the object. The object will move. Some of the train's kinetic energy is transferred to the stationary object. If one removes all sources working on and against a moving train on perfectly flat tracks--engine, friction and objects in its path-the kinetic energy of the train will never change. The train will continue to move forever at a constant velocity.

Other kinds of motion energy include heat, sound and electromagnetic radiation.

Heat occurs from the motion of molecules. The faster the molecules move, the more heat generated. A common source of heat is friction. In our previous example, friction performed work against the train. The kinetic energy of the train transformed itself to frictional heat in its wheels, axles and tracks (to a lesser degree, the air molecules). The train eventually stopped because its kinetic energy was entirely transformed into frictional heat. In most energy exchanges in nature, heat is part of the transaction.

Sound is another form of motion energy that occurs when a disturbance in a medium (commonly air) produces molecules to vibrate back and forth creating "sound waves." Each molecule receives the wave, vibrates back and forth and returns to its original position, but not before imposing a similar disturbance on its neighbor. The neighboring molecule repeats the same maneuver, as does each successor, thus creating a chain of disturbances that allows sound energy to propagate through the medium. Eventually, the sound waves hit an eardrum, causing it to vibrate. The vibrating eardrum creates signals to the brain that enable us to "hear" the sound energy.

One of the most important forms of motion energy is **electromagnetic radiation**. It exists everywhere throughout the universe and comes in many forms. Radio and television waves can be transmitted hundreds of miles through the air enabling music, images and conversation to magically appear in our homes. Microwaves, used in radar and modem cooking devices, ensure safe travel and a fast meal. Infrared radiation warms our skin and other vital regions of the universe. Visible light, the only form of electromagnetic radiation that we can see, enables our world to have definition and beauty. Ultraviolet radiation bums our skin and cures our plastics. X-rays help doctors diagnose problems in our bodies while gamma rays are found in many forms of radioactive decay.

Although we perceive and apply them differently, all forms of electromagnetic radiation are essentially the same phenomenon. Only the amount of energy per fundamental unit distinguishes a microwave from a beam of light.

The fundamental unit of electromagnetic radiation is called the **photon**, an infinitesimally small "packet" of energy. Radio waves have relatively low energy per photon. Microwaves have more energy per photon than radio waves but not as much as infrared radiation. A photon of visible light has more energy than a photon of infrared radiation but less than ultraviolet radiation. X-rays and gamma rays carry the most energy of all.

Electromagnetic radiation is created by accelerating or decelerating an electric charge. The greater the acceleration (or deceleration) of an electric charge, the more energy it will produce. It would take a much greater deceleration of an electric charge to create an X-ray photon than a microwave photon. Electric charges that are stationary, or those moving at a constant velocity, do not create electromagnetic radiation.

An electron, the most fundamental unit of negative charge, is the most common vehicle for creating electromagnetic radiation. For example, a radio station produces radio waves by accelerating electrons up and down a transmission antenna in a process called **oscillation**. An antenna is limited in its ability to rapidly accelerate and decelerate electrons, however. This is why antennas do not create visible light. Electron activity in atoms is the most prolific manufacturer of visible light. As explained later, this activity will be the basis from which lasers are created.

PROPERTIES OF ELECTROMAGNETIC WAVES

Water waves make a good model for the study of electromagnetic waves because they are commonplace, exhibit comparable properties, and move slowly enough to be carefully observed. There are a few profound differences between the two (for example, water waves must propagate *in a medium*-while electromagnetic waves need not), but not enough to impugn our comparison.

A wave is created by a disturbance in a medium (for electromagnetic waves, a disturbance may be created in empty space). In a swimming pool, a swimmer resting in the shallow end of the pool slaps his hand on the water. The disturbance creates a wave that moves from the shallow end to the deep end (for this example and all fictitious pools in this section, allow the edges to absorb all waves that hit it, thus eliminating the effect of reflected waves). The wave has a "crest" (high point) and a "trough" (low point).

A closer look at the wave would reveal that the water molecules do not travel with the wave. In fact, if you measured the net movement of all the water molecules due to the wave, it would total zero. One may ask, "If the water molecules aren't moving in the direction of the wave, what is?" The answer is energy.

If the swimmer slaps his hand many times in regular intervals, a wave with a series of crests and troughs is created. Each pair of one crest and one trough is called a **cycle** due to its tendency to repeat. If the intervals are fast, the crests and troughs (or cycles) will appear to be closer together. If a sunbather sitting halfway between the deep and shallow ends of the pool had a watch, she could count the number of cycles that pass by her each second. This value, the number of cycles per second, is called the **frequency** (for space economy, we use the letter **f** in equations) of a wave. The unit of one cycle per second is more commonly called a **hertz** (abbreviated **Hz**) in commemoration of the German physicist Heinrich Hertz. Because the frequencies of electromagnetic waves are quite high, larger units such as **megahertz** (one million hertz, abbreviated **MHz**) and **gigahertz** (one billion hertz, abbreviated **GHz**) are commonly used.

The velocity (v-measured in meters per second) of a wave is directly related to the medium in which the wave is travelling. If the pool was drained and filled with molasses, the velocity of the waves would be less than those moving in water. For the remainder of this section, it will be assumed that all waves generated in our fictitious pool are traveling at identical velocities.

The distance between two successive crests (or two troughs) on a wave is called the **wavelength** (l) which is measured in meters or subunits of meters. The most common subunit for wavelength measurement of electromagnetic waves is the **nanometer** (abbreviated **nm**) which is one-billionth (1×10^{-9}) of a meter.

When the swimmer slaps the water with slow intervals, the distances between the crests and troughs of the waves are large, hence the wavelength is long. The sunbather times very long cycles per second on such waves. The frequency is small. However, when the swimmer rapidly slaps the water, the crest and troughs seem to bunch together. The wavelengths are short. The sunbather counts many cycles per second. Waves with higher frequencies, therefore, have shorter wavelengths and waves with lower frequencies have longer wavelengths. The relationship between the two can be summarized in the equation:

$$f = v/l \text{ or } l = v/f$$

It is important to remember that, as long as you adjust their numerical values per the above equation, frequency and wavelength are interchangeable. In the study of light it is common to use either term.

The swimmer may also notice that when he slaps the water with more force, the crests become taller and the troughs become deeper. The height of the crests (or in many cases, the depth of the trough) is called the **amplitude** of the wave. If the swimmer continues producing waves, one long, unbroken string of crests and troughs will span the entire length of the pool. This is called **continuous wave** transmission (often called just **CW** transmission).

But, the swimmer could decide to produce one wave, rest (thus saving his energy), and then produce another wave followed by another rest period. By saving his energy between waves, the swimmer could slap the water harder and produce waves of greater amplitude. This is called **pulse** transmission. Lasers transmit in a similar manner; they are either continuous wave or pulse.

In a V-shaped swimming pool with two shallow ends converging to one deep end, two swimmers at rest (swimmer A and swimmer B, equal distance from the deep end) start slapping the water at exactly the same time and with exactly the same intervals. Not only would both waves (wave A and wave B) have the same frequency but, as they passed the sunbather, all the crests of wave A would pass exactly at the same time as all the crests of wave B. Similarly, all the troughs of wave A would pass at exactly the same time as the troughs of wave B. The two waves are said to be **in phase**.

Two waves being **in phase** or **out of phase** refers to a comparison of the two waves at a given point. In the example above, the given point is the exact spot where the two waves pass the sunbather. Two waves can be out of phase at a given point for a variety of reasons. The disturbances could have-started at different times. The frequencies of the two waves could be different, or the distance travelled to the given point (called the path length) could be more for one wave than the other. If the two waves were in different media, they could have different velocities.

Two waves having identical frequencies and velocities that are out of phase, will be so to the same degree at all points. For example, two waves at 60 Hz that are 80° out of phase at the sunbather will be 80° out of phase at the deep end of the pool and all points in between.

Two waves with equal frequencies and velocities starting at the same time can be out of phase if their path lengths are different, e.g., if swimmer A was slightly further from the sunbather than swimmer B. Holographers use this fact to create their holograms.

When wave A and wave B meet at the deep end, they will join together and this phenomenon is called **interference**. How they interfere depends on to what degree the two waves are in or out of phase (called their **phase relationship**). If wave A and wave B are in phase, the crests of the two waves will meet and combine to form one large crest for each cycle whose amplitude is the sum of the two individual crests. The troughs of wave A and wave B would also combine to form one large trough in each cycle whose amplitude is the sum of the two individual troughs. This is called constructive interference. (See diagram A.)

If the two waves are 180° out of phase, the crests of wave A would merge with the troughs of wave B (and vice

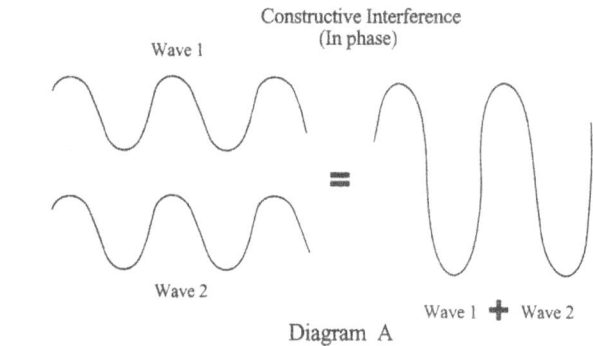

Constructive Interference
(In phase)

Diagram A

versa), cancelling each other's amplitudes. The result is a combined wave with little or no amplitude. This is called destructive interference. (See diagram B.)

Because the phase relationship of two waves can change from one point to another, the two waves can be in phase when they pass the sunbather, but out of phase when they hit the deep end of the pool. The ability of the two waves to stay in phase while they travel the length of the pool is called **coherence**. Waves that stay in phase for a long time are said to be very coherent.

Suppose the two swimmers agreed to create continual

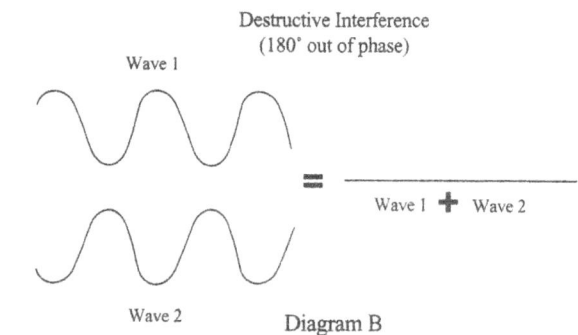

Destructive Interference
(180° out of phase)

Diagram B

constructive interference at the deep end of the pool. They would start slapping the pool at the same time while trying to maintain exactly the same frequency. If both swimmers have extremely good timing, they can keep the two waves coherent for a long period. But the swimmers, like other producers of waves (lasers, for example), aren't perfect. Their frequency may be slightly off.

The difference in frequency of any source (source is a common term for a device or system that produces waves) is called its bandwidth (abbreviated **Df**, and measured in hertz). The coherence of a source can be determined by measuring how long the waves stay in phase. Be it swimmers or a laser, the distance that the source can guarantee the waves will stay in phase is called its coherence length (abbreviated **L**; it is measured in meters or subunits of meters). The coherence length of a source is of great importance to holographers. It is directly related to bandwidth by the equation:

L = v/Df

Lasers are used to produce holograms primarily because no other source offers enough coherence.

Although electromagnetic waves exhibit exactly the same wave properties as the water waves described above, there are some notable differences between the two. Water waves propagate in two dimensions on a plane. Electromagnetic waves tend to propagate in three dimensions. As mentioned earlier, electromagnetic waves can propagate with or without a medium. Electromagnetic waves move extremely fast; water waves move relatively slowly. The interference of a water wave is determined by its amplitude-with electromagnetic waves, it is a function of its intensity.

In empty space, electromagnetic waves move at a velocity of 300 million meters per second (3 x 108 mis, also known as the speed of light-it is abbreviated in equations with the letter c). In his 1905 paper on special relativity, Albert Einstein correctly defined the speed of light as the absolute fastest velocity possible-a cosmic speed limit, so to speak. In air, the velocity of electromagnetic waves is just slightly less than "c". In most applications where electromagnetic waves are travelling through air, it is acceptable to use "c" as the velocity of the wave. Therefore, for electromagnetic waves travelling in free space or air:

f = c/l or l = c/f

and

L = c/Df

Electromagnetic waves are a union of electric and magnetic fields that are at right angles (90°) to both each other and the direction of their movement. (See diagram C.) When electromagnetic waves propagate, there is an infinite amount of directions in which they can travel. A laser is designed to channel waves such that their propagation is substantially in one direction.

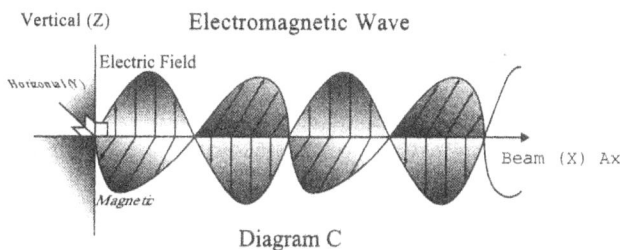

Diagram C

This unidirectional propagation of electromagnetic radiation is generally referred to as a laser beam.

Even while moving in a common direction, each wave can have its electric field and magnetic field oriented differently. The electric field can point straight down on the first wave, sideways on the second. There is an infinite amount of directions the electric field (or magnetic field, which stays exactly at a right angle to the electric field) can be pointing on the beam "axis" for each wave moving in the same direction.

Many properties of waves are more consistent if their electric and magnetic fields are properly aligned. The ability of electromagnetic waves to be aligned in the same orientation on the beam axis is called polarization. The human eye is not sensitive to polarization and cannot distinguish between polarized or unpolarized light waves. Some insects, like bees, are more sensitive and use polarization to determine direction. In holography, where consistency of the source's waves is critical, polarization is essential.

Because a reference point is needed in defining polarization, the electrical field is used to identify the position. If the electric field is travelling directly on the xz plane, the wave is defined as vertically polarized. How close the beam is to being polarized in the vertical position can be described by its polarization ratio. A laser beam with a 100: 1 polarization ratio is very close to being polarized in the vertical plane: a 500: 1 ratio is closer still.

Polarization can be achieved by several means, including reflection, transmission, scattering and birefringence. Birefringence is a phenomenon that occurs in certain crystals-such as calcite-and other materials. Such materials limit the absorption of waves to those with specific electric field orientations. Sunglasses use this effect, reducing the amount of glare received by the eyes. Crystals with maximum birefringence allow only one orientation to be absorbed and transmitted. The beam exiting the crystal is polarized.

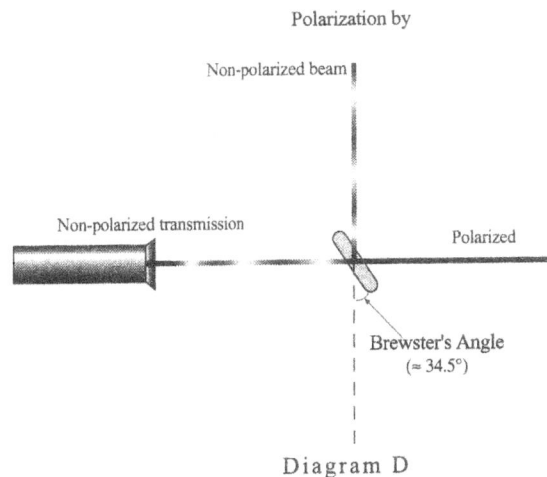

Diagram D

Scattering occurs when an atom deflects a photon away from it. Scattering of electromagnetic radiation in the earth's atmosphere produces partial polarization. Bees use this type of polarization for navigation. Polarization by scattering is not an effective source for most photonic applications. For lasers, polarization by transmission is more relevant.

If an unpolarized beam strikes a nonconductive, transparent target at a specific angle (called Brewster's angle; see diagram D), a polarized beam will pass through the target. This process is called polarization by transmission. Waves in the beam that do not have the selected electric field orientation are reflected from the target. Brewster's angle varies for different wavelengths and materials, but for most lasers it is in the neighborhood of 34.5 degrees.

PARTICLE PROPERTIES OF LIGHT

On October 19, 1900, Max Planck introduced a concept that was to revolutionize science. In an effort to resolve conflicts between scientific theory and experimental evidence, Planck suggested that **energy** (abbreviated in equations with the letter **E**) is not continuous, but instead comes in discrete little "packets" called **quanta**. Further, an "energy packet" of light was directly related to its frequency by the equation:

E = hlf

where **h** is defined as **Planck's Constant** and is equal to 6.63 x 10-34 joule-seconds.

Although the mathematics seemed to work, and it did resolve current conflicts between theory and experimental data, the implications of Planck's hypothesis were rather hard to accept. Clearly, light exhibited wave properties such as diffraction (bending of a wave around a comer), interference and polarization. Waves are inherently continuous and not discrete. Frequency, for example, used in Planck's equation is a wave phenomenon. Yet the energy in the same equation describes light as discrete packages of hlf. How could light possibly consist of particles and demonstrate properties of waves?

The numerous experiments and the profound mathematics that followed are extremely significant and detailed. The final result of two decades of scientific fervor was a new definition of the laws of physics now known as **quantum mechanics.**

At the core of quantum theory is the concept of **duality** which states that light, electromagnetic radiation, energy, and even matter are both a wave and a particle. Electromagnetic radiation itself is composed of minute "wave packets" called **photons** that demonstrate properties of both continuous waves and discrete particles.

In terms of wavelengths, the energy of one photon is expressed as:

E =hlc/l or l =hlc/E

As stated earlier, all electromagnetic radiation is the same phenomenon. Only the energy per photon is different. The fundamental unit of electromagnetic radiation is the photon. One can classify all forms of electromagnetic radiation by the wavelength (or frequency) of the photon. Because the wavelength involved in the common forms of electromagnetic radiation is small, it is usually measured in nanometers. In visible light, the wavelength of a photon determines its color. Red had the longest wavelength (740- 622 nm), followed by orange, yellow, green (577-490 nm), blue (489-430 nm), and violet (429-390 nm). White light is a mixture of all colors. Photons with wavelengths greater than 740 nm produce infrared (below red) radiation. Photons with wavelengths less than 390 nm create ultraviolet (beyond violet) radiation.

Photons can be created by transferring energy to an atom. In an atom, electrons reside in various positions and energy states. If the atom is stable, the electrons are defined

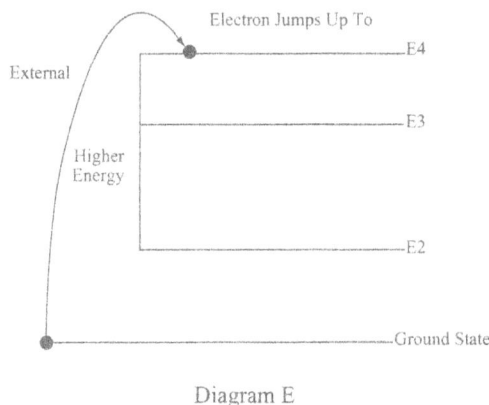

Electron Jumping to a Higher Energy State

Diagram E

as being -in their **ground** (lowest level of energy) state. When external energy is transferred to the atom, the electrons get "excited" and respond by jumping up to higher, or excited, **energy states**. (See diagram E.)

Quantum mechanics dictate that an electron cannot re-

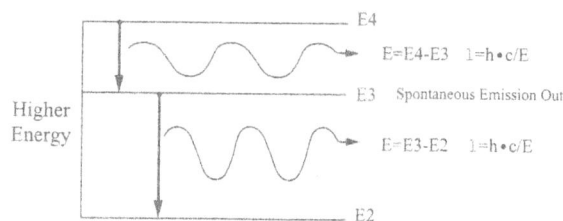

Spontaneous Emission

Diagram F

side between two energy states. It has to jump up all the way to a higher state or not jump at all. The electron will stay in the excited state for a very short period and then spontaneously drop back down to a lower energy state.

When the electron drops one or more energy steps (also called **transitions**), it releases the amount of energy difference between the two states (see diagram F) in the form of a photon. The wavelength of the photon is determined by the amount of energy released-the energy difference of the two energy states-through the formula l = hlC/E. If the energy released is small, the wavelength of the photon will be large and vice versa. This process is called **spontaneous emission.**

There are two principal mechanisms that enable energy to be transferred to an atom: absorption and collision. **Absorption** occurs when a photon bumps into an atom. If the photon's energy (determined by the wavelength of the photon using E = hlc/l) matches the energy difference between a lower energy state where an electron resides and a higher energy state (called an **energy band gap**), the atom will absorb the photon. The photon energy is transferred to the atom, kicking the electron up to the higher energy state. If the photon's energy does not match any of the energy differences in two excited states of the atom, scattering occurs, redirecting the photon without otherwise altering it.

Collision occurs when a moving particle (an electron, ion, atom or molecule) smashes into the atom with the proper amount of momentum. Some or all of the particle's kinetic energy is transferred to the atom, again raising the electrons of the atom to a higher energy state.

LASER THEORY

Albert Einstein was the first to recognize the significance of Planck's concept of quantized energy. He used Planck's E=hlf equation to derive his explanation of the photoelectric effect in 1905, for which he later received the Nobel Prize in Physics. In 1916, Einstein predicted another phenomenon now known as **stimulated emission**-the basis for all laser technology. The same year, Einstein also released his most prized work, the general theory of relativity. The theory of stimulated emission went unnoticed until the late 1940s.

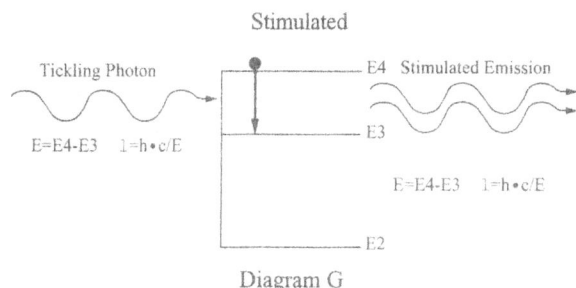

Diagram G

An excited electron in a higher energy state will spontaneously drop to a lower energy state and emit a photon. The wavelength of that photon is determined by the difference of the energy between the two energy states ($1 = hlc/E$). If, before the electron drops to the lower energy state, a photon with a wavelength identical to that which is about be produced by spontaneous emission passes by the exited atom, it will stimulate the electron to drop. This stimulation (also called **tickling**-see diagram G) will force an emission of a photon that is identical to one passing. This process is stimulated emission. Both photons have the same wavelength (and therefore the same frequency), are in phase, coherent, and travelling in the same direction. It is important to note that the "tickling" photon is not absorbed by the atom; it must only pass closely.

In an environment with many identical excited atoms, photons can multiply rapidly through stimulated emission. Two photons quickly become four which quickly become eight. Since atoms and photons are extremely small, eight photons can become billions in a reasonably short distance. This process of multiplying photons is called **amplification**; it is the essence of a laser. The term LASER itself is an acronym for **L**ight **A**mplification from **S**timulated **E**mission of (electromagnetic) **R**adiation. Billions of photons, with the same wavelength, in phase, coherent, and travelling in the same direction can be a very useful tool.

For lasing to occur, it is essential that there are more atoms with electrons in a higher energy state than those at the lower energy state-a condition called a **population inversion.** An atom with its electrons in lower energy states will absorb a passing photon instead of duplicating it. If there is more absorption than stimulated emission, amplification will not occur. How can a baker increase his inventory if he has five hungry children and only four cookies in the oven? A population inversion assures continuous multiplication of photons.

To create a laser (see diagram H), many atoms of one type must be contained in a given space. The atoms used for lasing are called the **active medium** and are housed in a container. Since the energy states of the active medium create most or all of the stimulated emissions. it is the active medium that determines the possible wavelengths produced by the laser ($/= hZc/E$). **Solid state** lasers have active media made from matter that is solid at room temperature. Neodymium and chromium ions (an ion is just an atom with either an excess or deficit of electrons) are common active media used in solid state lasers. The active medium in **liquid lasers** is, of course, liquid and that in **gas lasers** is gas. Common active media used in gas lasers are neon, argon, and ionized cadmium.

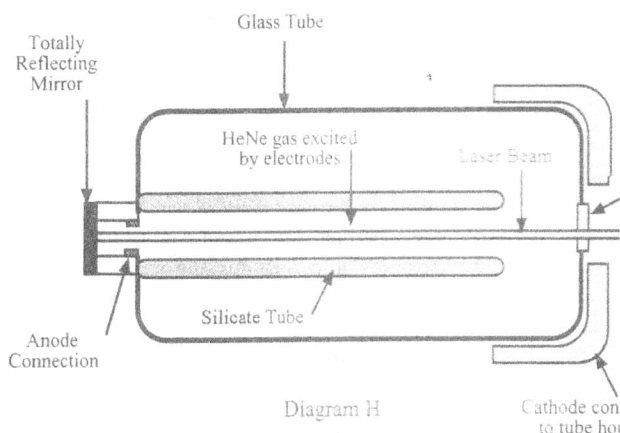

Diagram H

A **pump** transfers energy to the active medium, raising their electrons to an excited (higher energy) state. Brisk and continuous pumping will create and maintain a population inversion. As stated earlier. energy can be transferred by either absorption or collisions. In solid state lasers, an optical pump produces a flood of photons with energies that are easily absorbed by the active medium. Common optical pumps are flash lamps. laser diode arrays, and other gas, liquid and solid state lasers.

The vast majority of gas lasers use electron collisions to pump energy to the active medium. Normally, electrons will not flow through a neutral gas. If a large voltage (from 1,000 to 20,000 volts, depending on the gas) is applied to the gas, the gas will "break down" and allow a **discharge** of electrons to rush through it. This initial voltage is called a **spike**, and is applied to the gas through two electrodes (an **anode** and a **cathode**) on opposite ends of the container. Once electron current is flowing through the tube, the voltage is automatically reduced (by 90 to 4,000 volts, depending on the gas). The discharge of electrons and other charged particles collide with the atoms in the active medium, enabling energy to be transferred. Both the spike voltage and the operating voltage necessary to operate the laser are furnished by a power supply, usually an external box that transforms either 117VAC or 220VAC to the required voltage.

Many active media are not efficient at receiving energy from the pump. In such cases, the active medium must be combined with a transfer medium-a substance that compensates by efficiently collecting energy from the pump and then passing it on to the active medium. Solid state transfer media should be good absorbers; gas transfer media require the ability to efficiently receive energy from collisions.

The transfer medium must be compatible with the active medium in two ways. First, it must have some common higher energy states that provide a channel to efficiently pass energy to the active medium. Second, it must be chemically compatible, allowing peaceful coexistence with the active medium as well as the other components inside the tube.

There are typically 5-20 transfer medium atoms for every active medium atom. Inert helium makes a good transfer medium and is used in helium-neon (**HeNe**, pronounced Hee-Nee) and helium-cadmium (**HeCd**, pronounced Hee-Cad) gas lasers. Common transfer media in solid state lasers are yittrium atrium garnet (**YAG**), yittrium lithium fluoride (**YLF**, pronounced "Yelf"), sapphire, and ruby.

The container that holds the active medium (and transfer medium, if applicable) must protect it from elements that may interfere with the lasing process. In solid state lasers, the crystalline transfer medium encompasses the active medium atoms, forming a strong durable solid structure that serves as the container. Because the transfer media in solid state lasers house and transfer energy to the active medium, they are called **hosts**.

For gas lasers, the container is almost always a long cylindrical **tube**. Tubes are made of various materials; ceramic and glass are the most common. Inside the tube is an equally-long, yet very narrow (3 millimeters) **bore** that allows lasing to occur in a straight, usually horizontal path.

During lasing, a wide variety of spontaneous and stimulated emissions occur throughout the tube with photons propagating in every conceivable direction. All emissions except those travelling down the bore exit from all sides of the tube with limited or no amplification. Those photons travelling down the length of the bore continue to multiply through stimulated emission. By making the bore long enough, one could continue the lasing process until adequate amplification was achieved. At this point, a billion or so photons exit the bore in the form of a laser beam.

Unfortunately, in order to achieve sufficient amplification, the bore would have to be forty feet long. A forty-foot laser would be extremely awkward to both transport and operate. A shorter bore is needed to make the device more practical.

By passing photons back and forth (called **optical feedback**) several times through a shorter bore, adequate amplification can be attained. This is accomplished simply by placing mirrors on both sides of the bore. If both mirrors are 100% reflective, extremely high amplification is achieved. However, no photons exit the laser (photon exiting the laser is called **transmission**). This, of course, has no value whatsoever.

However, if one mirror could reflect some of the photons back into the bore while allowing the remainder to exit, both amplification and transmission could occur. Such a mirror found on lasers is called the **output coupler**, or **OC**. The second mirror is known as the **high-reflector**, or **HR**. A perfect high-reflector will provide 100% reflection. In actual lasers, there is a small amount of light transmitted from the HR, referred to as **leakage**.

Power is the amount of energy a source produces each second. It is measured in units of joules per second, more commonly called **watts**-in honor of the British scientist James Watt. The power in a laser reflects the amount of photons per second exiting the laser.

Laser designers strive to maximize the power produced by the laser. However, if the OC allows too many photons to exit, the number of photons returning to the bore may not be adequate to provide significant amplification. This, in turn, limits the amount of transmission. Therefore, the amount of transmission and reflection provided by the OC must be properly balanced to complement both the amplification process in the laser **cavity** (area between the two mirrors) and the transmission from the cavity. In most lasers, output couplers allow 1-3% of the photons to be transmitted .

It is common to have multiple wavelengths lasing inside the cavity. Because there is a variety of paths of energy states for which an excited electron can travel back to the ground, there is a variety of energies (and therefore, wavelengths) that can be emitted throughout its descent. The electron can also bounce down two steps at a time-maybe three---each time emitting photons of higher energy and lower wavelengths.

Certain transitions are more dominant, however. The more spontaneous emissions produced at a given wavelength, the greater probability of stimulated emission. The more stimulated emissions, the more amplification. The most dominant wavelength in a given laser is called the **primary wavelength**. The second most dominant wavelength is called the **secondary wavelength**.

Photons with undesirable wavelengths can be eliminated from the cavity by putting special thin film coatings on the mirrors. Such coatings only allow a specific range of wavelengths (for example 430-460 nm) to reflect back into the cavity. Thus photons with undesirable wavelengths will not multiply.

The lasing process described above is continuous wave transmission. Lasers that produce this kind of transmission are called, expectably, **continuous wave lasers** (or just **CW lasers**). Lasers that provide pulse transmission are called **pulse lasers** by some and **pulsed lasers** by others.

A continuous wave laser can be converted to a high energy pulse laser by installing a **Q-switch** in the laser cavity. Q-switches are devices that enable the active or transfer medium in the cavity to collect maximum pump energy before beginning the process of stimulated emissions. The Q in Q-switches is an abbreviation for "quality factor" to represent the quality of a feedback system. In the "low Q mode," a Q-switch limits optical feedback in the cavity for a very short period. If optical feedback is blocked, continuous stimulated emission will not occur.

During this period, the active medium continues to collect energy from either the pump or the transfer medium. A high percentage of electrons is elevated to higher energy states creating a large population inversion. When the Q-switch opens (high Q mode), a rapid and powerful episode of stimulated emission occurs in the cavity until the Q-switch is again closed. The active medium begins receiving energy, and again its electrons begin to elevate in preparation for the next high Q mode.

The repetitive bursts of lasing in the cavity result in a string of powerful energy pulses departing from the output coupler. The average length of time of a pulse is referred to as its **pulse length** (measured in seconds and subunits of seconds). The number of pulses per second is called its repetition rate (or more commonly, **rep rate**, measured in hertz).

There are four types of Q-switches: chemical, electro-optical, acousto-optical, and mechanical. The first three are common and found in a variety of applications. Mechanical Q-switches are seldom used because they tend to be slow, noisy and produce unwanted vibrations.

The laser tube and mirrors are held by a mechanical support structure almost unanimously referred to as a **resonator**. The title, nevertheless, is inaccurate. A resonator is a system consisting of a laser cavity and mirrors that enables rapid bidirectional optical feedback. It oscillate's. Most physicists will readily admit that the term is imprecise; however, there is no other term available other than "mechanical support structure." Conforming to the majority, the term "resonator" will be used in this paper with the knowledge it is incorrect. The resonator has the task of keeping the bore straight and aligned with the mirrors. This is not an easy assignment. The lasing process produces an ample amount of heat. The heat creates thermal expansion, which tends to shift the mechanisms that hold the mirrors (**mirror mounts**) and bore. The bore itself, being long and quite narrow, is extremely susceptible to thermal distortions.

When the laser cools down, the components of the laser tend to contract. Even the best resonators will sometimes fail to keep the bore and mirrors in line. Because it is easier to align the mirrors than the bore, alignment devices are placed on the mirror mounts.

Such devices, called **tilt plates** or **xy plates**, enable the operator to change the positioning of the mirrors either sideways (horizontal, or "x" position) or up/down (vertical, or "y" position) without changing the "z" position (frontwards and backwards-this would change the cavity length which can only hurt the laser's performance). Prop-

er adjustment of the tilt plates enables maximum amplification inside the cavity, producing maximum laser power. The resonator must also provide mechanical protection from the routine bumps and bruises that may occur.

The tilt plates and mirror mounts are secured in the resonator by three or four resonator rods, which span the length of the laser. Resonator rods are the backbone of the resonator. They give mechanical support to the entire laser head and the hardware that holds the tube.

On large lasers, the resonator rods are made from carbon graphite and are generally one to two inches in diameter. Carbon graphite has a very low **coefficient of expansion,** which means it will have minimal movement (expansions and contractions) when temperatures fluctuate. In smaller lasers, invar is generally used. Invar has a larger coefficient of expansion than carbon graphite, but provides equal strength at one-fourth the thickness.

In polarized lasers, **Brewster windows** are attached to the ends of the bore. The windows, nonconductive and transparent panels placed at Brewster's angle, seal the bore and polarize the photons that pass through it. One of three methods may be used for sealing the windows to the bore: epoxy, frit, or optical contact. Sealing the bore and securing the Brewster window by epoxy was one of the first methods used on gas lasers. A space-grade epoxy glue is evenly distributed on a quartz **Brewster stub** and meticulously fastened on the bore. Frit sealing involves heating glass between the Brewster stub and the bore.

Perhaps the most effective method of sealing the windows, yet hardest to do properly, is optical contact. Optical contact requires a precise mechanical fit between the Brewster stub and the bore. The bore is heated, microscopically melting the Brewster stub directly to the bore.

In low-powered lasers, such as air-cooled ion lasers, HeNe, and HeCd lasers, the Brewster windows are almost always made from fused silica. Fused silica is preferred due to its low absorption of electromagnetic radiation, which enables the highest possible transmission. In higher powered lasers, such as large-frame ion lasers, fused silica is susceptible to solarization. **Solarization** occurs when an excessive amount of the transmitted photon energy is absorbed by a Brewster window, changing its optical properties. Two (of the more common effects of solarization are thermal lensing and color centering.

Thermal lensing is caused when photon energy absorbed by the Brewster window is converted into heat. Heat circulating in the window changes its optical properties. It also warms the air surrounding the Brewster window, distorting the optical properties of the air. Both the window and the surrounding air act as a randomly shifting lens that causes a slight variation in the direction of the beam. When the shifted beam hits the mirror, it may not reflect precisely down the center of the bore. Part of the beam may "clip" upon entering the bore, causing a significant reduction in power. The effects of thermal lensing are very similar to those of a misaligned mirror. It is common for an unaware operator to try to correct the malfunction by adjusting the tilt plates. Often, it will work-temporarily.

Unfortunately, heat energy is not stationary. The distortions in the lens and its surrounding air can change, causing the beam to shift again. Or, if the operator turns the laser off, thermal lensing may not occur again until hours after restarting it. All previous adjustments of the tilt plates are no longer valid. The laser is now legitimately misaligned.

Thermal lensing is extremely frustrating if not detected. An operator who finds the large frame ion laser constantly out of alignment may find it necessary to inspect the Brewster windows.

Color centering is the result of extreme solarization. In color centering, the Brewster window absorbs enough energy from the laser beam to change its molecular structure. When this occurs, the Brewster window will lose its transparentness.

Because most of the energy of a beam is in its core, the molecular restructuring is generally restricted to the center of the Brewster window. Photons will no longer pass through the damaged region, producing a beam that has no light in its center. This "donut" shaped beam is unusable in most applications. To help reduce the effects of solarization, thermal lensing, and color centering, Brewster windows on large frame ion lasers are generally made from crystal quartz.

ADDITIONAL PROPERTIES OF LASERS

Any light source that delivers exactly one wavelength is said to be **monochromatic**.

Ideally, stimulated emissions from a group of identical fuel atoms should produce very distinct wavelengths (or frequencies) lasing within the cavity; for example, a HeCd laser would produce amplification at 325.0 and 441.6 nm. By using proper coatings on the mirrors, the 325.0 nm wavelength can be removed from the lasing process, thus producing a monochromatic beam at 441.6 nm. In actuality, however, this is not the case.

In the same manner the two swimmers discussed earlier produced slightly different wave frequencies, the lasing inside of the cavity produces minute variations of photon frequencies (or wavelengths) in the transmitted beam defined as the laser's **bandwidth** (Df).

Several factors contribute to variation of photon frequencies inside the cavity, including the motion of an atom at the time it emits a photon.

Variation of photon wavelengths in a laser, called its **linewidth** (Dl, measured in nanometers) is exactly the same phenomenon as a laser's bandwidth (Df). Quantitatively, however, the two will not have the same values. Linewidth can be numerically converted to bandwidth (and vice versa) by the following equations:

$$Dl/l = Df/f \text{ or } Dl = Dfl l/f \text{ or } Df = Dl lf/l$$

Often, it is helpful to eliminate frequency entirely from our equations. This may be done by inserting the expression f = c/l into the above equations and applying some basic algebra. The results are:

$$Dl = Dfl l^2 \text{ and } Df = Dl l c/l^2$$

Coherence length (L) can also be expressed in terms of wavelength and linewidth:

$$L = c/Df = l^2/Dl$$

Waves moving back and forth in a confined region create constructive and destructive interference. If the frequency of the waves matches the resonant frequency of the region, constructive interference will occur throughout the length of the region. A set of non-moving waves, complete with crests and troughs, forms in the region. These "standing" waves now dominate.

By changing the length of the region (or the frequency of the waves), the two frequencies no longer match. Destructive interference is introduced, and the standing waves disappear. If you continue to change the length, you will find other discrete distances (called harmonics) that will enable the frequency of the region to match those of the waves. Destructive interference is again replaced by constructive interference, and the standing waves reappear.

By increasing the length of the region, you enable more standing waves to exist within its boundaries. By decreasing the length of the region, fewer standing waves exist.

A laser has waves (photons are waves) travelling back and forth in a confined region (the laser cavity). Because many frequencies exist within the laser's bandwidth (Df), some of them will match the resonant frequency of the structure. Standing waves will form within the cavity. Only those frequencies creating standing waves will continue to lase. These frequencies are called longitudinal modes.

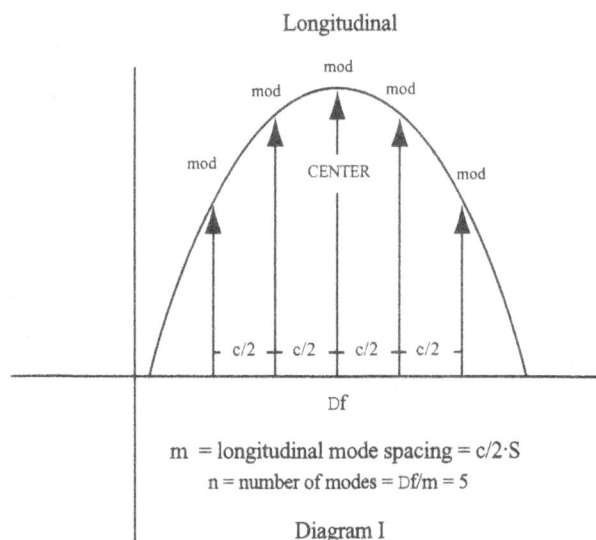

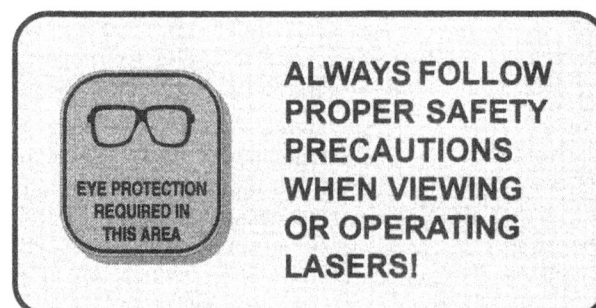

Longitudinal

mod
mod   mod
mod   CENTER   mod

c/2   c/2   c/2   c/2

Df

m = longitudinal mode spacing = c/2·S
n = number of modes = Df/m = 5

Diagram I

EYE PROTECTION
REQUIRED IN
THIS AREA

ALWAYS FOLLOW PROPER SAFETY PRECAUTIONS WHEN VIEWING OR OPERATING LASERS!

Within the bandwidth is a set of distinct longitudinal modes spaced equally apart. There are no frequencies lasing in between them. The distance between each mode is called the **longitudinal mode spacing** (abbreviated m in terms of frequency, measured in hertz- see Diagram I).

The longitudinal mode spacing is determined by the separation of the cavity mirrors (abbreviated S, also called the **cavity length**, measured in meters) and can be calculated with the following equation:

$m = c/2lS$

In terms of wavelengths, longitudinal mode spacing (**M**, in nanometers) can be calculated:

$M = ml l^2/c = l^2/2lS$

The number of longitudinal modes (n) in the bandwidth can be determined by the equation:

$n = Df/m$

In terms of bandwidth or linewidth, the number of longitudinal modes should be identical. The equation in terms of linewidth is:

$n = (2lSlDl)/l^2$

Because the mode spacings are very close together (generally less than 1/1000 of a nm), very small changes in the cavity length can cause the modes to move. In argon-ion and other lasers that generate a significant amount of heat, the cavity can expand or contract enough to cause the longitudinal modes to literally jump over each other. This phenomenon is called **mode hopping**. In many laser applications, such as holography, mode hopping can cause undesirable effects.

Another mode that manifests itself inside the cavity is the **transverse electric and magnetic mode**, more commonly called the TEM mode. Light will propagate with distinct and defined geometrical paths. The most fundamental path will produce a clear, uninterrupted spot when projected on a target. Other paths, or "modes," will have dark irregularities (called **nodal lines**) separating the spot. The nodal lines can be either vertical or horizontal. (See diagram J.)

These patterns are classified with subscript using the form TEM_{vh} (where "v" designates the number of nodal lines in the vertical direction and "h" designates the number of nodal lines in the horizontal direction)–for example TEM_{10}, TEM_{20}, TEM_{11}. One of the more interesting patterns is the TEM_{01*}, a mode that produces a large circular node in the middle of the beam. Because of its distinct pattern, TEM_{01}, mode is often referred to as the **donut mode.**

Lasers can be built to produce only the fundamental mode TEM_{00} by reducing the ratio of bore diameter to the diameter of the TEM_{00} beam, usually to less than 3:1. This can be achieved by properly selecting mirror combination- that encourage TEM_{00} transmission.

Generally, long thin bores produce TEM_{00} mode more readily than fatter ones. In holographic applications, TEM_{00} transmission is essential.

TEM$_{00}$ Gaussian Energy

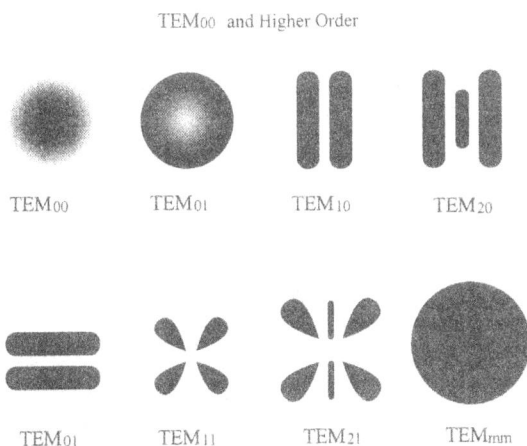

Diagram K

TEM_{00} beams can be focused down to the smallest possible size spot known as the **diffraction limit**. In theory, the diffraction limit of a beam is its wavelength. In real applications, the smallest possible size of a focused spot is slightly higher. In holography, the ability to focus down to a "tight" spot is not an advantage since the beam is expanded.

The beam of a **multimode laser** has a combination of (many modes resulting in an uneven energy distribution in its cross section. Multimode beams (commonly written **TEM$_{mm}$**) deliver more energy than, but have none of the advantages of, those modes listed above. A TEM$_{mm}$ beam can also be focused down, but not nearly as tightly as TEM$_{00}$, Although multimode lasers are used in a variety of applications where laser power is the primary concern, they cannot be used for holography.

Mechanical instabilities in the optics and tube can cause the beam to wander. A laser's beam-pointing stability (also called angular drift, measured in microradians) measures how much the laser beam drifts from the beam axis. Beam pointing stabilities of 60 microradians or less are considered good.

TEM$_{00}$ and Higher Order

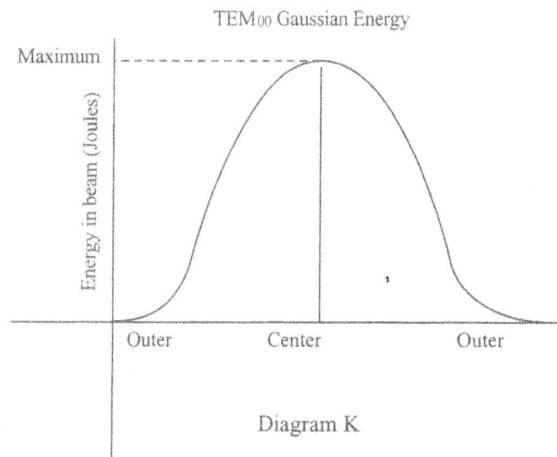

TEM$_{00}$ TEM$_{01}$ TEM$_{10}$ TEM$_{20}$

TEM$_{01}$ TEM$_{11}$ TEM$_{21}$ TEM$_{mm}$

Diagram J

Primary Types of CW Lasers Used for Holography

The three primary types of CW lasers used in holography are argon-ion, HeCd, and HeNe lasers. Each has distinct advantages that are related to the holographer's needs. Typically, the type of recording material, size of the hologram, and operator budget determines which laser is best suited.

In embossed holography, photoresist is the primary medium used for recording images. The use of photoresist enables mass-produced holograms. Photoresist chemically etches the holographic image onto a glass plate. The optically-engraved glass plate (called a **master**) is made conductive, and then electroplated-which produces a **shim**. The shim is placed on an embossing machine for mass stamping of embossed holograms.

Because photoresist is extremely sensitive to wavelengths between 420-450 nm, HeCd lasers (which lase at 441.6 nm) are most often chosen. HeCd lasers are much more cost-effective to operate than comparable lasers, enabling originators of holograms to keep their operating costs down.

Artistic holography isn't constrained by the necessity to mass-produce. This gives the holographer freedom to choose from a variety of emulsions to produce holograms. In such cases, most holographers prefer to use emulsions sensitive to the primary wavelength (514.5 nm) of an argon laser. Argon lasers provide an attractive combination of high power and long coherence length, enabling holograms that are both large and visually striking.

In most forms of holography, the coherence length of the laser determines the size of the hologram. Generally, a 10 cm coherence length laser produces holograms that are 10 cm x 10 cm (4" x 4"). Seasoned holographers routinely extend the size of their holograms without increasing their coherence length, some being able to shoot a 6" x 6" while using a source with 10 cm coherence length.

Novice holographers typically can't justify the hefty expense of an argon laser. A HeCd laser, though much more attractively priced, still costs over $10,000. For "weekend shooters" on a budget, the HeNe laser provides a cost- effective alternative.

Certain laser parameters are essential to all forms of continuous wave holography. The beam must be TEM_{00} mode. A polarized beam (at least 100: 1) is necessary in all forms of holography except dot matrix; the low exposure times of dot matrix holography, typically 5 milliseconds per dot, reduce the need for polarization. The laser cannot generate excessive heat or vibration, two mortal enemies of the holographer.

With the exception of dot matrix holography, high power is very desirable. Less photon power requires longer exposure times. Not only do longer exposure times raise the cost of labor (professional holographers do not work for free), but they increase the odds of something going wrong. As every holographer will gladly tell you, things do go wrong.

Argon-Ion Lasers

Before the advancement of the helium-cadmium laser, argon-ion lasers were the preferred choice for almost all forms of CW holography. The argon laser provides a generous amount of power, polarization and coherence-three of the more prized parameters in holography.

Argon-ion lasers produce lasing at many wavelengths between 454.5-528.7 nm-usually eight to ten-and can be equipped with a **prism wavelength selector** in the cavity to allow the operator to select a specific wavelength. Of primary interest in holography are the 514.5, 488.0 and 457.9 nm wavelengths. The 514 nm line is the most powerful, followed closely by the 488.0 nm line. Large-frame argon lasers can provide nearly 10 W of power at 514.5 nm and over 6 W at 488.0 nm. Most argon lasers have a polarized beam with a polarization ratio of 100: 1 or greater. The combination of high power and high coherence length make argon-ion lasers the best laser for the serious artist who can afford its substantial price.

In embossed holography, the powerful 514.5 and 488.0 nm wavelengths of an argon laser have little effect on the photoresist used to record the holograms. Because photoresist is extremely sensitive to wavelengths between 420 and 450 nm, only the 457.9 nm line can effectively expose it. Unfortunately, the 457.9 nm wavelength is relatively weak in comparison with the more powerful 514.5 and 488.0 nm lines-less than 20% of the relative power. Large-frame ion lasers produce approximately 1 W of power at 457.9 nm.

Argon-ion lasers can be equipped with another intracavity device called an **etalon**. An etalon is a wedge-shaped piece of high quality optical glass which, by means of constructive and destructive interference, can eliminate longitudinal modes from the laser cavity. The etalon acts as a separate laser cavity inside of the main laser cavity. When the beam enters the etalon, it is reflected inside the wedge to form its own optical feedback before exiting. By adjusting the angle where the beam enters and the temperature (which changes its index of refraction and cavity length) of the glass, the etalon can choose which longitudinal modes will survive (through constructive interference) and which ones will not (through destructive interference). The beam exits the etalon, plus or minus a few longitudinal modes, to resume amplification in the main cavity.

By reducing the number of longitudinal modes, the etalon reduces the bandwidth. Lowering the bandwidth raises the coherence length. In essence, the etalon allows the holographer to control how much coherence length the laser beam will have. While most lasers refer to their coherence length in centimeters, an argon laser with an etalon can attain coherence lengths of many meters. The trade-off, however, is power. An etalon will produce losses of 30% or more, depending on degree of bandwidth reduction.

The active medium in an argon laser is a pure argon gas. Because argon is a good energy absorber, it needs no transfer medium. Argon atoms are excited by passing a high-current density discharge through a ceramic tube. An initial spike of a few thousand volts breaks down the low pressure gas (approximately 1 torr), then the voltage drops to 90-400 volts while the current jumps to 10-70 amps (dc). The discharge current is concentrated in a small-diameter bore in which stimulated emission takes place and must be high enough to ionize the argon gas (hence, the title argon-ion laser). An external magnet placed immediately outside the tube produces a magnetic field parallel to the bore axis that helps confine the discharge to the bore.

The conditions inside an ion-laser plasma tube are extremely harsh. Highly charged electrons and ions violently collide with the tube and bore, eroding the surfaces and contaminating the gas. A sizeable amount of UV radiation is generated in the cavity which, over time, tends to damage the Brewster windows and other optical surfaces.

Many times, microscopic particles from the walls of the bore can peel off and fall into the beam path. These pesky flakes can cause minute losses in power called drop offs. More advanced materials used in the inner cavity of argon-ion laser tubes have reduced the amount of lasers experiencing drop off problems.

The average lifetime of an argon-ion laser tube is directly related to how high the operator sets his tube current. Those trying to achieve maximum power will find an average tube lifetime of around 2,400 hours. More prudent operators can expect approximately 3,000 hours per tube.

Although argon lasers are capable of producing a significant amount of optical power, the energy efficiency of the argon laser is actually quite poor. Because stimulated emission takes place in energy transitions far above the ground state, much energy is required to pump the electrons up to an excited state.

The inefficient conversion of energy creates a substantial amount of heat. To remove this heat, metal disks are brazen inside the tube to allow it to transfer from the bore to the outside of the tube. A metal duct outside of the tube circulates water which transports excessive heat away from the tube. To keep the temperature of the tube low enough to operate, large-frame ion lasers require a water flow rate greater than three gallons per minute.

Once the water exits the laser head, it must either be recirculated or dumped down the drain. To recirculate water, a heat exchanger is placed in the water flow loop. The exchanger works much like a radiator, extracting heat from water, exiting the head and transferring it to a different medium (usually air or other water). Cooler water exits the heat exchanger and is sent back to the tube.

Although heat exchangers can be relatively effective, their high vibration makes them unappealing in sensitive applications like holography. Most holographers prefer an open-cycle system in which the water flows from the tap to the laser, and then down the drain.

The power supply must also be water cooled, although water circulating in the middle of high-voltage circuitry is not a favorable combination. Condensation and leakage jeopardize the performance and safety of the unit.

Due to the high-current requirements needed to produce stimulated emissions, argon lasers operate from 208VAC, three-phase line voltage. The extreme conditions of the laser tube tend to induce nontrivial expansions and contractions in the laser cavity. making argon lasers susceptible to mode hopping and poor beam pointing stability.

Though not perfect, argon-ion lasers offer me best combination of power and coherence length for artistic holography. The price of a large-frame ion laser generally starts at $30,000. Re-tubing costs range from $13,000 to $15,000.

HELIUM-CADMIUM (HECD) LASERS

The 441.6 nm line of a HeCd laser exposes photoresist used in embossed holography ten times more effectively than the 457.9 nm line of an argon laser. Unfortunately. for many years HeCd laser were only capable of delivering a maximum of 40 mW TEM$_{00}$, The marginal power resulted in unusually long exposure times for originators of 2D/3D (also called multiple plane), 3D and composite holograms. A l W argon laser at 457.9 nm, comparable to a 100 m\Y HeCd laser in its effective exposure rate, could expose photoresist 2 112 times faster than the most powerful HeCd laser.

Further, an HeCd laser could only provide 10 cm of coherence length, adequate enough for a standard 4" x -l" master. As noted earlier, only the more resourceful holographers could use a HeCd for the more illustrious 6" x 6" image. In comparison, Argon lasers can provide an unlimited amount of coherence length-albeit at the expense of power-by having an etalon installed.

As for originators of dot matrix holograms, in which the exposure time is trivial, HeCd lasers were readily adopted. A HeCd laser costs less to buy and operate, is easier to use, has less maintenance, and lasts longer than a comparable argon laser. HeCd lasers do not suffer from common argon laser malfunctions such as thermal lensing, color centering, power supply leakage and mode hopping. Further, HeCd lasers operate off of standard 117VAC, need no water to cool the tube and, therefore, no special plumbing is required in the facility.

In the last four years, HeCd lasers have quadrupled their power output. A typical HeCd laser can deliver more than 150 mW TEM$_{00}$ and provide 30 cm coherence length- more than enough for a 6" x 6" hologram. The effective exposure on photoresist for HeCd lasers now meets or exceeds large-frame argon lasers, while saving the average (holographer in excess of $800 per month per laser on electricity and water bills. This improvement has created a profound change in the embossed holography marketplace. Today, seven out of every eight lasers bought for commercial embossed holography are HeCd's.

The active medium used in a HeCd laser is cadmium. Standard HeCd lasers use naturally reoccurring cadmium that consists of a blend of three isotopes: Cd_{112}, Cd_{114}, and Cd_{116}, It is abundant in nature, and therefore costs pennies per gram. The stimulated emissions of cadmium provide a primary lasing wavelength of 441.6 nm, with a secondary wavelength of 325.0 nm. One manufacturer has patented a 353.6 nm HeCd laser. With naturally reoccurring cadmium. the spectral bandwidth is 3.0 gigahertz, which translates into 10 cm coherence length. Standard HeCd lasers can produce up to 120 mW TEMoo at 441.6 nm.

By processing naturally reoccurring cadmium in a manner similar to processing nuclear grade uranium, anyone of the three isotopes can be isolated, producing isotopically enriched cadmium-more commonly called single isotope cadmium. Because the technology and equipment required to produce single isotope cadmium is restricted, the processing of it is extremely expensive. This significantly raises its price. Currently, single isotope cadmium sells for over $1,600 per gram.

An average HeCd laser consumes five to eight grams of cadmium. Using single isotope cadmium can increase the price of the laser many thousands of dollars. The highest powered HeCd laser with naturally occurring cadmium is priced around $18,000. An equivalent laser with isotopically enriched cadmium sells at $25,000.

Having only one isotope lasing in the cavity produces a third of the spectral bandwidth (1.0 GHz). One third of the bandwidth nets three times the coherence length. In essence, the use of single isotope cadmium raises the coherence length from 10 to 30 cm. As an added benefit, single isotope cadmium also produces more power-about 30% more at 441.6 nm. Single isotope HeCd lasers deliver up to 170 mW, TEMoo at 441.6 nm. This translates into 1.7 W of equivalent power when compared to the 457.9 nm line of a large-frame argon-ion laser in effectively exposing photoresist.

Similar to argon-ion lasers, the pumping process in HeCd lasers is accomplished by a high density discharge that breaks down helium gas (the transfer medium). The discharge is produced by an anode and a cathode and confined to a long, narrow glass bore. The spike voltage is typically 13,000 volts and after breaking down the gas (ionization) is reduced to approximately 2,000 volts.

After breakdown of the helium gas, cadmium placed near the anode in a cadmium reservoir is heated to about 2500 C by a heater wrapped around the reservoir. In approximately five minutes, the cadmium starts to vaporize. Through a natural process called catephoresis, the cadmium vapor migrates uniformly from the anode, through the bore, and towards the cathode. Because catephoresis exists, a uniform distribution through the bore is possible. Otherwise, the lasing process would be impossible to control.

It is critical that the cadmium is properly "trapped" before reaching the cathode. Cadmium ions being deposited on the cathode would drastically alter the electrical properties of the laser, affecting the laser's performance. A cold trap is placed centimeters from the cathode to stop the advancing cadmium. Another cold trap is placed in front of the Brewster window, to eliminate the deposit of cadmium on the window.

The strict regulation of both the helium and cadmium vapor pressures is vital to the performance of the tube. The amount of cadmium in the bore can be detected easily by measuring the tube's voltage. A feedback circuit is placed in the head that adjusts the cadmium heater should the tube voltage read too high or too low.

Helium must also be kept at a proper pressure. Helium, being a very small atom, can diffuse out of the tube; the amount is seldom of consequence. During operation, though, the cold trapping process tends to trap the smaller helium atoms under the larger and more massive cadmium atoms. The helium atoms get buried, and the tube pressure eventually lowers. To replenish the lost helium, a refill bottle is placed in the tube with a supply of high pressure helium. When the tube senses low helium pressure, a heater wrapped around the refill bottle is switched on. The elevated temperature raises the kinetic energy in the bottle, creating even higher helium pressure in the refill bottle. The higher temperature also widens the atomic spacing of the container. The combined effect creates an increased diffusion rate of helium from the refill bottle to the tube.

Helium-cadmium lasers that are stored risk subtle migration of helium atoms from refill bottle into the tube. This creates an excess of pressure in the tube, which can make the laser difficult to start. HeCd lasers, when stored, should be started at least once a month and operated 2-4 hours to stabilize the system.

HELIUM-CADIUM lASER

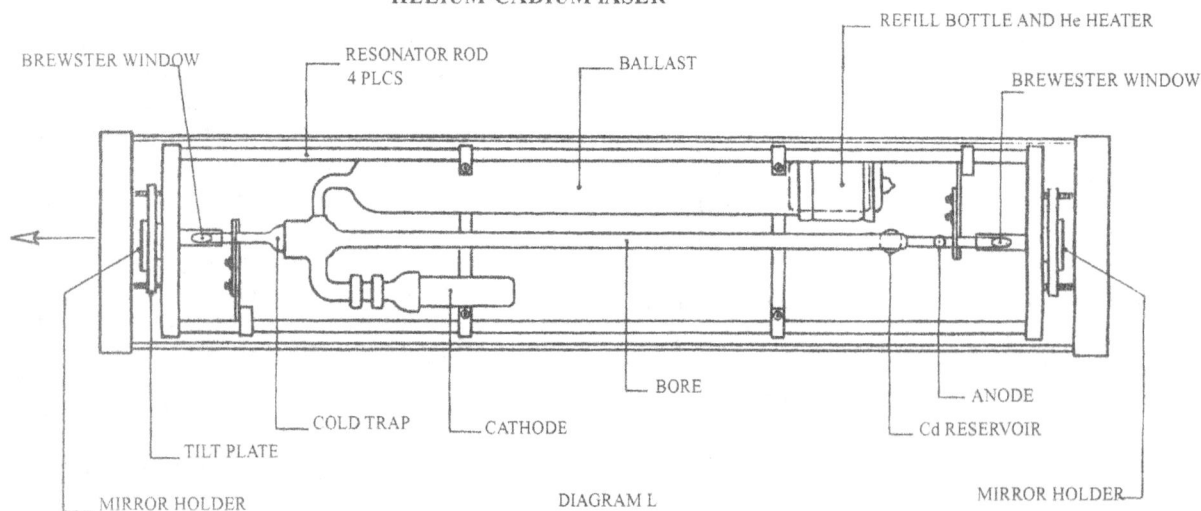

DIAGRAM L

A typical HeCd laser tube typically lasts 4,000 hours. Single isotope cadmium laser tubes, because manufacturers tend to limit the amount of cadmium in the reservoir, last about 3,500 hours.

HeCd lasers have quickly reached a position of dominance in the embossed holography marketplace. The combination of high power, effective exposure rate, low cost of operation, and ease of use make it the laser of choice for embossed holography.

Helium-Neon (HeNe) Lasers

The HeNe laser was the first gas laser to be commercially available, brought to market in 1961. Over 30 years later, the HeNe laser is still the most commonly used laser. Supermarkets use HeNe lasers to scan the bar codes on packages for quick and efficient customer check out. High schools and universities find their low price, ease-of-use, good beam pointing stability, and long tube lifetimes extremely attractive. HeNe lasers operate from a 117VAC source and are air-cooled.

An average HeNe laser costs a few hundred dollars, making it an affordable tool for those who normally could not afford the expense of an argon-ion or HeCd laser. HeNe lasers are low powered, typically delivering between 0.5 and 1 mW, TEM_{00} at 632.8 run. More expensive models are available, delivering up to 35 mW, TEM_{00} at 632.8 run. The beam is generally polarized with a coherence length between 20 and 30 cm. An intracavity etalon may be installed for greater coherence, but the corresponding loss of power tends to create extremely long exposure times. The average lifetime of an HeNe laser tube is about 15,000 hours.

The first HeNe laser ever demonstrated emitted a 1,153 run wavelength, but almost all HeNe lasers are utilized at the 632.8 run line. HeNe lasers emitting green, yellow, and orange wavelengths are also available, but their low power makes them ineffective in most commercial applications.

The active medium in an HeNe laser is neon. As in HeCd lasers, helium is the transfer medium. The laser tube (see diagram H) consists of a glass envelope (bulb containing the cathode) with a narrow bore through its center. The bore can be anywhere from 10 to 100 cm in length, depending on how powerful the laser is.

A 10,000 volt (dc) spike breaks down the two gases in a narrow capillary tube. The voltage drops to between 1,000 and 2,000 V with a current of a few milliamperes. Electrons in the discharge pump both helium and neon atoms to excited states. The more abundant helium atoms collect most of the energy, then transfer it to lower energy or ground-state neon atoms through a series of inelastic collisions.

This transfer of energy is very efficient for two reasons. First, both the helium and the neon gases have two higher energy states with comparable energy values. Second, both pairs of matching higher energy states are characterized by prolonged delays before allowing their electrons to drop. Energy states that hold their electrons longer (up to many milliseconds) are called metastable states.

When the abundant helium atoms in the tube are excited by the discharge current, more of them will have electrons in one of the two metastable states than in other higher energy states. A leaky bucket in a rain storm that holds water longer than a comparably sized peach basket, will more likely have rainwater in it the next day.

A sizeable population of helium atoms with electrons in the metastable state is built up. The excited helium atoms wander in the tube, collide with non-excited (ground state) neon atoms and transfer their energy to them. Since the higher energy levels of the two gases match, the amount of energy transferred to the neon atom is just enough to raise its electron to a metastable state. Soon, most neon atoms have their electrons in metastable state. A population in- version soon develops and lasing begins.

HeNe lasers are being challenged by low-powered laser diodes that emit at the 650 and 670 run wavelengths. Laser diodes are extremely small, light, use very little current, need extremely low voltages, and are less than one fourth the price of an HeNe laser. Laser diodes have inherent properties that are detrimental to holography, however. Low coherence lengths (less than 2 cm), and wavelength instability make the use of laser diodes in holography unfeasible.

The low-cost HeNe will always be the perfect laser for novice holographers. The combination of affordability, high-reliability, and ease of use makes this laser perfect for the production of budget holograms.

CW Lasers

9

This chapter first examines the "holographer-friendly" features that define the latest generation of ion lasers. It then introduces DPSS laser technology, including its advantages and current limitations. An overview of some of the lasers currently available from three of the world 5- major laser manufacturers follows. The chapter concludes by discussing alternative sources for laser equipment and important considerations for those contemplating refurbishing their existing equipment.

ION LASERS IMPROVE; DIODE-PUMPED SOLID-STATE LASERS INTRODUCED

by Paul Ginouves (Coherent)

Until recently, successfully performing holography in most R&D or production applications usually required the use of an ion laser. The ion laser, in turn, required hands-on operator attention to maintain optimum alignment and achieve single-frequency operation. Fortunately, recent developments in laser technology have eliminated both these requirements. Specifically, ion lasers have undergone significant maturation in terms of ease of use and implementation, resulting in true "hands-free" operation, even for applications as demanding as holography.

The first argon-ion laser was made in 1964. Ion lasers opened the door to practical holography and have dominated this application because of their low-noise, high-power output characteristics. Indeed, photopolymers and other materials have been developed with sensitivities that specifically match ion laser wavelengths. As listed in Table 1, argon lasers can produce several watts of narrow-line output at several wavelengths in the UV through green. Krypton lasers have slightly lower output power, but can produce over 2 watts of single-frequency output in the red (647.1 nm) as well as lower powers at other visible wavelengths. Mixed gas (krypton/argon) lasers produce a combination of these output wavelengths.

The basic elements of a single-frequency ion laser head for holography are shown in figure 1. A prism is used as a wavelength filter to select the wavelength line of choice (e.g., 488 nm). Each laser output line has a typical width of 0.004-0.01 nm. This means that the coherence length of the laser is less than 50 mm. Unfortunately, successful holography requires a coherence length many times longer than the optical path difference. For most holographic applications, the linewidth of the laser output must be narrowed by using an intracavity etalon - a simple optic that serves as a narrow-bandpass filter.

Output Line (nm) Argon Ion	Small Frame	Large Frame
All lines, multimode	5W	20W
457.9	0.2W	0.8W
488.0	0.9W	4.2W
514.5	1.2W	5.4W
Output Line (nm) Krypton Ion	Small Frame	Large Frame
All lines, multimode	1.0W	4.6W
413.1	0.15W	1.1W
647.1	0.5W	2.1W

Table 1. Typical single frequency outputs for the higher power visible wavelengths of small and large frame ion lasers.

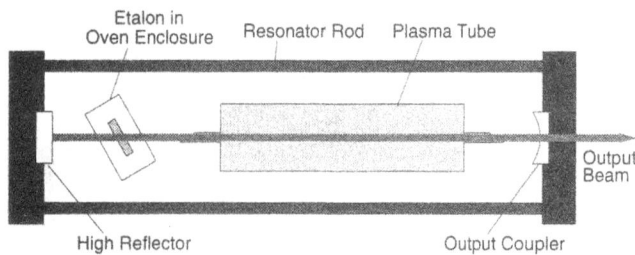

Figure 1- Schematic of key elements of a single-frequency ion laser head.

Figure 2 illustrates how the etalon achieves narrow line operation. Each laser line actually consists of many discrete wavelengths, or longitudinal modes. These modes satisfy the equation, $n\lambda = 2L$ where n is a very large integer, λ is the wavelength, and L is the cavity length, i.e., the distance between the rear mirror and the output coupler. As shown in Figure 2, the loss due to the etalon restricts laser oscillation to a single longitudinal mode.

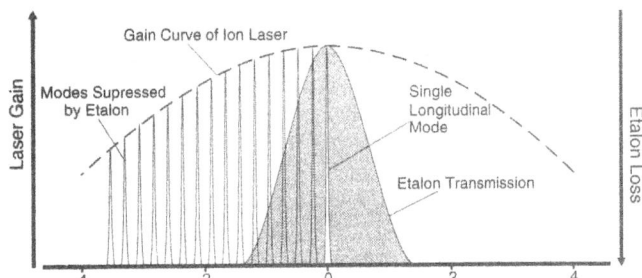

Figure 2 - Frequency Offset From Line Center (GHz)

Ion Lasers Continue to Evolve

In the past five years, manufacturers have made tremendous progress in the area of active stabilization. Early ion lasers had serious stability problems, particularly when operating narrow-line, i.e., with an etalon.

Waste energy from the lasing process generates many kilowatts of heat. Consequently, minor changes in cooling-water flow or tube current can cause measurable changes in the overall head temperature. For this reason, in high-end products, the cavity mirrors are supported on a low-expansion alloy (SuperInvar) structure. But the distance between the cavity mirrors is typically greater than 1 meter. Thus, even with a low-expansion resonator, small changes in temperature have the potential to change the alignment of, and distance between, the cavity mirrors.

Angular misalignment results in a reduction in output power, poor mode structure, and higher noise. For these reasons, the laser required constant attention and adjustment in order to maintain optimum performance.

In 1989, one laser manufacturer, Coherent, eliminated this alignment problem using an approach known as PowerTrack™. With this system, internal photo diode sensors detect changes in the beam power. Signals from these sensors are interpreted by a microprocessor that directs actuators to make compensating angular adjustments on the mirror mounts. Alignment is thus continuously maintained with no operator intervention--ensuring maximum power at a given current, highest transverse mode quality, and lowest optical noise.

Mode Hopping

With single-frequency operation, however, there is still the problem of mode-hopping: the sudden shift from one longitudinal mode to another. Shifts are accompanied by a drop in output power and fluctuations in coherence length. This phenomenon occurs because temperature affects both the etalon thickness and the laser cavity length, thus changing their wavelength characteristics.

If the head temperature drifts, the cavity mode may drift off the center of the etalon transmission peak. (See figure 2.) This causes a drop in laser power. As the drift becomes larger, the next cavity mode will eventually come close to the etalon center and the laser output will suddenly "jump" to this mode.

Because of the associated power and wavelength fluctuation, mode-hops during hologram mastering, or H1 to H2 transfer, have the potential to completely ruin the hologram. This has long been cited as the leading drawback of ion lasers in holography.

To deal with this problem, a number of stabilization schemes have been employed by laser manufacturers. In 1995, Coherent introduced the v-Track™ system on its Sabre[R] ion lasers. In this system, the etalon is enclosed in a stabilized oven to maintain constant temperature. The cavity mirrors are supported on mounts that permit both angular and translational adjustments . To maintain single-mode operation, the laser's microprocessor senses any temperature-induced mode drift and directly compensates for cavity expansion or contraction by lengthening or shortening the distance between the mirrors. This stabilizes the length of the cavity.

This automated system virtually eliminates the effects of temperature drift in the laser head, ensuring hours, and even days, of mode-hop-free operation. Also, warm-up time is reduced to a few minutes. The benefits are greatly increased throughput for any holography application, as well as enabling longer-exposure holograms, such as rainbow holograms, to be recorded with consistently high quality.

To offer additional operational simplicity, these Sabre lasers can be operated by a hand-held remote module. Besides convenience, the use of a remote control can help reduce unwanted motion in the holography studio.

DPSS Lasers

Like many other areas of photonics, laser technology is inexorably moving towards all-solid-state solutions. For holographers, the area of most current interest is diode-pumped solid-state (DPSS) technology. These lasers have been under development since the mid-1980s and are now delivering suitable performance for a number of holographic applications, most notably for nondestructive testing (NDT) and HI to H2 transfers of masters.

A DPSS laser uses a gain medium of Nd:YAG (neodymium yttrium aluminum garnet) or Nd:YV04 (neodymium yttrium aluminum vanadate). These are efficient lasing materials that produce near-infrared (1,064 nm) output. This near-IR light can be frequency doubled to the green (532 nm) by a nonlinear crystal such as KDP (potassium dihydrogen phosphate).

Many lasers incorporating Nd:YAG or Nd:YV04 use high-energy flashlamps to optically pump these mate-rials. Such lasers are used in holography for generating pulsed masters of moving objects. In DPSS lasers, the pump energy is supplied by laser diodes. These are more powerful versions of the same types of laser found in CD players. They efficiently convert electricity to light and do not require water cooling. Also, their output can be tuned to match the absorption profile of the crystal gain medium so that very little of their pump light is wasted.

In multiwatt lasers, these laser diodes can be located in a compact power supply and the pump light supplied to the head via fiber optics. This results in an extremely compact, high-power laser head and minimal heat loading, requiring no external cooling.

DPSS Lasers for Holography

Until very recently, no DPSS laser offered the appropriate characteristics for the holography marketplace. Some companies concentrated on developing high powers (lip to 1 watt by 1995) but with multimode output and, hence, poor coherence. Other lasers were designed to capitalize on the potentially long coherence length of Nd:YAG and Nd:YV04. Originally producing tens of milliwatts of green output, these single-mode lasers had reached the hundreds of milliwatts by 1995. In 1996, however, there was a quantum leap in performance with the development of a laser (Coherent's Verdi ™) which produces 5 watts of CW, single-mode output.

As shown in figure 3, this laser uses a ring configuration for efficient single-mode operation and overall compactness. The compact cavity results in favorable mode-spacing, making it easier to select a single mode using the internal etalon. The probability of temperature-induced mode-hopping also is greatly reduced. Nonetheless, to ensure mode-hop-free operation, active stabilization is employed. This closed electronic system is completely transparent to the user. In fact, the only control on this laser is the power switch.

This laser offers several important benefits for holographers because of its all-solid-state construction. It is rugged, reliable, compact, requires no cooling-water, operates from a standard 110 V supply and is highly portable. Fur-

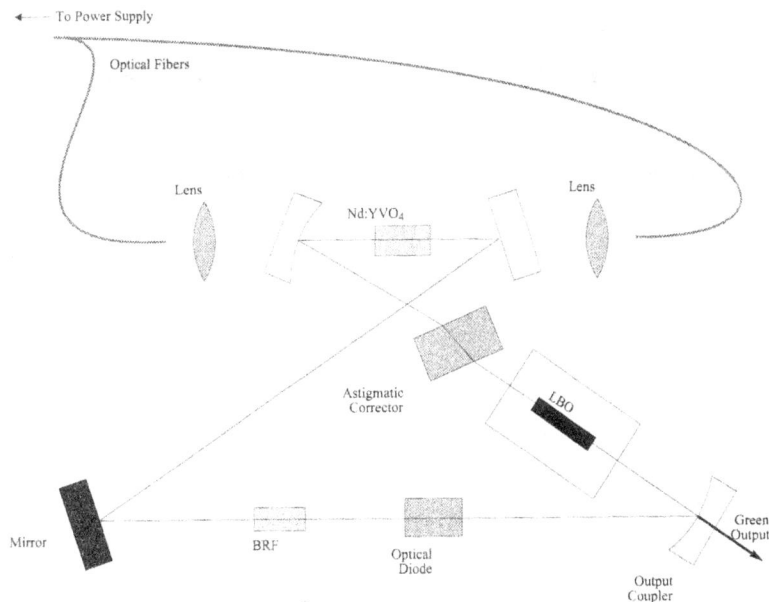

Figure 3

Figure 3. The active medium in a DPSS laser is a neodymium-doped crystal that is optically pumped by light from laser diodes.

thermore, the spectral linewidth of the output is much narrower than an ion laser-making for a much longer coherence length.

At this time, the only significant limitation of DPSS lasers is that they are restricted to a single visible output wavelength: 532 nm. Ion lasers will therefore continue to be the preferred lasers in applications requiring other visible wavelengths.

CONCLUSION

To summarize, it is unlikely that ion lasers for holography will ever be as simple and reliable to operate as a light bulb. However, recent progress in automated operation and active cavity stabilization systems has certainly moved ion lasers a long way in the direction of that ideal goal. At the same time, a new laser technology that does have the potential for light-bulb simplicity and reliability has become available to holographers. With these tools, expect continued expansion in the applications of holography.

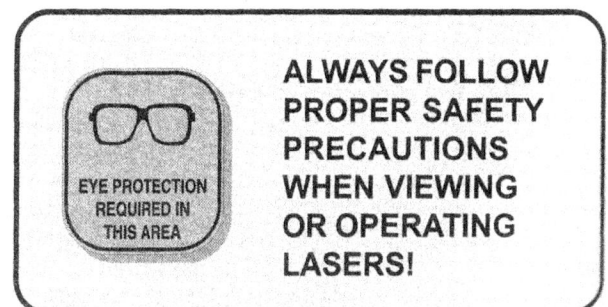

ALWAYS FOLLOW PROPER SAFETY PRECAUTIONS WHEN VIEWING OR OPERATING LASERS!

EYE PROTECTION REQUIRED IN THIS AREA

COHERENT, INC. LASER GROUP

MANY CHOICES FOR HOLOGRAPHERS

Coherent, Inc., is a major manufacturer of lasers and laser systems for scientific, medical, micromachining and entertainment applications. The Laser Group, one of four divisions within the company, designs and manufactures ion, CW, YAG, YLF, ultrafast, CO2, tunable-dye, diode, and diode-pumped solid-state lasers. Headquartered in Santa Clara, California, the Laser Group was the foundation upon which Coherent was built and today is the most diverse business group within the company, providing laser products targeted at many different electro-optical applications, including holography. Below is a partial list of its lasers that are useful to holographers.

VERDI™ DPSS LASER

Coherent's Verdi™ is a single-frequency ring laser design producing a compact and efficient source of CW green (532 nm) light, suitable for exposing photopolymer. As of fall '98, several photopolymer hologram manufacturing facilities had integrated it into their mass replication equipment.

The Verdi™ is also particularly well-suited to NDT applications, as its high power and long coherence enable faster testing of larger parts. In addition, because of the laser's portability, there is the potential to test these larger parts in the field instead of hauling them into a laboratory. The laser is capable of transferring (copying) HI holograms made with high-power pulsed Nd:YAG lasers. Transferring a hologram with the same wavelength used in the original recording greatly simplifies the transfer process, and can increase production yields.

The 2 W Verdi™ uses the same ring laser cavity as the existing 5 W Verdi, resulting in a single-frequency output with exceptionally low noise. In the 2 W version, a single, fiber-delivered diode bar is used as the pump source, rather than the two diode bars used in the 5 W version. Both the 2 W and 5 W versions require only standard 110/220 volt single-phase power and no external cooling. Numerous other laser systems, based on the Verdi™ platform and utilizing these technologies, are under development.

SUITABLE WATER-COOLED ION LASERS

INNOVA 300 Series Ion Lasers The INNOVA 300 Series Ion Laser System is a full-feature small-frame ion laser, offering high power, low RMS and peak-to-peak noise, power and mode stability, and actively stabilized single-frequency operation. The 300 Series is available in either argon or krypton. Multiline visible power up to 10W, TEM_{00} and multiline UV power up to 1W are available. To enhance productivity and performance, PowerTrack™ Coherent's actively stabilized optical cavity-is a standard on the INNOVA 300. And, for single-frequency applications, ModeTune simplifies etalon optimization and ModeTrack eliminates mode-hops.

Type	Model	Power (W)
Argon	INNOVA 304	4
Argon	INNOVA 305	5
Argon	INNOVA 306	6
Argon	INNOVA 307	7
Argon	INNOVA 308	8
Argon	INNOVA 310	10
Krypton	INNOVA 301	1
Krypton	INNOVA 302	1

The INNOVA Sabre Ion Lasers

The INNOVA Sabre Ion Laser System is a large-frame laser that combines a very stable basic design with performance-enhancing active components. Sabre's Sentry system will automatically acquire lasing, tune to a specific wavelength, and peak for maximum power, TEM_{00} mode, and minimum noise. PowerTrack™ and v-Track™ provide immediate warm-up in multiline, single-line, or single-frequency applications, and ensure stable performance even in changing environments.

Type	Model	Power (W)
Argon	Sabre TSM 10	10
Argon	Sabre TSM 15	15
Argon	Sabre TSM 20	20
Argon	Sabre TSM 25	25
Argon	Sabre DBW 10	10
Argon	Sabre DBW 15	15
Argon	Sabre DBW 20	20
Argon	Sabre DBW 25	25
Krypton	Sabre DBW Kr	4.6

INFINITY ND:YAG

In addition to the CW lasers listed above, the company has recently produced a pulsed Nd:Yag laser (the Infinity Series) capable of >20 W output at 532 nm, which could interest researchers doing pulsed holography.

SPECTRA-PHYSICS

PIONEERING COMPANY STILL SUPPLIES INDUSTRY

Spectra-Physics, based in Mountain View, California, is the world's largest and oldest designer and manufacturer of helium-neon lasers. (Their first HeNe was introduced in 1963.) Over the years, numerous silver-halide holograms have been mastered and copied using the company's red emitting lasers. These lasers have proven reliable and extremely durable. Consequently, the company has established a very solid reputation within the holography industry. The company's line has since expanded to include other CW lasers that are suitable for exposing photoresist and photopolymer emulsions, including water cooled ion lasers and DPSS CW lasers. The company also manufactures Nd:Yag pulsed lasers that can be used for some holographic applications.

HENE LASERS

Currently the company offers two versions of high power helium-neon lasers with a long (approx. 1m) coherence length: Model 127 (a complete unit) and Model 107B (for OEM applications). Both versions deliver either 25 or 35 mW of TEM_{00} polarized 632.8 nm light, though most working holographers will order the former.

According to company literature, Model 127 features an improved mirror configuration, mounted on adjustable plates, that enhances stability. The resonator design minimizes the effect of temperature changes on output power. Hard-sealed plasma tube Brewster windows make the plasma tube impervious to contamination and provide unlimited shelf life. A large cathode and gas reservoir greatly increase operating lifetime, which typically exceeds 20,000 hours. The plasma tube and the power supply are integrated into a single compact unit. The laser comes equipped with a threaded mounting bezel for various optical accessories.

STABILIZED WATER-COOLED ION LASERS

As new recording materials became available to the holography industry, other types of lasers become necessary for commercial production. However, unlike the simple air cooled HeNe's, the more powerful argon-ion and krypton lasers were water-cooled and a bit more finicky. Beam stability and output reliability became even more crucial in manufacturing environments.

Beam motion, or wander, is caused by changes in the optical cavity alignment of a laser. These changes result primarily from temperature variances as the laser warms up, alterations in the ambient environment or cooling water temperature, and changes in the operating current to the laser. Beam motion obviously hinders hologram production. The engineers at Spectra-Physics designed several

methods to ensure reliable single frequency operation, called BeamLok®, Z-Lok®, and J-Lok®.

The BeamLok design is centered around a quadrant cell detector and an actively steered piezo-mounted output coupler. A small portion of the beam is split off outside the laser cavity and is used as a reference beam into the quad cell. An error signal is generated for the piezo-mounted output coupler. With the output mirror moving in the X and Y axes, beam pointing is maintained and wandering eliminated.

Z-Lok was developed to address the common problems that occur in single frequency operation: mode hops. Mode-hops occur when the cavity and etalon lengths change relative to each other, primarily as a result of temperature variations. This can plague the holographer during critical single frequency exposures. To eliminate mode-hops, Z-Lok uses a temperature stabilized etalon as a reference, locking the laser's cavity to it. The end result is minimal absolute frequency drift and mode hop elimination.

J-Lok: In addition to mode hops, short term jitter caused, for example, by cooling water flow can also affect single frequency exposures. Reduction of this jitter is especially important for holography, with exposure times in the microseconds-to-minutes time scale. Using the inherent response capability of the piezo electric crystal employed in all BeamLok lasers, J-Lok reduces the low-frequency jitter in the 10-500 Hz range.

The BeamLok® Series

BeamLok is the trade name of Spectra-Physics' family of argon, krypton, and white light ion lasers that employ the aforementioned stabilization methods. The BeamLok series offers a range of output power options in visible wavelengths ranging from 454 .5 to 514.5 nm and TEM_{00} operation. The small frame 2060/65 systems provide up to 10 W multi-line visible output. The large-frame 2080/85 systems are configured for 12 to 30 W multi-line argon, up to 5 W multi-line krypton, and 14 W multi-line white light outputs.

DIODE-PUMPED SOLID-STATE LASERS

Spectra-Physics also makes a line of solid-state lasers (Millennia) that can be used to expose green-sensitive photopolymer emulsions. These lasers could be very effective for contact copying H1s to IDs, though the coherence length probably makes mastering impractical. These lasers are air-cooled, which makes them inherently more stable and more convenient to operate than their water-cooled counterparts. The solid-state technology offers higher efficiency, lower utility requirements and operating costs, smaller size and weight, and longer lifetime. The Millennia technology platform has been extended to include 2 W (Millennia II) and 10W (Millennia X) versions.

MELLES GRIOT

MELLES GRIOT ACQUIRES LiCONiX

Melles Griot, a worldwide supplier of a wide range of photonic products including lasers, announced that on September 25, 1998, an agreement was reached to acquire LiCONiX. Located in Santa Clara, CA, LiCONiX is a leading supplier of low noise and high power helium-cadmium lasers used for automated optical inspection, precision stereo lithography, holography and research applications. The company will continue to manufacture LiCONiX lasers at the Santa Clara facility.

In 1997, Melles Griot also acquired the business of Omnichrome Corporation. Omnichrome was a leading supplier of helium-cadmium and ion lasers to scientific and original equipment manufacture (OEM) markets. As a result of these acquisitions, the Melles Griot product line now includes a full spectrum of helium-cadmium, helium-neon, argon-ion, and krypton-argon lasers with wavelengths from 325 nm to 3.39 mm and power output to 500 mW Many are appropriate for holographic applications.

HeNe Lasers

Of special interest to display holographers is a new line of economical 35 mW+ helium-neon lasers featuring the 05 LHP 928 linearly polarized and 05 LHR 928 randomly polarized models. These lasers use the Melles Griot hard-sealed internal cavity mirror construction for long lifetime and long-term mirror alignment. As with all Melles Griot lasers, beam delivery systems and other accessories are available.

Here are some frequently asked questions regarding HeNe lasers:

How do you determine the coherence length of a helium-neon laser?

MG - Coherence length is defined as the length over which energy in two separate waves remains constant. With respect to the laser, it is the greatest distance between two arms of an interferometric system for which sufficient interferometric effects can be obtained: Lc = c / D n L. Coherence length will vary from laser to laser as a function of the Doppler broadened gain width; however, for an HeNe, 20-30 cm is typical.

What is the Doppler broadened gain width of a helium-neon laser?

MG - helium-neon lasers can range from 800 MHz to 1600 MHz full-width-at-half-maximum (FWHM) depending on the design. The typical red HeNe is 1400 MHz. The width of a single mode located under the gain curve is approximately 1 MHz.

Is it practical to consider re-gassing a HeNe laser?

MG - While re-gassing can provide some extension of the output performance in some gas lasers like the CO_2, Argon and the higher powered side arm HeNe's (which have external optics), it is not recommended or provided for smaller internal mirror coaxial tubes. Typical end-of-life failure for an HeNe is cathode sputtering. This occurs when the protective oxide layer on the cathode is expended through continuous bombardment by the laser discharge. There is no cost effective way of regenerating this layer. When the oxide layer is expended, the discharge itself vaporizes the "raw" aluminum and deposits this material, in its vapor state, on other surfaces such as the optics and bore.

Other Suitable CW Lasers

In addition to their helium-neon lasers, the company carries a range of other CW lasers that are suitable for commercial display holographic applications, including helium-cadmium (200 mW/442 nm), argon-ion (100 mW/457-514 nm and 300 mW/454-514 nm) and krypton-argon-ion (up to 150 mW/467-752 nm) lasers. Contact Melles Griot for a full product catalog.

Diode Laser for Interferometry

Of special interest to industrial holographers is a new diode laser with long coherence length and stable power output. For instance, model 56 IMS 663 is a completely self-contained, temperature-stabilized diode laser system with long coherence length and stable output, especially designed for holographic interferometry. It delivers 18 MHz direct modulation capability and high-power output in a small footprint, and requires only 5 VDC to operate. A front panel C-mount is included for beam expanders and spatial filters.

Features:

- 685 ± 10 nm
- >3 m coherence length
- <100 MHz linewidth
- Power stability < 0.05% over 60 minutes
- >25 mW delivered output power
- dl/dT is 0.6 GHz/°< C
- 8 MHz (-3dB) modulation bandwidth

LiCONiX

LiCONiX is now a Melles Griot line of UV and blue lasers, based on helium-cadmium (HeCd) technologies and includes HeCd lasers with all 325 nm, 442 nm, and 354 nm wavelength options. (The 354 nm line was discovered and patented by LiCONiX.) Company engineers claim that holographers mastering embossed holograms will greatly benefit from using their 442 nm output HeCd lasers rather than using comparable argon-ion lasers (458 nm) due to the fact that standard photoresist emulsions are ten times more sensitive to the former wavelengths.

Also, according to company literature, the bandwidth of HeCd lasers using natural Cd vapor is approximately 3 GHz, corresponding to a coherence length of 10 cm. Using isotopic-enriched Cd, the bandwidth is reduced to 1 GHz, corresponding to a 30 cm coherence length. Both the single wavelength output and the relatively long coherence length make the HeCd laser a practical choice for embossed holography mastering (the company's "NX" models employ isotopically enriched cadmium for enhanced coherence length and higher power).

HeCd lasers range in power from a few mW to almost a quarter of a watt. As with all CW lasers, the output power is determined by the gain length-the size of the laser. LiCONiX offers HeCd lasers in three sizes: small frame 61 cm (24 inch), medium frame 102 cm (40 inch), and large-frame 140 cm (55 inch). The company's medium and large-frame Embosser Series was designed to give holographers making embossed holograms and diffraction gratings the high power they require.

There is, as always, a trade-off between power and mode. LiCONiX offers all frame sizes in TEM_{00} and TEM_{mm} configurations. For holographers, the company states that their TEM_{00} mode lasers are "true TEM_{00}," rather than the less exact "visual Gaussian." The M-squared value is typically less than 1.10. For more information on laser beam characterization, contact a company representative.

Recently LiCONiX released a new CADLine series of lasers to comply with current and projected CE mark regulations for sale into the European community. Their performance is identical to the older lasers listed below, but European model numbers may be different than those listed.

LARGE-FRAME HECD LASERS

Model: Embosser II Series

          325 nm, 354 nm, 442 nm

Application: Embossed Holography

Features: World's highest power HeCd laser.

          Available in two different coherence lengths: 10 cm and 30 cm.

          Expected Life: >4,000 Hrs

          Polarization, plane vertical >55:1

Expected Life: >4,000 Hrs

          Polarization, plane vertical >55:1.

Large-Frame HeCd Lasers (all TEM_{00} Mode)

Model	A	Power	Dia.	Coh. Len.
Emboss II	(nm)	(mW)	(mm)	(cm)
3620N	325	20	1.2	10
3630NX	325	30	1.2	30
46120N	442	120	1.3	10
46150NX	442	150	1.3	30
46170NX	442	170	1.3	30

MID-FRAME HECD LASERS

Model: Embosser I & Series 200

          325 nm, 354 nm, 442 nm

Application: Embossed Holography

Features: Convection cooling and carbon-fiber composite resonator rods provide outstanding stability.

          Excellent mode quality.

          Low Noise, 200 series incorporates LiCONiX "Noise Lock" feature.

          Expected Life   >4000 Hrs.

          Polarization, plane vertical >500:1

Mid-Frame HeCd Lasers (all TEMoo Mode)

Model	A	Power	Dia.	Coh.Len.
Emboss I	(nm)	(mW)	(mm)	(cm)
3210N	325	10	1.0	10
3216N	325	16	1.0	10
3220NX	325	20	1.0	30
7205	325	5	1.0	10
4270N	442	70	1.2	10
4290NX	442	90	1.2	30
Series 200				
3207N	325	7	1.0	10
321 IN	325	11	1.0	10
3215NX	325	15	1.0	30
7203N	354	3	1.0	10
4320N	442	30	1.2	10
4240N	442	40	1.2	10
4250N	442	50	1.2	10
4260NX	442	60	1.2	30

The company also produces a range of optical equipment for the holography studio, including a power meter.

ALTERNATIVE SOURCES FOR LASERS

For those who cannot afford to purchase a new laser from the original equipment manufacturer (or an authorized distributor), there are alternative sources for lasers suitable for holography. Specifically, there are companies that specialize in selling never-been-used "surplus" equipment and others that resell used and/or refurbished hardware, at prices far below retail. Besides offering low prices, these companies often have access to "obsolete" lasers and related equipment which holographers still find desirable.

It is also possible, but often less practical, for knowledgeable buyers to find decent deals on lasers from various other sources-such as optical laboratories, universities, manufacturing plants, and corporate auctions. It is worthwhile to hunt for bargains at these places if you have the time and the expertise. Unfortunately, these sources do not provide the technical support, selection, and service needed by most professional holographers and hobbyists.

For these reasons, we recommend that you contact those companies whose primary business is the resale and/or repair of electro-optical equipment (see the Business Directory in this book for a listing of such companies). These companies have developed a close working relationship with the laser industry and thus can provide a high level of customer service.

Since many of these companies are staffed by experienced laser aficionados, they are often able and willing to answer technical questions before and after the sale. Salespeople are typically encouraged to spend some extra time with novice shoppers in order to ascertain their specific needs and steer them to the right equipment (as in most industries, the factory's sales force is typically geared to service the larger corporate accounts and don't have the resources to answer a lot of basic questions from small businesses and hobbyists interested in purchasing single units). In addition, most of the companies we know about in this business have one or two people on staff who are especially familiar with the lasers required by holographers.

Another good reason to shop with resale businesses is product availability and selection. Most of these companies carry a wide selection of lasers from different manufacturers in various price and power ranges, rather than only a single product line. Prices usually range from a few hundred dollars for low powered units to thousands of dollars for more industrial gear. Although most holographers are hunting for ReNe and argon lasers, it is also possible to find an assortment of more esoteric hardware, such as Reed, Nd:Yag and ruby lasers.

Most notably, these companies often stock models which the manufacturer has discontinued, even though they have proven extremely useful to holographers in the past. These companies also sell components (laser heads, power supplies, optics, etc.) for customers interested in assembling their own units and spare parts that might otherwise be unobtainable. Several U.S. companies mentioned that they routinely customize equipment to make it suitable for countries with different electrical systems.

Finally, it is important to consider warranty protection. Gas lasers do have a finite life expectancy and are rather delicate pieces of equipment. Reputable companies should provide some sort of guarantee, especially on used equipment (some buyers might prefer buying refurbished lasers, as the life expectancy can be more easily ascertained). Every company we surveyed offered a warranty, ranging from 30 days to a full factory equivalent. Obviously, it benefits the customer to have one, especially if the laser is being shipped.

Where does this surplus and used laser equipment come from? One company we interviewed specializes in purchasing large quantities of surplus components directly from the OEM and resells both pre-assembled packages or individual parts. This "factory fresh" equipment is typically obtained from excess inventories of discontinued models, spare parts, and production overruns. Another company we asked acquires large lots of used equipment from corporate users who are upgrading their equipment or switching technologies. For example, many inexpensive ReNe gas lasers became available when supermarket chains purchased new bar-code scanners built around solid-state diode lasers. Other businesses purchase older units from a variety of sources and refurbish them to factory specs. All these companies are potential sources of good equipment at discounted prices.

Since the primary reason for buying surplus, used, or repaired lasers is price, we surveyed several businesses to find out how much a customer could reasonably expect to save. Steve Garret, owner of Midwest Laser Products, replies, "The amount of money saved depends on the cost of the laser; however, one can often save (an average of) 50% off the price of new equipment."

Martin Rasa of MWK Industries concurs: "The clear advantage of purchasing a surplus or used laser is cost. A new ReNe laser purchased directly from a manufacturer or one of their distributors will usually cost about three to six times more than a similar unit purchased from us. For example, in a recent catalog we listed a surplus lO mW ReNe and power supply for one-third the cost of a similar system offered by a major distributor." Substantial savings are available for refurbished equipment, too. Rasa concludes, "Though the price of new laser equipment can be far beyond the means of the average individual, used (and surplus) equipment makes holography an affordable pursuit for people, companies and institutions on a tight budget."

LASER REFURBISHING AND REPAIR

REFURBISHING GAS LASERS

an interview with Don Gillespie (El Don Engineering)

A holographer's livelihood depends upon having a laser that functions properly. Educators and hobbyists also rely on their lasers to perform well, though many have "inherited" aging hardware that has been questionably maintained. Fortunately for professionals and aficionados alike, if problems do arise with this essential piece of equipment, there are companies around the world that specialize in fixing a wide variety of lasers.

In this article, we present an interview between Holography MarketPlace (HMP) and Don Gillespie (DG) of El Don Engineering concerning the repair of lasers, especially the merits and economics of refurbishing or reprocessing the laser tubes of gas lasers. Mr. Gillespie is highly experienced and is very familiar with the needs of holographers.

HMP: What type and power of lasers can you repair and/or refurbish? Do you specialize in a particular type?

DG: We specialize in the repair of Spectra-Physics helium-neon and argon-ion lasers of all ages and models. And in particular, the HeNe models 120, 124, 125, 127, and the industrial models 107 and 907. Holographers commonly use all of these. We also repair the popular S-P 165 and 171 argon-ion high power laser.

HMP: For the less technical laser users, please list some obvious signs that their laser needs to be refurbished.

DG: Failing HeNe lasers usually begin to blink on and off a short while after they are turned on. They generally lase while they are on and may produce a usable power output. However, the blinking is a pretty sure sign the gas is used up. Another indication of failure is the tube color turns essentially blue or white. Tubes that fail in this mode generally do not blink, do not lase at all, or lase poorly.

HMP: What's involved in re-gassing a laser? How much will performance improve? How much can a customer expect to pay for such repairs?

DG: Re-gassing is a misnomer. One does not simply replace the gas. Generally speaking, the tube is removed from the frame, opened up, and inspected. Those parts that may require replacing, including windows, cathode, and getter, are replaced and those parts that may not need replacing, such as the glass, are inspected, cleaned, and reused. The tube is then re-gassed to the manufacturer's original specification using the same (nonradioactive) isotopes it used. It is an involved process. A properly refurbished or reprocessed laser tube will perform to the same specifications and have a service life equal to that of a newly fabricated tube. The cost savings is obviously in the reuse of the component parts of the tube that are not "used up" in the original service life of the tube.

The cost of these repairs depends, of course, on the size and type of laser tube. Also, HeNe laser repairs are less expensive than argon laser repairs. HeNe repairs cost as little as S500 for the 6 milliwatt S-P Model 120, and as much as $2,500 for the S-P Model 125 that generally produces 65 to 70 milliwatts.

Tube repairs to argon lasers such as the popular S-P model 165 (5 watt output and water-cooled type) generally cost between $3,500 and $5,000 depending on the nature of the tube failure. These prices include the simple field modifications recommended by the manufacturer, and minor repairs.

HMP: How many times can a laser be re-gassed?

DG: Generally, several.

HMP: What electrical/mechanical problems do you typically encounter? How much can a customer expect to pay for such repairs?

DG: The more common electrical failures include the loss of current regulation to the tube, and also transformer burnout. With the exception of transformers, the cost of parts is small: often a few tens of dollars. The bench rate to troubleshoot and replace the parts is $75 per hour. Most HeNe's can be repaired in a half day or less. Most argons take a half day or more. A whole lot has to have gone wrong to take more than a day to repair. We have a complete electronics shop and machine shop and do all our work in-house. The electronics repairs are done by an electrical engineer.

HMP: How frequently do you run across alignment or other "optical problems"? How much can a customer expect to pay for such repairs?

DG: Not too often. These problems generally occur when the laser user is a "knob twister" and can't let the alignment alone. I strongly recommend that all alignment be done using the screws on one end of the laser, only. This way the user has a chance of getting the alignment back in order if he/she loses the light beam. Another problem with alignment comes up when a laser is shipped to the customer. Although well-packed, lasers sometimes lose their alignment in shipping. This is usually very upsetting to the customer. However, on the whole, I have been able to talk the user through the often trivial adjustments that are necessary to correct the problem over the telephone. To be successful, it does require that the user be one who can follow instructions and be a good communicator. It can be nerve rattling. The cost of non-warranty optical alignment work is the same as the bench repair rate: $75 per hour.

HMP: Is it possible or feasible to "upgrade" lasers (i.e., install new or better components)?

DG: I presume the intent of the question is directed toward the purpose of improving the power output of the laser. The answer is generally no. The manufacturers are really pretty sharp at their engineering and work right to the limits set by the physics of the device. By the same merit, we are able to restore the laser to equal or exceed the original power output specification, as the specification is usually conservative. The manufacturer doesn't want to get caught short either, so he leaves a little comfort space at the top when he sets the specifications. Regarding component failure: if a part shows a propensity to fail regularly, we replace it with a more hefty capacity part, or if required we re-engineer the circuit so it will take the load.

HMP: What's the best way to ship equipment to you?

DG: I recommend either UPS or FedEx for those lasers small enough to be shipped by these carriers. The laser head and the power supply should be shipped in separate boxes. Each unit should be wrapped in bubble cell packing, with the bubbles out so they will engage the foam peanuts that should be used as the balance of the packing. Don't skimp on packing, including the ends. The carrier may refuse claims of damage if the packing is insufficient. The box you send it in with is the box it will come back in. If the box is insufficient or of poor quality, we may replace it. There may be a small charge for the box and materials if we are required to do this. Do not use- wooden crates for any but the largest lasers. They are expensive to make, expensive to ship, and contrary to common judgment, often result in the contents being damaged. They don't land as lightly as a cardboard carton when thrown or dropped. It is important to insure the content of each box.

HMP: Do you warrant your work?

DG: Yes, workmanship and materials that are provided in the repairs are warranted for one year from the date of shipment FOB, Ann Arbor, Michigan. There are some exceptions on electronic repairs, which may be warranted for 90 days.

HMP: How long do repairs typically take? Do you do rush jobs?

DG: Repairs usually take two or three days to two weeks. The time required depends on what the nature of the failure is. On the whole, ReNe's are repaired faster than argons. Electronic work goes quickly unless we are delayed by parts availability, which is not often, as we stock most replacement parts. Rush jobs are usually accommodated, as not everyone is concerned about being "first in, first out." They just want the work done in a reasonable time.

HMP: Is there something else you would like to add to this interview?

DG: Yes. I have been at this business a long time, about 30 years. I have enjoyed the work and the people I have met doing it. I welcome calls from those who may have a question to ask and I do try to fix the problems "long distance" if the conversation is not too long. Also, we do have lasers in stock for sale, and purchase selected used lasers. I wish to thank Ross Books for the opportunity to talk with you and your readers.

For further information, contact:
Don Gillespie
EI Don Engineering 4629 Platt Rd.
Ann Arbor, MI 48108
Phone: 734 973 0330
Fax: 734 973 0330
Email: eldonlaser@aol.com

10
Pulsed Lasers

Pulsed lasers seem to be undergoing a slight renaissance in the holography industry, mainly because of a few individuals who tirelessly promote the advantages of using these systems. This chapter will describe the newest systems available from manufacturers that specialize in this technology.

PULSED, NEODYMIUM GLASS LASERS FOR HOLOGRAPHY

by David Ratcliffe (Geola)

Pulsed lasers offer one very large advantage over conventional CW systems for display holography: a complete indifference to vibration and motion. In no uncertain terms, these lasers free the holographer from his normal rule book. With the nanosecond exposure times characteristic of the Q-switched pulsed laser, human and animal portraits become possible: unstable models, falling leaves, water jets, smoke, and even speeding bullets and fast impacts: all definitely out of the question with CW but possible with pulsed....

Of course, until now there have been several catches. First, there have been few available pulsed laser systems designed especially for holography-and this is pretty important unless you are a trained laser engineer and you don't mind redesigning a laser intended for another application. Second, the holography laser systems available until now have operated exclusively in the very deep red, a wavelength characteristic of the ruby crystal and which can at best be described as "inconvenient" for display holography.

A new ruby laser with enough energy to be useful usually comes with a price tag approaching six digits. Second-hand systems admittedly do fetch a more moderate sum. But put all this together with the fact that no one has ever really offered a professional commercial camera system based on an available laser, and the fact that the nontrivial task of designing and putting together the optical scheme outside the laser has always been the holographer's job,

and then one can perhaps understand why pulsed lasers have remained until now a somewhat untouchable dream to all but an elite few.

To further exacerbate this situation, Agfa pulled out of the (red) holographic emulsions market in 1997, leaving no clear producer for film or plates suitable for ruby laser use.

This rather depressing situation is however not the end of the story. In a somewhat serendipitous fashion, there have recently appeared several manufacturers of holography laser systems whose products are based on the more advanced neodymium technology and who have done what no ruby laser manufacturer ever did: supply entire camera systems-not only for mastering but also for transfer copying.

RUBY VS. NEODYMIUM

Notwithstanding the rather crucial issue of emulsion availability for ruby lasers that has recently arisen, there are several important reasons why today's holography laser manufacturers have opted for solutions based on neodymium technology rather than ruby.

Ironically, the ruby laser has come to be regarded amongst the holographic community as the only seriously viable laser for pulsed holography. However, this has happened mostly for the reasons of availability discussed above and appears not to be based on a sound appraisal of the relative merits of the different possible systems.

We must first observe that ruby is actually a three-level quantum system. Lasing in neodymium is, by contrast, governed by a four-level scheme, making the neodymium laser inherently more efficient than ruby. Practically, this difference becomes visually evident when you compare typical neodymium glass laser power supply units with those of ruby lasers. The neodymium glass units are usually small single phase modules that can be plugged into any household supply. Ruby units are large industrial-type racks requiring hefty three-phase outlets. And because ruby lasing is fundamentally less efficient than neodymium, optical pumping in a ruby laser must be much higher. This leads to complications that do not arise in a weakly pumped system.

But perhaps even more important than these size and power consumption issues is the fact that the green frequency-doubled neodymium light is far more appropriate as a lighting source for display holography than the far-red light of ruby. For portraiture applications, it has been known for many years that ruby light penetrates the outer skin layer and is reflected only by the relatively uniform subcutaneous layer. This is what is responsible for those waxlike images you see in ruby portraits devoid of any surface detail. Neodymium, of course, doesn't do this. Light at around 530 nm behaves as we are used to in our normal everyday experience. And, in addition, our eyes are much more sensitive to this part of the spectrum.

Finally, every CW holographer knows how useful spatial filtering is when you are making transfer copies and even when mastering. It is extremely difficult and highly discouraged to use spatial filtering techniques with a ruby laser' as back-reflection from an accidental optically induced spark may do serious damage to the laser. With neodymium lasers, the frequency conversion crystal that converts the infrared lasing light to the visible green acts as a kind of very efficient optical protector, effectively preventing back, reflected light from reentering the laser and causing damage. This difference basically means that neodymium laser beams can be cleaned up a lot easier than ruby beams and hence systems based on neodymium can do most things a CW laser can, including the production of transfer copies.

CHOOSING A PULSED LASER

Neodymium lasers come in various shapes and forms. Those most useful for holography applications are based on the neodymium-doped yttrium aluminium garnet crystal (Nd:YAG) or the neodymium-doped yttrium lithium fluoride crystal (Nd:YLF). Both crystals are optically pumped by flash lamp (as in the ruby laser) and emit in the infrared. YAG lases at 1064 nm and YLF a little lower at 1053 nm". A nonlinear frequency-doubling crystal such as KDP, KTP, or DKDP is used to convert the emission into a green output at 532 nm (YAG) or 526.5 nm (YLF). Essential features for any holography laser are a linearly polarized TEMoo transverse mode with a high quality quasi-gaussian beam profile and true single longitudinal mode operation.

YAG Lasers

Most of the large laser manufacturers such as Continuum, Coherent, Spectra, Spectron and Quantel make standard model neodymium YAG lasers giving energies in the green of up to around 1 to 1.5 J. Such lasers use a neodymium YAG oscillator followed by one or more YAG amplifiers. Depending on whom you talk to and when, YAG lasers are or are not useful for display holography.

The true answer is that such lasers can be used for display holography with some effort but are not designed for this and therefore they usually represent an inappropriate choice for such an application.

All standard YAG lasers achieve "Single-Longitudinal Mode" operation-which is the mode you need to ensure long coherence-by the use of a complicated and costly technique known as injection seeding which requires continual firing of the oscillator and which necessitates pulse shuttering for standard holography use. Injection-seeded systems are moreover difficult to realign and require the use of expensive scopes with nanosecond resolution. In addition, YAG lasers are usually designed to pump doubling or tripling crystals whose conversion efficiency is critically dependent on the beam profile. Hence it is unusual to find such a laser (having the appropriate amount of energy per pulse) with the quasi-gaussian beam profile that one requires for display holography. Of course one can filter a non-optimal profile but then your useful energy decreases. And let's face it, even without filtering, one joule of energy for an approximate figure of $150,000 (including laser, injection-seeding system, and scope) is just too much.

Hybrid Neodymium Glass Lasers

Hybrid neodymium glass lasers are undoubtedly the best solution for display holography available today. Those that may be considered for this application are lasers in which the oscillator is either neodymium YAG or YLF and which employ one or more glass amplifiers. Usually a ring cavity oscillator is used with passive Q-switching. Single longitudinal mode is thus produced without the need for injection seeding. The advantage of a glass amplifier is

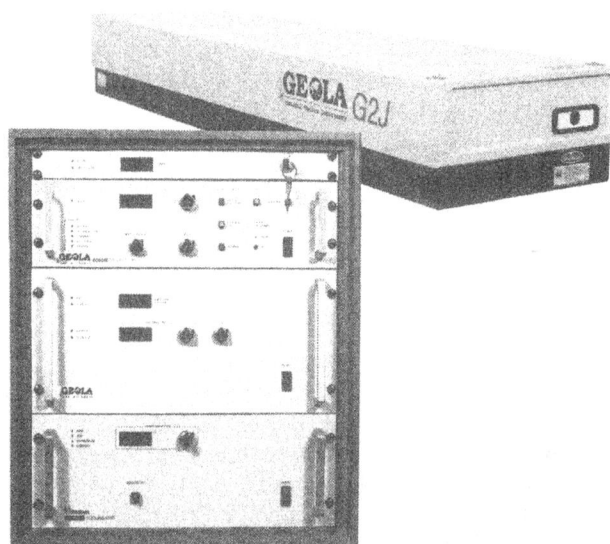

Fig. 1. Geola's compact G2J holography laser based on modern neodymium technology.

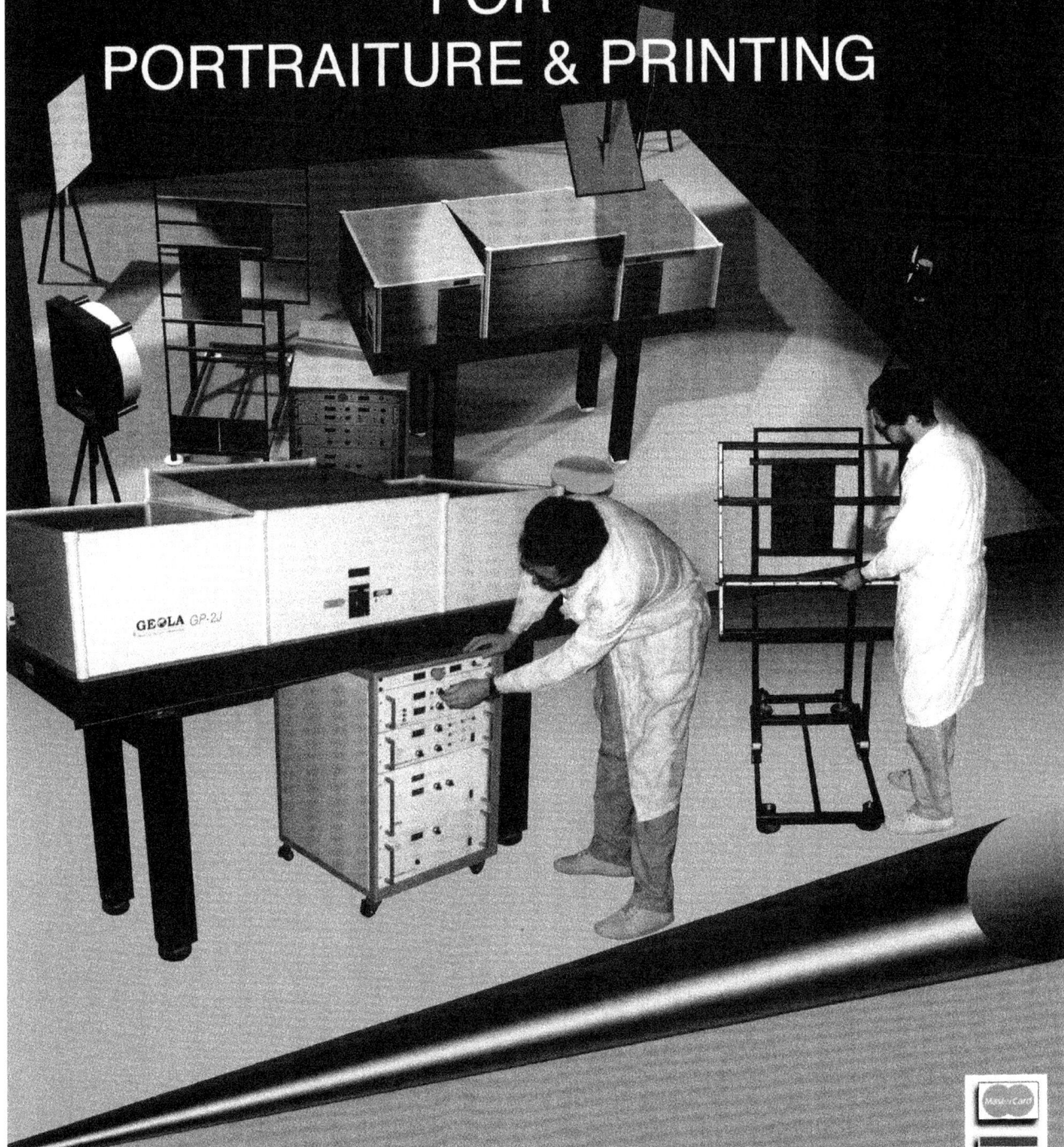

AUTOMATIC HOLOGRAPHY
CAMERA SYSTEMS
FOR
PORTRAITURE & PRINTING

GEOLA GP-2J

GEOLA

General Optics Laboratory
P.O. Box 343, Vilnius 2006, Lithuania

For more information please
contact our office or check out
our home page at:
www.geola.com

Lock Out Mode-hops Drift and Jitter

With Z-Lok and J-Lok

Lock in the ultimate in single-frequency performance with a BeamLok® ion laser equipped with Z-Lok® and J-Lok®.

This is the only commercial system to actively stabilize the three important parameters for single-frequency operation: mode hops, long-term frequency drift and short-term frequency jitter.

Mode-hops, drift, and jitter are all stabilized with the same compact PZT device which is integrated into every BeamLok ion laser.

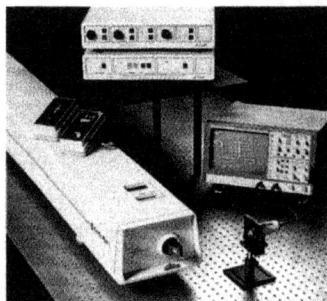

The Ultimate in Single-Frequency Performance.

❏ Mode-hop free operation
❏ Drift ≤ 30 MHz/°C
❏ Jitter ≤ 2 MHz

So, if you're performing holography, interferometry, LDV studies, or other applications that require the most stable single-frequency performance available, specify a BeamLok ion laser with Z-Lok and J-Lok. And lock in stability.

In North America, call your local Spectra-Physics office, or dial 1-800-775-5273.

⑤ Spectra-Physics
Spectra-Physics Lasers

Call: 1-800-SPL-LASER (775-5273)   Web: http://www.splasers.com   E-mail: sales@splasers.com

Ion King.

BeamLok

BeamLok Ion Lasers:
The Pride of the Industry.

BeamLok[s] ion lasers have been the leaders in power and beam position stability since they were introduced in 1990. They *still* are. Here's why:

1. **Ultra-stable resonator structure** provides *passive* control of output power, mode, and noise.
2. **BeamLok**[®] technology ensures unsurpassed beam position and power stability from a few seconds after turn-on to turn-off, every time.
3. **Z-Lok**[®] eliminates mode hops and makes drift negligible during single-frequency operation.
4. **J-Lok**[®] drastically reduces short-term single-frequency jitter to the lowest level in the business.

5. **SilentLite**[™] cuts *multiline* noise by up to 10:1.
6. **Patented Q-M endbells and window coatings** ensure exceptional mode performance and eliminate optical degradation to prolong operating life.
7. **Ultra-clean manufacturing technology** delivers the highest performance plasma tubes with maximum lifetime.

Maximum beam stability means maximum simplicity for you. No tweaking, adjusting, or worrying. Just turn it on and forget it.

For more information, in North America call the ion trainer at your local Spectra-Physics office or 1-800-SPL-LASER (1-800-775-5273).

ⓢ Spectra-Physics
Spectra-Physics Lasers

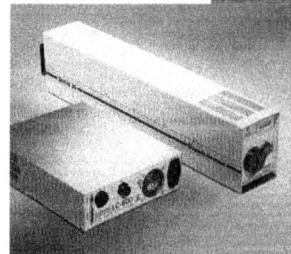

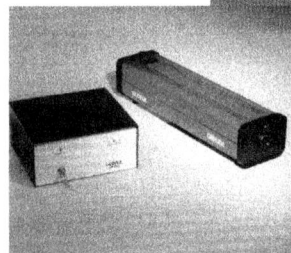

that it is effectively more efficient than a YLF or YAG crystal and additionally the glass material can be produced cheaper and easier in the large sizes and high optical standards required for amplification. This allows one to make a smaller and cheaper laser with higher output energy. The disadvantage, which in fact is not a disadvantage for holography at all, is that this type of laser can typically only be fired once every few minutes. This is due to the lower thermal conductivity of glass.

One practical solution for the design of an Nd:glass holography laser is to pair a neodymium YAG oscillator with

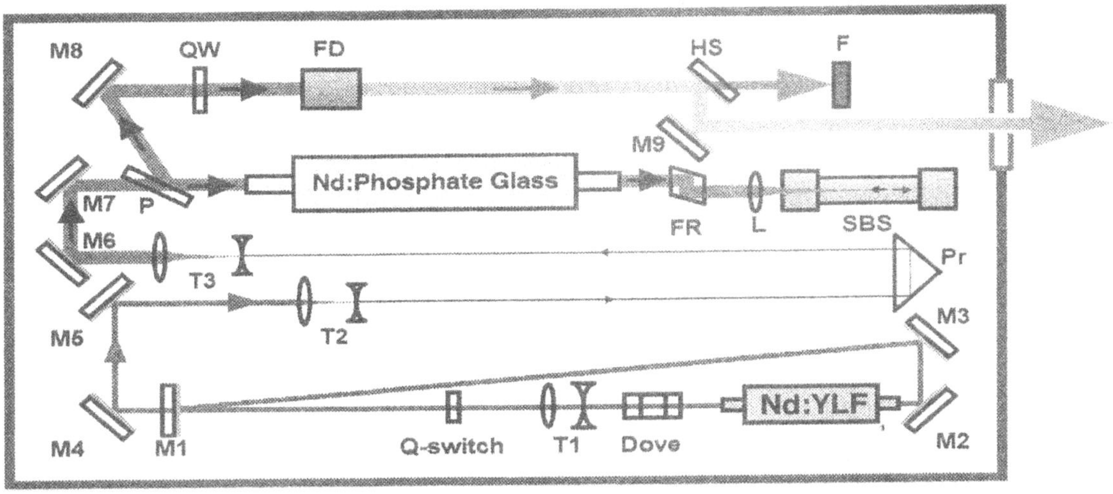

Other Lasers Appropriate for Holography

In the author's opinion, there are currently no serious candidates for monochromatic display pulsed holography applications that can compete with neodymium glass lasers.

Fig. 2. Optical Scheme of the GlJ and G2J lasers. Notable features include a simple high energy ring oscillator, double-pass glass amplifier with SBS mirror, and frequency doubling by DKDP.

silicate glass amplifiers. Such a system has been successfully constructed by Ron Olson of Laser Reflections and has been reported on in the '96'97 edition of this publication. Ron has been producing and continues to produce many high quality holograms using this laser. He· uses his system both for mastering and copying.

Another solution to the problem is to pair neodymium YLF with phosphate glass amplifiers. Phosphate glass has significantly more gain than silicate glass and this is a distinct advantage. In addition, YLF, with lower gain than YAG, allows more energy to be stored in the oscillator without the risk of superluminescence, an effect that can be a problem when designing a YAG/silicate system. YLF has weaker thermal lensing than YAG and hence, even though it has higher thermal conductivity, YLF can tolerate more optical pumping without propagation modification. This is important when you want to design a laser that has a high-thermal-stress, low-energy-alignment mode in addition to the normal high-energy, low-frequency mode.

* In fact it is technically feasible to modify a ruby laser such that the output beam transits a Faraday isolator. However, this modification is expensive and brings other disadvantages .*

* It should be noted that YLF actually has a more powerful line at 1047 nm but this line does not match well with an emission line of glass.*

GEOLA AND HOLOR

New Pulsed Laser Systems

by David Ratcliffe (Geola)

Geola, short for General Optics Laboratory, is a high-tech, European company located in Vilnius, Lithuania. In addition to producing custom lasers for the science community, this company has recently started manufacturing a family of advanced hybrid Nd: YLFlNd: Phosphate glass lasers designed specifically for display holography, called the "GxJ" series. All GEOLA's "GxJ" lasers are based on a high-stability single transverse and longitudinal mode ring-cavity master-oscillator, and are commercially available as models G11, G2J, G5J, and G8J with respective output energies of 1,2, 5, and 8 J at the second harmonic wavelength of 526.5 nm. For higher energy applications Geola offers multichannel holography laser systems such as the GM32J model, which produces a total energy of 34 J in four object channels (32 1), plus one reference channel (21).

Inside the GxJ Lasers

All GxJ lasers offer a fast repetition mode that can be used for the alignment of external optical elements. This mode energizes only the oscillator and produces a lower energy green beam that has propagation characteristics identical to the main high energy beam. Such a feature seems to be a significant improvement over the more common technique of using an additional CW alignment laser that inevitably is of a different wavelength and has different beam parameters compared with the main laser emission. The G11 and G2J lasers employ a single YLF ring oscillator incorporating passive Q-switching by $Cr+4$:GSGG and a single double-pass phosphate glass amplifier with SBS mirror. The general optical scheme and external appearance of the lasers is reproduced in figures 1 and 2.

There are several features worth noting in the Geola oscillator. The choice of $Cr+4$:GSGG over such better known materials as LiF not only ensures reliable Q-switching but also aids single-mode operation. The use of very high purity YLF produces an almost unheard of oscillator output energy-more characteristic in fact of a YAG laser with one amplifier than a simple oscillator. This makes the design of the rest of the laser much easier and less critical when it comes to setting the transverse beam profile. Since the oscillator is what actually determines the coherence length and basic mode of the final beam, one can appreciate just how vitally important it is that the oscillator does not respond in a sensitive fashion to external conditions. Besides using high-stability precision component holders for all optics, Geola has opted for a three-mirror vertically mounted cavity design that is very insensitive to the usual bending moments felt by the optical base.

All GxJ lasers use the standard technique known as "diffraction cleaning" (optical elements M4, M5, T2, PR, T3) to improve the spatial distribution of the oscillator output before driving the Nd:phosphate glass amplifier. Geola uses a two-pass phosphate-glass amplifier incorporating phase-conjugation by Stimulated Brillouin Scattering (SBS) in order to guarantee high quality spatial distributions. The use of a Brillouin cell is very important for two reasons. First, SBS allows the formation of a diffraction-limited beam by compensation of the aberrations and distortions in the wavefront' which are produced by the hot Nd:phosphate glass rod. This assures identical beam divergence and propagation direction in both the high repetition low energy alignment mode and in the usual high energy low repetition mode. The second reason is the greater energy extraction possible with a double-pass scheme without self- excitation. Here the Brillouin mirror serves as a selective reflector which reflects only a coherent signal and not the noise from any amplified spontaneous emission. The choice of SBS fluid used in the Brillouin cuvette is crucial in such high energy applications and here Geola has drawn on its considerable experience in the high energy laser-fusion domain.

An important general point to make is that Geola has paid special attention to the mechanical stability of their laser heads. First, they have used a floating three-point suspension system for the honeycomb superinvar/stainless optical base. In addition they have designed all optical mounts to the exact precision and function required. Finally, an optical scheme has been designed that cancels out temperature-induced expansion making the GxJ highly stable devices.

Since YLF lases in the infrared at 1053 nm, a harmonic generator must be used to double the frequency of the emission output. For this, the GxJ use a DKDP crystal sealed in a temperature-controlled dry cell as the harmonic generator. A precision engineered angle tuning mechanism based on a simple mechanical design ensures excellent long-term stability of the harmonic output.

The technique permits an energy conversion efficiency to 526.5 nm of up to 60%. Harmonic separation is achieved by pairs of dichroic mirrors giving >99.7 % separation.

As I mentioned above, two of the most important parameters for a holography laser are beam distribution and coherence. The GxJ lasers have been constructed to have output pulse durations of no more than 25 ns giving coherence lengths in excess of 3 m. The spatial profiles have been optimized for holography and all lasers may hence be used for both mastering and copying.

The main differences between the G 11 and G2J lasers are a larger diameter phosphate glass rod in the G2J amplifier (d = 12 mm instead of 10 mm for the G11) and special higher-damage-threshold dielectric coatings on the optical elements. The higher energy models, G5J and G8J, additionally utilize a final-stage large diameter amplifier. This amplifier is single pass and uses focal plane translation incorporating vacuum spatial filtering. Vacuum filtering ensures effective decoupling between amplifiers, thus preventing unwanted lasing, whilst also improving beam quality'

It is in fact more accurate to say that the beam quality is improved by the actual process of image translation which is of course integral to the vacuum filtering.

Power Supplies and Control Electronics

The main features of the GxJ control electronics are remote operation by wireless push button, an LCD shot counter, digital countdown to next pulse, digital programming of the amplifier and oscillator voltages, push-button selection of "Alignment" and "High-Energy" modes, digital laser-coolant temperature stabilization and readout, and a variety of interlocks against over-voltage, flash tube activation prior to rod cooling and accidental shots. The wireless push button has proved to be a very successful feature in holography applications where control of the laser pulse from awkward positions is often essential when working alone. All laser models are produced with a variety of parameters that are controlled by 10 tum potentiometers and displayed by LED digital readouts. These include the flash lamp voltage, pulse repetition rate and triggering delay (internal and external triggering is provided). Panel indicators signal interlock input and various status indications.

Multichannel High Energy Laser Systems

There are some unique applications in large format display holography that require even more energy than an 8 joule laser can give. However, in nearly all cases most of the energy is required for object illumination and hence the spatial distribution is relatively unimportant for the majority of the laser light. Geola's GMxJ (multichannel) series of custom ultra-high-energy-holography lasers has been designed to take advantage of this fact by producing separate beams for the reference and objects. The reference beam in all GMxJ lasers is adjustable up to 2 J of energy in the green and has near-perfect beam parameters. One or more object beams produce up to 8 joules each. Since the technology used to produce these lasers is scalable, systems with any number of object beams may be produced. The GMxJ series of lasers appears ideal for the creation of ultra-large-format rainbow, master-reflection, or transmission holograms.

High Repetition Rate Lasers

Occasionally the holographer will need a pulsed laser that is capable of high frequency repetition. As we have already mentioned, glass lasers can only be fired every few minutes due to the thermal properties of the glass amplifier rods. Hence, in addition to its GxJ series, Geola now offers a series of YLF-YLF lasers where the glass amplifier is replaced by one or more YLF rods. Currently, Geola has one standard model available named "Quickfire." This laser is housed in a box identical to a G5J and produces 0.5 J per pulse at up to 5 Hz. All other principal beam parameters follow the GxJ range of glass lasers. Custom models having higher output energies are available on request.

THE GREENSTAR LASERS

In addition to Geola there is another interesting source for holography lasers. For many years the S.1. Vavilov State Optical Institute (SOl) in St. Petersburg, Russia, has been producing holography lasers based on the neodymium technology for mostly Russian clients. Recently the institute has spawned a commercial company called Holor that markets specifically this type of laser. The S.O.1. also makes entire holography camera systems based on the Greenstar series of lasers.

Holor's Greenstar lasers use optical designs similar to Geola's. Two models are offered, the GS-2 and the GS-4, which respectively produce 2 and 4 joules of output energy in the green. The lasers use a ring oscillator based on the YLF material and passive Q-switching. Amplifiers are based on phosphate glass as in the Geola systems. Table 1 lists a summary of essential parameters.

Parameter	GS-2	GS-4
Wavelength (nm)	526.5	526.5
Pulse Duration (ns)	30	30
Pulse Energy (J)	2	4
Rep. Rate (pulse/min)	1	1
Beam Divergence	Lowest	Mode
Coherence Length (m)	>3	>3
Beam Diameter (mm)	10	10
Production Time (months)	3	5

Table 1. Main technical parameters of the Greenstar lasers manufactured by Holor.

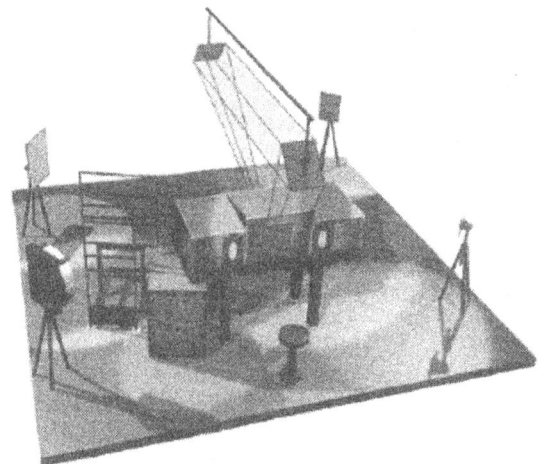

Fig. 4. The GP-2J Camera in Mastering Mode.
The subject or model to be holographed is placed on the red stool in front of the camera. Illumination of the subject is verified by using the laser's pilot mode. When a satisfactory illumination has been set, an unexposed holographic plate is then loaded into the main plateholder. The control unit is now set for exposure and when the holographer is ready he simply presses a button on a small remote control unit to activate the main flash. Laser light exits from the front two illumination ports and from a rear reference port to combine at the holographic plate where a hologram is recorded by interference of the two beams.

HOLOGRAPHY CAMERA SYSTEMS

S.O.l. was really the first to offer a compact integrated holography camera system based on a neodymium laser. Their system, now named GREEF and based on the Greenstar lasers, is a compact unit containing laser and all optics and accessories that are required for shooting a 30- by-40-cm master hologram suitable for the production of a reflection copy by CW transfer. Due to its size and construction, the camera can be transported easily from location to location.

GREEF is a concept that is vitally important to the holographer as the unit is complete. You no longer need to design your own external optical scheme because it is done. for you. All the display holographer needs to do is to adjust the lighting and press the button to take the portrait. Perhaps if one imagines where photography would be today if photographers had to design and make their own lens and shuttering systems, one can start to understand the importance of the availability of such integrated holography cameras-particularly when one also understands that pulsed laser light is a lot harder and more dangerous to play with than the light from your usual HeNe or argon laser.

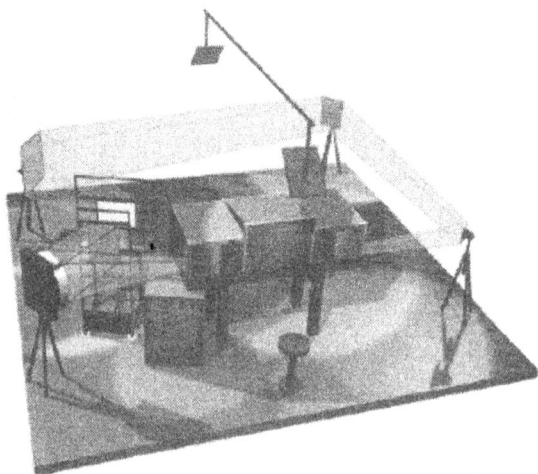

Fig.. 5. The Gp-2J Camera in Copying. Mode
After the master hologram is processed and dried it is inserted into the transfer rig. The laser is set to pilot mode and a white screen is placed opposite the master hologram in the transfer hologram holder. The holographer can align the master hologram by observing the image projected onto the screen. An unexposed holographic plate is now inserted in place of the screen. The laser and control unit are then reset for exposure and the flash activated by remote control as before. The hologram must then be processed and dried after which it is ready for viewing in white light.

In 1998, S.O.l. and Holor have developed and refined GREEF and have succeeded in making the power supply 1.5 times less than before. Current dimensions are 14 by 21 by 75 cm (12 kg) for the GS-21aser and optics box, 28 by 45 by 50 cm (38 Kg) for the power supply and 17 by 20 by 25 cm (2 kg) for the cooling unit. Holor sees applications for its new device in art, education, scientific research, holographic interferometry, spectroscopy, medicine, and archeology. Clearly its small size, which makes it currently the world's smallest pulsed holography camera, makes it ideal for the holographer who is dealing

with subjects and artifacts that cannot be brought to his studio. To further facilitate the concept of a truly portable holography camera, S.O.l. is now working on a version capable of operation in white light.

INTEGRATED MASTER/COPY SYSTEMS

In 1997, Geola took the holography camera idea one stage further: to produce an integrated compact machine that does both the mastering and the transfer copying to reflection or rainbow format.

One of the big problems in the past has been how to make the copies. If you used a pulsed laser for mastering then you ended up with a transmission master that had to be transferred somehow to produce the white light viewable reflection hologram. To do this one usually used a krypton laser on a large conventional optical table. Usually the pulsed laser that was used was a ruby which really ruled out HeNe copies. Even with krypton, color control was a problem and the transfer process would take far longer than the mastering. When you realize that the transfer setup could cost nearly as much as the pulsed laser you start to understand why it might be a good idea to use the same pulsed laser to do the copy. In addition, using the same frequency of laser light for the copy and the master is, of course, ideal.

As I have pointed out above such a system is a practical impossibility with ruby lasers and has been the natural outcome of the newer neodymium technology.

The GP2J System

The GP2J is Geola's first Integrated Master/Copy Camera system. Commercial units are currently available for sale with a four to six month construction delay depending on factory workload.

The GP2J is based on a G2J laser and comprises a central unit containing laser, optics and the master plate holder, an integrated power/control unit, several mirrors, and a small transfer rig . The entire machine fits into a room 5 m-by-5 m with a low to medium height ceiling. This space leaves ample room for scene construction and for walking between the various components.

The GP2J is highly automated. No manual adjustments are required either to the laser or any of the high energy optics during mastering, copying, or switching between the two modes. Sophisticated electronic controls activate servomotors in all crucial components meaning that practically all the holographer has to do is load plates, press buttons, and look at digital displays. The GP2J is capable of producing up to 50 30-by-40 or 40-by-60 cm reflection holograms per day with an operating staff of two technicians.

Geola is now in the process of introducing a range of GP- xJ Master/Copy systems covering the needs of both conventional holographers and a new breed of client that the company foresees. These newcomers to the market will start to use holography without actually understanding how it works, much like how the guy at the pharmacy who processes photos with an automated 1 hour machine hasn't got a clue what actually happens inside this device.

Importantly, during the course of 1999, Geola is planning to introduce the following machine options:

White Light Operation

Normally GP-xJ cameras should be operated in moderate intensity ambient red light. However, with the white light option, special light-tight loading boxes with large aperture shutters permit normal operation in almost any type of ambient light.

Automatic Latensification System

Some holographic materials require a procedure known as latensification as the first step of processing after laser exposure. This process works by illumination of the exposed holographic plate by uniform white-light for a specific time.

The ALS option provides an automatic post-exposure latensification while the holographic plate is still located in the plate-holder by means of a special programmable incoherent light source. This option is targeted at customers who will be using the camera system in conjunction with an automatic processing machine.

Shot Image Control System

In a standard GP-xJ system, alignment of the master hologram is effected by observation of the image produced by the pilot mode onto a special white screen. Manual adjustment of the transfer hologram holder in the x, y, and z planes then determines the final image framing and the image projection. Default settings of the reference/copy ratio and energy density are then usually sufficient to produce an excellent final hologram.

The SICS option replaces this manual alignment and setting procedure with a computerized system that allows the operator to select different image projections and alignments from a computer terminal. In addition the system is capable of calculating the optimized reference/copy ratio and energy density for any given setting as well as warning against over-and-under exposure when manual settings are adopted.

Color Control System

Although color control is usually effected by processing, more accurate results may be obtained by preheating in a special oven.

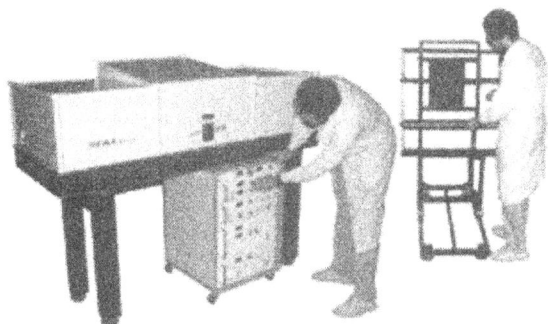

Fig. 6 The GP-2J in operaztion.

Copy Control System

In printing applications, one master hologram must be transferred many times onto a roll of film. The CCS option consists of a motorized film box that fits onto the transfer rig. The film advance is controlled automatically by the camera control unit. This system is compatible with ALS.

High Frequency Exposure

Standard GP-xJ camera systems are capable of producing one exposure every two minutes. By installation of a Geola Quickfire laser this rate may be dramatically increased up to five exposures per second. This option is particularly useful for high-volume printing applications.

All Geola's camera systems are compatible with leading makes of automatic chemical processing machines.

COLOR SYSTEMS

Both Geola and Holor are currently working on color lasers and color camera systems. Neither of the companies currently has a finished commercial product so don't hold your breath too long. However, it will come. Geola has been concentrating on the concept of Raman scattering of doubled Nd radiation (RSDN). The company's published work currently appears to show results significantly ahead of any other group regarding hydrogen conversion efficiencies. Geola is now turning to studies of Deuterium Raman conversion in the hope to realize a two-color Green/Red laser within the next 12-24 months. This laser would be used in Geola's planned next stage two-color master/ copy cameras which would look and work almost exactly as the GP2J except for the fact that the final reflection hologram would have two colors. For most portrait applications, the lack of the final blue color is of far lesser importance than the inclusion of the red.

An important point to note here is that color laser cameras are not much use unless you have the proper holographic emulsions. I have already mentioned that there is a current problem with the availability of ruby emulsions. Concerning this Geola has been collaborating closely with the Russian emulsion producers Slavich in the development of both green and near-red sensitive materials. Excellent emulsions now exist for both neodymium green (526.5 nm) and for hydrogen RSDN at around 620 nm, both in the form of glass and film.

Holor is hoping to produce a three-color laser by a rather different mechanism to Geola's with output energies of 0.4 J at 440 nm, 1 J at 526 nm and I J at 660 nm. Their estimated time to market is 12 months.

DIGITAL MASTERING

As we have seen, pulsed lasers were conventionally thought of as only being useful for the manufacture of the initial master transmission hologram. However, with the advent of neodymium lasers we see that pulsed lasers can make a vital difference in the production efficiency of the copy hologram. This is particularly evident in the case of Geola's Copy Control and High Frequency options. A GP-xJ camera fitted with such options can print up to three hundred 40-by-60-cm film holograms per minute, a task which would have to be regarded as totally impossible for a CW laser and yet which is surely required if holography is to find a use in the printing industry. Hence, it is rather clear that a major role of the pulsed laser must now be appreciated to be in the efficient production of the copy hologram.

Mastering by any kind of laser, whether CW or pulsed, has one big disadvantage. The image information in the hologram is not compatible or controllable by today's computer software. If display holography is going to make it to real-life mass production, this situation must be changed.

It is for this reason that Geola is now turning its attention for the future to a multicolor system comprising digital mastering and a normal pulsed analogue transfer.

CONCLUSION

Pulsed laser holography is really in the process of a quantum jump: not because fundamental new discoveries have occurred in laser physics or optics, but because a few well-known concepts have finally been carefully and systematically applied to what must be regarded as a pretty simple problem for laser physics: display holography.

The ruby laser has been largely responsible for the rather pessimistic situation up until now. Sure, you can make holograms with this laser but the color is wrong, spatial filtering requires at best undesirable and costly modifications, alignment is difficult, and the laser is (sizewise) a monster. Perhaps it is not so surprising after all then that most people have become disenchanted with pulsed holography?

But introduce neodymium technology, which by the way has been around for ages in Russia in the form needed for holography, and things get easier. Cost and size come down, beam quality goes up. The color is naturally optimum. Alignment is easy. And spatial filtering lets you re-

ally do anything you like. The GxJ series of lasers presents the first real commercial pulsed holography lasers to become available on the market and others will follow. This type of lasers is highly competitive with CW neodymium or Argon lasers from both a dollar point of view and in what the laser can offer. It will also make portraits....

With reliable neodymium glass lasers finally here, it has been an obvious step to make commercial integrated systems for people who don't like playing with laser beams, and to make a single automated device for mastering and copying. With neodymium, things can be made small. And with computer automation, thermal color control and automated processing, perhaps that day when you will be able to walk down the street to your local holomat and get your 1 hour 3D portrait for fifty bucks has gotten just that bit closer.

Finally, for those who are skeptical about holographic emulsion availability, I should stress that neodymium lasers allow you to continue using available Agfa products. After all, Agfa has never indicated that it would stop producing photoplates for the lucrative microelectronics market, and of course these plates are just what the new lasers need. As for film substrates, the Russian producer Slavich is now producing an equivalent product to the Agfa 8 E 5 6 - which I might add is significantly cheaper!

For further information, contact
Geola
P.O. Box 343
Vilinus
Lithuania
Phone: 370-2 232737
Fax: 370-2 232838
Email: sales@geola.com
Web site: www.geola.com

The Ronald Reagan holographic portrait.
The first and only Presidential "pulsed" portrait hologram.

Left - Former President Ronald Reagan poses for a holographic portrait at Brooks Institute of Photography on May 24, 1991. The studio was set up by Holicon Corporation and the portrait (right) was produced by Hans Bjelkhagen (also pictured), Michel Marhic, Ernest Brooks II, Dr. John Landry and Fred Unterseher. Prints are in the collections of the National Portrait Gallery and the Ronald Reagan Library. Additional limited-edition copies available. For more information, please email: hologram@well.com.

Photograph of one copy.
Size - 12.5" x 16.5";
Color - orange/gold;
Created by a team of
master holographers.

LASER REFLECTIONS

A CUSTOM-BUILT PULSED LASER SYSTEM

by Ron Olson (Laser Reflections)

In our holography studio, we use a passively Q-switched Nd:YAG oscillator operating in the TEMoo mode. Two etalons and a passive Q-switch (BDN dye) define a single longitudinal mode. We have a 500 MHz scope and a fast photodiode to look at the temporal profile, but seldom use it as a diagnostic tool.

A preamplifier of Nd:YAG provides-100 mJ at 1 Hzasan input to the Nd:Silicate Glass amplifier, which we use in a double-pass configuration to supply us with just over 2 J at 1,064 nm. A 40 mm long KD*P doubling crystal is - 50% efficient and we work routinely with 1 J at 532 nm. The frequency-doubling without the final amplifier stage in operation is extremely inefficient-allowing us to work at a manageable level (1 mJ) for image composition-we do not have an alignment laser within the system.

We take great care to underfill our final amplifier, ensuring a beam spatial profile which is "very nearly Gaussian." Much of the criticism regarding Nd:YAG lasers concerned beam spatial profiles-but the dark days of Nd:Y AG crystal growing are long past-to the point that except for fanatics like myself, commercial Q-switched Nd:YAG lasers as they leave the production floor would be adequate for most pictorial holography.

We split - 10% of the energy for the reference beam (variable via a half-wave plate and a dielectric polarizer) which is delivered to the plate by a large float glass mirror at Brewster's angle. The remaining 90% of the energy goes to the two ground- glass diffusers which ensure that the illumination beams are eye-safe.

We work under red safelights and to maximize the studio session, we often put aside half of the transmission masters for processing the day following a shoot-at this point we have that much confidence in our technique. In the last two years we have produced more than 100 large format transmission masters and more than 300 reflection copies- all on glass.

SAFETY CONCERNS

When asked if a holographic portrait system such as ours is truly franchisable-I caution people immediately. I have almost 20 years experience with high-energy, high-power lasers, and what is natural and instinctive for me is not necessarily the case for people not familiar with 1 joule/ 100 megawatt lasers. The laser company in which I am a founding partner produces more custom-designed high power lasers than any in the U.S. and it cannot be overstated that such lasers should be operated only by trained personnel who are thoroughly familiar with laser optics, laser power supplies, and laser safety issues.

We would be happy to design a custom Q-switched neodymium based laser for anyone with the requisite $150,000- but a technician trained in the operation and maintenance of a Class IV laser is a requirement at the time of installation and training. No laser system in this power class is truly hands-off. Our experience, with some 500 laser systems shipped worldwide, is that unskilled operators who attempt to learn their craft on-the-job jeopardize the laser system, their own safety, and the safety of others in the studio/lab. In addition, it is extremely unlikely that any- one would be issued studio liability insurance such as we carry without a certifiable history in Class IV Laser System operation.

For further information, contact:
Laser Reflections, Inc.
589 Howard Street San Francisco, CA 94105
Phone: (415) 896-5958 Fax: (415) 896-5171
Email: hologram@laser-reflections.com
Web site: www.laser-reflections.com

Ron Olson of Laser Reflections with YAG laser (Photography by Brendan Beirne)

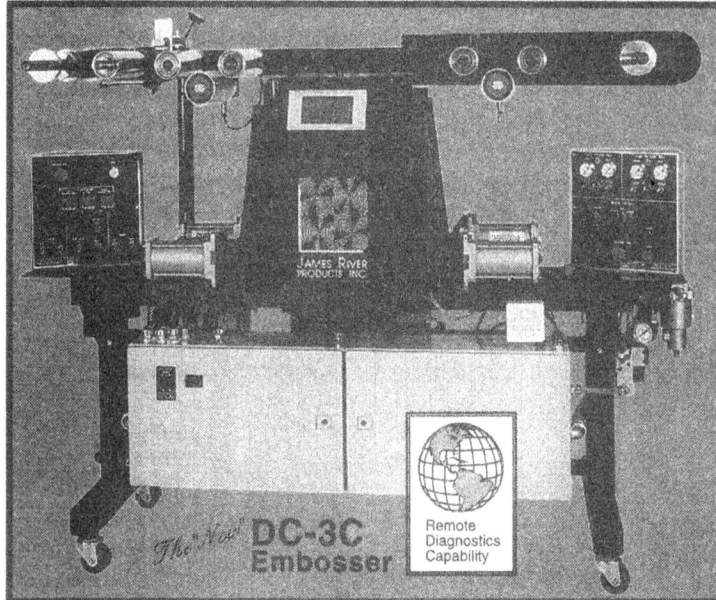
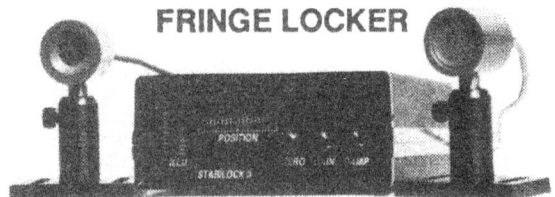

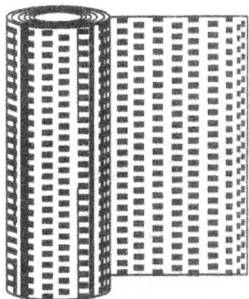

11

Production Equipment

This chapter describes a type of hologram production equipment that is increasingly being used for security and packaging applications-dot matrix hologram machines. As embossed hologram production grows, more and more factories require the ability to originate holograms without the expense of a maintaining and staffing a complete holography studio. These machines offer these facilities a solution to their problem, as well as provide existing studios a new way to create unique visual effects.

DOT MATRIX HOLOGRAM MACHINES

A dot matrix hologram is, as the name implies, a hologram created from an array of smaller units (dots). Each dot is actually a separate hologram that produces a predetermined result, usually a simple diffraction grating. When illuminated with a white light source and viewed at the correct angle, the diffraction grating produces a tiny spot of color. The microscopic gratings can be arranged to produce a single composite picture, a sequence of related images, and/or a variety of "special effects."

Dot matrix technology allows for computer control of each individual pixel in a holographic image. Systems that generate dot matrix holograms currently have resolutions that range up to 2,000 dots per square inch (each dot measuring 16 microns), though resolutions of 25-600 dpi are more common. This gives the holographer design options that are not available with traditional holographic originations. In addition, the process creates holograms that have some very useful properties, especially for commercial applications.

MAJOR ATTRIBUTES

Brightness Control of each pixel in a holographic image allows the holographer to optimize the brightness of each and every image component in relation to a viewer. This results in images that are easily recognized under the less-than-ideal lighting conditions which are present in most commercial environments.

Viewing angle Dot matrix holograms, unlike most other embossed holograms, have a very wide potential viewing angle. This is due to the fact that the diffraction gratings that make up the hologram can be positioned to redirect light in a variety of directions.

Kinetic effects Dot matrix holograms can easily incorporate dynamic effects that make the image appear to move or change as the viewer shifts position or the hologram is tilted. For example, designs can be made to "pinwheel" (radiate in and out when the hologram is rotated) or "flash" on and off.

Subject matter Any two-dimensional artwork that can be digitized can be used to make dot matrix holograms. Corporate logos are often reproduced. More often, repeating geometric patterns are generated. Many commercially available graphic design software programs (such as Adobe's Photoshop™) can be used to generate the required digital files.

One-step Mastering Aside from enhanced visual effects, dot matrix technology allows a wider variety of companies to originate holograms. Although minimal training on the currently available systems is required and an extensive knowledge of embossed holography is certainly beneficial, these machines are intended to eliminate the need for a highly experienced holographer on staff. As of today, many dot matrix systems have been purchased by existing hologram manufacturing facilities, but these machines can be operated by less-specialized production departments. Operators should have experience with computer controlled mechanical devices and a knowledge of lasers (along with laser safety procedures).

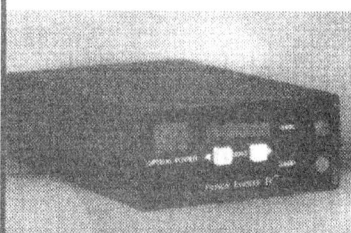

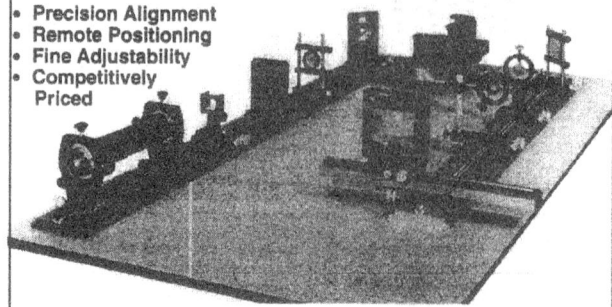

APPLICATIONS

Currently, most dot matrix hologram production systems are being used to create masters for holographic security labels or holographic packaging materials. The complex dot matrix designs that can be produced aid in document authentication. To further enhance anti-counterfeiting efforts, the proprietary computer software which translates the original artwork into a matrix of dots can encrypt security information in a way that is invisible to the casual observer. In addition, with some systems the hardware which creates the gratings can be set up to produce holograms with other signature effects (fingerprints) which can be traced back to the machine which made them.

Dot matrix holograms incorporated into embossed packaging materials help overcome many limitations of holograms created by other means. Dot matrix holograms are extremely eye-catching. They are bright and colorful even under fluorescent lighting. The animated designs attract attention. The repeating geometric patterns typically produced can be seen if the package is right-side up or upside-down. Many hologram manufacturers combine dot matrix holography with conventional optical mastering techniques in order to utilize the advantages of both systems-resulting in fully 3D images with striking 2D effects.

PRODUCING DOT MATRIX HOLOGRAMS Producing dot matrix holograms is a four-step process: artwork origination, digital-to-dot transfer, mastering and replication. In brief, a digitized image is downloaded into a dot matrix transfer system which converts digital parameters into a corresponding set of mechanical positioning instructions. These instructions guide a laser-based machine that uses holographic techniques to produce thousands of separate diffraction gratings. These diffraction gratings are recorded onto a photoresist plate, creating a master hologram. The master hologram is plated, and embossing shims are produced. The shims are used to emboss the hologram into a chosen substrate by commonly used methods .   .

It is important to note that there are several other digital-to-optical transfer systems that create holographic images from an assemblage of smaller image components. However, they are not dot matrix systems. Many employ LCD screens and are based on holographic stereogram theory. Although they have some properties that are similar to dot matrix systems, they produce very different results. (See Holography MarketPlace 5th Ed. for a discussion of these "LCD/HOP" systems.)

BASIC THEORY

Dot matrix holography was developed in the mid-1980s by Frank S. Davis at Advanced Digital Holographics, Inc. In 1992, the patent was assigned to Dimensional Arts, Inc., a U.S. company that builds and sells a line of production machinery based on the Davis patent.

(Editor's Note-Other companies, including Cfcl Applied Holographies and Toppan Printing, have patented methods that produce dot matrix holograms, but choose to keep their technology "in house." Ahead Optoelectronics claims to hold patents for dot matrix technology which is incorporated in the machines that they sell. Westmead Technology also sells dot matrix equipment. We encourage potential users of dot matrix technology to research this issue further, as patent protection may vary from country to country.)

A fundamental principle of the Davis method involves the use of a laser beam which is first split in two, and then recombined to create tiny interference patterns (holographic dots) on a recording material. This is the basic dot matrix patent. Changing the angle and orientation of the intersecting laser beams changes the interference pattern, which is the primary optical property this technology exploits.

Each interference pattern, once recorded on photoresist, becomes a surface relief diffraction grating. The diffraction grating operates like any split beam hologram- an illuminating beam coming in from the appropriate angle is redirected along a specific corresponding path (usually perpendicular to the surface of the hologram). It's like setting up an array of microscopic mirrors so that incoming light can be aimed back to the viewer in a precise way.

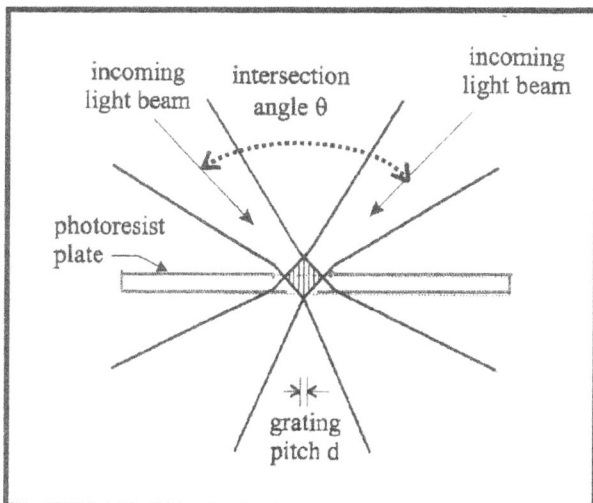

Fig. 1. Creating a grating with two beams.

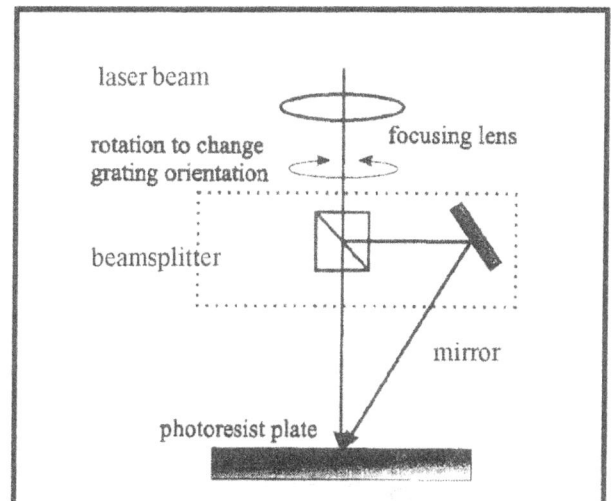

Fig. 2. A traditional dot matrix writer.

CREATING GRATINGS

There are two important parameters for each grating:

1) The first parameter is the grating pitch, which is deter- mined by the intersection angle of the two incoming coherent lightbeams. The larger the intersection angle, the smaller the grating pitch. More specifically, the grating pitch equals: the wavelength of the incoming laser beam/ sine of the intersection angle.

2) The second parameter is the grating orientation, which is determined by the direction of the two incoming laser beams. (See figure 1.)

Playback Control

Pitch and orientation of each grating pixel are the most fundamental parameters to consider when designing a diffraction grating. These two variables determine how the incident light beam can be diffracted precisely to where the observer is. The orientation of each grating pixel is a function of three variables:

1) The light source position,

2) The observing location, and

3) The position of each grating pixel.

The grating pitch of each grating pixel is influenced by another variable,

4) The color wavelength produced by each grating pixel in relation to other pixels in the same image.

Thus, a graphic designer needs first to decide the location of a fixed light source, the position of the observer, the color distribution and the desired viewing effect of every graphic pixel of the original graphic design. All of the data mentioned above must be entered into the design program. The software program automatically calculates the pitch and orientation of each dot.

VIEWING ANGLE

The viewing angle associated with each microscopic grating in a dot matrix hologram is limited by the optical geometries used to create it. Angles of view are very small

However, when the observer is located within the designated viewing angle for that specific pixel, all the redirected light is concentrated within that small angle. This creates very bright dots.

KINETIC EFFECTS

If many gratings are aimed at one viewing zone, very bright images result. If the gratings direct light away from the viewer, "dark" areas result. The ability to control a sequence of thousands of tiny "flashing lights" allows the designer to create a variety of special visual effects such as disappearing images, zooming-in, zooming-out or changing shape.

COLOR

Different colors can be produced by changing grating angles. Each dot matrix system has its own palette of colors available to the hologram designer. Like most embossed transmission holograms, colors are chosen in relation to a specific viewing zone. As the viewer moves out of this primary viewing zone (or the hologram is tilted) the intended color will shift to the next color in the spectrum. Prismatic effects result.

SOME "EXISTING DOT MATRIX SYSTEMS

Some of the technical developments investigated during the last few years are reviewed below.

For the traditional two-beam interference dot matrix writer shown in figure 2, the laser beam goes through the focusing lens, and is then split into two converged beams by the beamsplitter. The laser beam reflected by the mirror intersects with the other beam to form a small grating pixel on the photoresist plate. The beamsplitter and the mirror are fixed on a rotational stage. The orientation of gratings and the size of each grating pixel recorded on the photoresist plate can be altered by rotating the stage and by moving the lens respectively.

In figure 3, the beamsplitter, mirrors, and lens are all fixed inside a rotational stage. The laser beam source is first split into two beams by the beamsplitter. The two laser beams

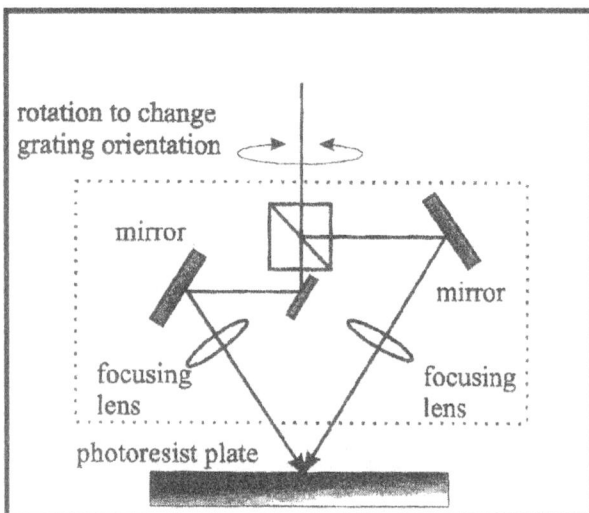

Fig. 3. Another configuration of a dot matrix writer.

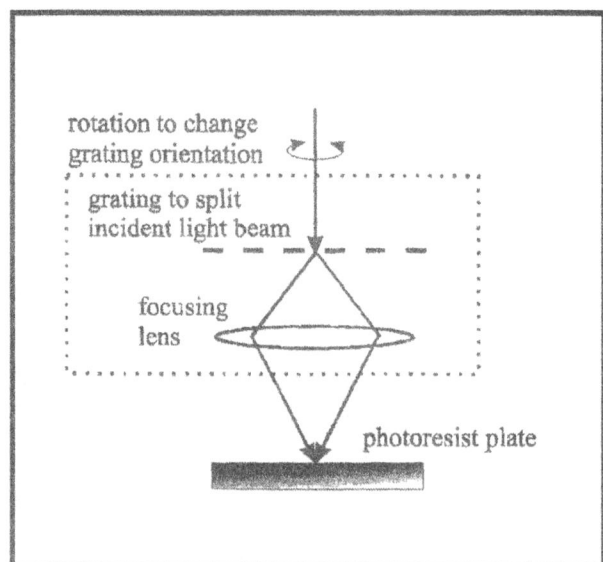

Fig. 4. A grating is used instead of a beamsplitter.

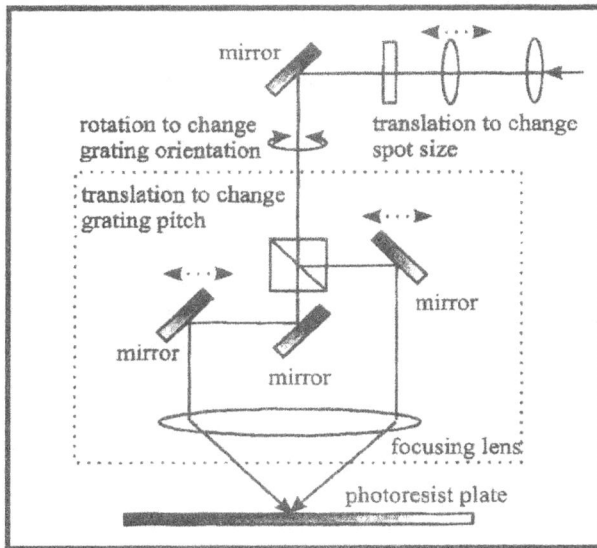

Figure 5. A dot matrix writer designed by AHEAD.

are reflected by mirrors and then converged by the lens. These two laser beams finally intersect and form a grating spot on the surface of the photoresist plate.

Figure 4 shows an optical configuration where the beam-splitting grating and lens are fixed inside a rotational stage. When the laser beam source goes through the beam-splitting grating, only a +1 order and -1 order of the diffraction beams will be selected to pass through the focusing lens. Finally, these two beams will converge and form an interference spot on the photoresist plate.

To achieve the goal of altering the grating pitch within each grating pixel, the intersecting angle between the two interfering laser beams must be changed. The main difficulty in designing and constructing a system that can change the grating pitch on the fly lies in the fact that a focusing spot will move in and out of the photoresist-plate whenever the intersection angle of the two recording laser light beams is changed.

The Dimensional Arts System

Dimensional Arts, now part of the HoloCom group of companies, was the first company to successfully market dot matrix technology. The company sells the "Light MachineTM," a basic dot matrix production system for tiled patterns and imaging, as well as an enhanced version specifically built for security applications. According to company president Ken Harris, the company sells the only patent protected dot matrix system commercially available in North America. In a notable development, Harris reports that his company recently has been awarded another, more extensive patent covering "holographic dot" technology. "Further patents pending will cover three-dimensional holograms made using stereo pairs of holographic dots and additional color-control methods," he adds.

Harris notes that dot matrix technology has proved reliable and popular since his company

placed its first machine over five years ago. He points to a list of major embossed hologram manufacturers that use his machines for everything from small security holograms to large poster-sized decorative holograms (including American Bank Note Holographics, Transfer Print Foils, Hologramas de Mexico, and Crown Roll Leaf, among others).

The AHEAD System

According to company literature from AHEAD Opto-electronics, a Taiwanese company that builds and sells dot matrix machines, the lack of grating pitch variation capability in other systems prevents the graphic designer from specifying colors in order to achieve more vivid effects. In addition, the absence of color specifying capabilities actually prevents an accurate preview function in such systems. "Many hologram manufacturers and designers are thus forced to create a look-up table to compensate for the incorrect results," states a company report.

Company spokesperson Julie Lee, elaborates: "One of the optical configurations that not only keeps the advantages of the traditional design but also overcomes drawbacks associated with other designs, is our Sparkle™ machine. (See figure 5.) In AHEAD's design, the incident laser beam is split by the beamsplitter to become two incident laser beams. The two coherent light beams are focused onto the photoresist plate by a specially designed focusing lens. The grating pitch of the grating pixel formed on the photoresist plate can be varied on the fly by translating a set of stages to change the distance of the two coherent light beams before impinging on the focusing lens.

As is shown in figure 6, the larger the distance between the two coherent light beams, the smaller the grating pitch. The focusing lens shown in figure 5 and 6 has been especially designed and fabricated in order to prevent the spherical aberration of the focusing lens to have the focusing dot location changed whenever the distance of the two incoming coherent laser beams is changed.

"The grating pixels generated by this new class of dot matrix writers are able to have spot sizes, grating orientation, and grating pitch completely specified by the designer. The images constructed by this type of dot matrix writer can exert many pre-specified colors at a specific viewing angle. In addition, all of the special effects that

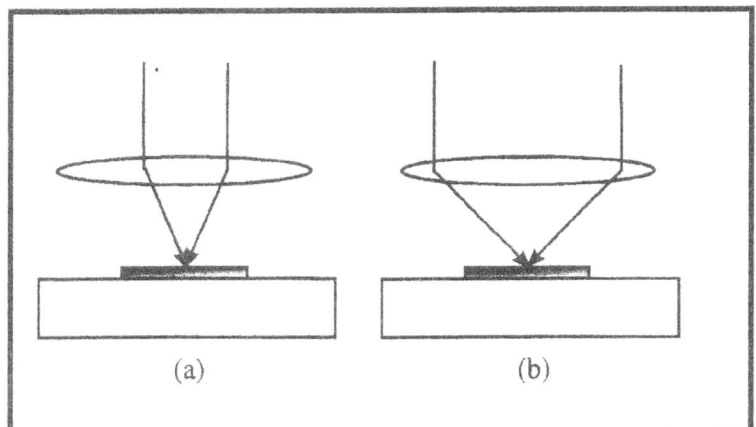

Figure 6. Changing the pitch of the grating.

can be created from other traditional two-beam interference dot matrix writers when an observer continuously varies the viewing angle, can certainly be generated."

CONCLUSION

Dot matrix technology is continuously being advanced and improved. Limits are being pushed towards higher resolution, different shapes other than dots, higher speed, more security features, better color selection, smaller overall machine size, and lower cost. Ease of use and cost effective production methods should result in more dot matrix holograms being used in a variety of applications.

For further information about the original, patented dot matrix machines and current dot matrix technologies, contact:
Dimensional Arts
401 Carver Rd.
Las Cruces, New Mexico 88005 USA
Phone: 505 527 9183 Fax: 505 527 9927
Email: arts@holo.com Website: holo.com/arts/

For information about other dot matrix systems, contact:
AHEAD Optoelectronics, Inc.
BI, No. 130, Sec. 3, Keelung Road Taipei, Taiwan
Phone: (+886)-2-2369-1520 Fax: (+886)-2-2362-0485
Email: sales@ahead.com.tw Web site: www.ahead.com.tw/

Westmead Technology Ltd
Unit 7 0150 - St. Georges Industrial Estate
Wilton Road CamberJey GUI5 2QW ENGLAND
Phone : 44 (0)1276685455 Fax 44 (0)127662810
Email: westmead@globalnet.co.uk
Web site: www.westmead-technology.com

Other companies that produce dot matrix holograms can be found in the International Business Directory.

DIMENSIONAL ARTS

The Dimensional Arts' Light Machine is the fifth generation computerized holographic printer based on the patented technology invented in 1986 by Frank S.

Davis. Images created on the Light Machine are unique in that they can be viewed under any lighting condition, and they are kinetic.

The Light Machine uses Dimensional Arts' patented technology to create unique patterns and images based on repeatable codes with infinite variations. Images created on the Light Machine cannot be duplicated by conventional holographic imaging technologies, or on a Light Machine without the proper codes.

The machine also has scanning capabilities and special drawing software for custom computerized imaging capabilities. Images can be scanned into the computer on the UMAX. High Resolution Color Scanner (included) or they can be created in Adobe Photoshop. The image file can also be used to create masks for mixing the dot matrix images and patterns with conventional 2D, 2D/3D, 3D, stereogram holograms, and white or controlled grating registration marks and lines. Custom software created by Dimensional Arts converts the image and allows the image to be previewed on the screen before it prints it out on the Light Machine.

Features

The Dimensional Arts' Light Machine operates as a stand alone unit offering computerized push button origination capabilities that require minimal training to operate. It is comprised of an optical bridge (which processes the beam from a Liconix 15 mW HeCd Laser) mounted on a Newport Isolation Workstation and high-precision stages which are driven by a digital interface (all controlled by an mM PSI I computer system.)

The Light Machine comes in two versions for standard commercial production. One has a resolution of over 200 dpi and operates at over 4 dots per second. It is 36" x 36" x 48" high, is not vibration sensitive under normal circumstances, and runs on standard line voltage from most any wall plug (110V). It requires a source of compressed air for periodic refills of the air tank for the isolation legs of the workstation. A higher resolution (thousands of dpi) model of the Light Machine is also available.

Security Model

For hard security applications, the security model Light Machine offers Dimensional Arts' patented Dot Shape Control. Dot Shape Control allows the predetermination of the actual shape of the dots comprising the image. The security model Light Machine also includes a selection of three registered angle writing heads for color mixing. Each registered angle is unique to the machine with which it is sold, which adds an additional level of security. Sale of the security model Light Machine is reserved only for legitimate security printers.

AHEAD

SPARKLE™ I DOT MATRIX MASTERING SYSTEMS

FEATURES

High speed (15 dots/sec).

High resolution (150-1300 dpi).

Fully computer keyboard controlled.

Multiple colors can be user specified.

Adjustable grating pitch allows for 6 color specification.

Powerful, friendly, easy-to-use GUI controlling program.

Handles .bmp files for easy downloading.

Realistic preview function.

Calculates an optimal exposure path to create master

Specifications
Light source:
HeCd laser, wavelength 442 nm

System dimensions:
900 mm (L) x 900 mm (W) x 1500 mm (H)

Grating pitch:
adjustable, specifies up to six colors

Dot resolution:
150 dpi to 1300 dpi.

Exposure rate:
15 dots per second.

Exposure size:
Standard 8" x 10" (300 mm x 300 mm maximum)

Yellow light space:
approximately 4 m squared

System contents
Pentium 166 or higher or equivalent computer.

MS Windows™ 95/ NT operating systems.

Kinetic vibration isolating optical table.

Optical bridge, XY stages and 5-axes controller.

Sparkle 97 GUI software (system control, pattern generation & preview).

Adobe Photoshop 4.0, CorelDraw 7.0 & KPT 3.02.

Electric power interrupt sensor and UPS.

WESTMEAD

LIGHTGATE 1270 DIFFRACTIVE PIXEL PRINTER

The Lightgate 1270 is a high-resolution diffractive pixel printer. With this system (patent pending), the resist plate is enclosed during exposure, enabling the Lightgate 1270 to be placed in any convenient location. No special room lighting is needed. The system is also designed so the resist plate is inverted, preventing dust problems during exposure. The computer interface ensures simple training and operation.

In addition, the laser can be situated remotely from the Lightgate 1270. This completely negates problems caused by heat and allows easy access to the laser head. No re-alignment is required when using variable thickness resist plates. The resist substrate can be of any thickness, facilitating the use of in-house plate spinning

Warranty: 12 months parts and labor subject to terms and conditions.

According to the manufacturer, new machines will be available shortly. Here are the features of the currently available mosel.

FEATURES

2000 dpi resolution.

25 exposures per second.

256 rotational angles.

16 cm x 16 cm exposure area.

High-precision translation table.

50 mW HeCd Laser.

Pentium class PC.

Lightgate 1270 EXEL
25 cm x 25 cm exposure area.

Real-time variable beam angles.

16.7 million color diffractive pixels.

SGI or Pentium MMX 300.

Sparkle™

Dot Matrix Mastering System

Applications:
- Packaging - multiple colors creat colorful packaging paper
- Promotional - dynamic images make eye-catching products
- Anti-counterfeiting - high resolution deters counterfeiting
- Security - adjustable grating pitch increases security elements

Features:

- High speed and high resolution
  (up to 15 dots/sec and up to 1300 dpi)

- Fully computer keyboard controlled ; powerful, friendly, easy-to-use GUI program

- Unique optical head allows user to specify multiple colors and 3D images

- Software allows easy combination of various dot resolutions in one image

- Handles any *.bmp file for easy downloading

- Allows user to incorporate dynamic or kinetic effects

- Preview function allows a rough or detailed preview

- Calculates optimal exposure path to create master

Brief Specifications:

Light source	He-Cd laser, wavelength 442 nm
System dimensions	900mm(L) × 900mm(W) × 1500mm(H)
Grating pitch	software adjustable, six colors
Dot resolution	150 dpi to 1300 dpi
Exposure rate	up to 15 dots per second
Exposure size	standard 8" × 10" ~ maximum 300mm × 300mm
Yellow light space	approx. 4m^2
System contents	Pentium 166MMX or equivalent computer MS Windows™ 95/98/NT operating systems Kinetic vibration isolating optical table Optical bridge, XY stage and 5-axes controller Sparkle™ GUI software (system control, pattern generation & preview) Adobe Photoshop 4.0, CorelDraw 7.0 & KPT 3.02 Electric power interrupt sensor and UPS

AHEAD OPTOELECTRONICS, INC.

Tel: (+886) 2- 2369-1520   Fax: (+886) 2- 2362-0485
B1, No. 130, Section 3, Keelung Road, Taipei, Taiwan, R.O.C.
http://www.ahead.com.tw
e-mail: sales@ ahead.com.tw

AHEAD Optoelectronics, Inc.

AHEAD is a high-tech company specializing in holographic products and systems, optoelectronic and optomechanical systems as well as piezoelectric systems.

AHEAD's holographic specialities include diffractive optical elements, lenticular lens, dot matrix mastering systems, embossed holograms and furrow overlay embossing. In addition, AHEAD also designs and produces high performance optical instruments including integrating sphere ellipsometry analyzers, optical wavefront metrology equipment, laser encoders, laser writers, differential laser Doppler interferometers, etc.

AHEAD, is part of the Wah Lee Industrial Corp. Group of Companies which has over 1000 employees worldwide with offices in 13 different countries.

AHEAD has ties to National Taiwan University, Industrial Technology Research Institute (ITRI), CSIST, as well as various other industrial worldwide high-tech firms in the semiconductor industry, DVD industry and hard disk industry.

AHEAD offers two dot matrix mastering systems (Twinkle at 150-1300 dpi and adjustable grating orientation, and Sparkle at 150-1300 dpi with both grating pitch and orientation adjustable).

✎ Twinkle ✐
Dot Matrix Mastering System
(Basic System of Sparkle)

Applications:
- Packaging: rainbow effect creates unique colorful packaging
- Promotional: adjustable grating orientation creates 2D/3D kaleidoscopic images
- Security: high resolution can be used as added security feature

Features:
- 150-1300 dpi and adjustable grating orientation
- Fully computer controlled, easy alignment
- Rainbow effect and dynamic effect
- Powerful, friendly, easy-to-use GUI program
- Handles any *.bmp file for easy downloading
- Calculates optimal exposure path to create master

Brief Specifications:

Dot resolution	● 150 dpi to 1300 dpi
Exposure rate	● up to 15 dots per second
Orientation	● yes; adjustable
Grating pitch	● no; Cannot specify center, rainbow effect only

AHEAD OPTOELECTRONICS, INC.
Tel: (+886) 2-2369-1520   Fax: (+886) 2- 2362-0485
B1, No. 130, Section 3, Keelung Road, Taipei, Taiwan, R.O.C.
http://www.ahead.com.tw
e-mail: sales@ ahead.com.tw

12

Recording Materials

This chapter will cover the four most commonly used recording materials for holography: silver-halide emulsions (glass plates and film) , photopolymer films, photoresist, and dichromated gelatin. Always use proper laboratory procedures when working with these materials and all related processing chemicals. A listing of suppliers for all these materials can be found in the Business Directory section of this book.

INTRODUCTION TO SILVER-HALIDE RECORDING MATERIALS

Silver-halide recording materials are commonly coated onto films or glass plates for use in traditional cameras, in laboratories, and in factories. However, they were not originally designed for holographic applications.

The important difference between photographic and holographic materials is resolving power, usually expressed in lines per millimeter. Whereas photographic films usually cannot resolve more than 50-100 lines per millimeter, holographic applications require 1,250-2,500 lines per millimeter.

Another difference is sensitivity, which is typically expressed as an ASA number. For example, popular photographic films are rated at ASA 120-400. Exposures are usually measured in hundredths of a second. Silver-halide holographic recording materials are so much less sensitive than standard films that their ASA would only be rated as fractions. Therefore, their sensitivities are usually expressed in micro-joules per square centimeter (or ergs per centimeter squared). Exposure times are typically measured in seconds, or even minutes, depending on the amount of laser light available.

The silver-halide emulsions are coated on either glass plates or film. Glass is preferred for most holographic applications, especially reflection holography, due to its ri-

gidity. The most popular size glass plates are 4" x 5" and 8" x 10", though display holographers often prefer to work with bigger sizes. It is most economical to purchase larger sizes and cut plates to your own specs, if this is feasible.

The same factories that coat glass plates usually offer their emulsions on film, too. A major advantage of film is that it is less expensive than glass, is easier to cut and curve, and is much more suitable for automated reproduction processes-as it is often available on rolls. The main difficulty faced by holographers using low powered CW lasers and film is keeping the film absolutely motionless during the exposure. Although the film can be sandwiched between clear pieces of glass, better results are obtained when the emulsion is left uncovered. Vacuum mounts and various other devices have been designed to accomplish this feat.

Some plates and films are supplied with an antihalation coating on the back, which can be useful when making transmission holograms, as it helps to cut down on unwanted internal reflections. These plates cannot, however, be used to make reflection holograms, so check product codes before ordering.

CURRENT AVAILABILITY

In response to anticipated demand, several major manufacturers of silver-halide emulsions adapted their existing production techniques and formulations to provide affordable recording materials for holographic applications. This spurred the growth of the holography industry, which needed a reliable supply of basic materials. These silver-halide emulsions were high quality, ready to use, and had a reasonably long shelf life.

Unfortunately, the worldwide demand for these particular photosensitive materials has leveled off, due in part to the increased use of electronic imaging, especially in the field of Non-Destructive Testing. Since the current combined needs of commercial holographers, educational facilities and hobbyists do not compare to other industrial and mass market customers, some major manufacturers (most notably, Agfa and Ilford) have ceased production entirely. However, the supply situation appears to be gradually improving as other factories expand their production capacity and establish new distribution networks. (To prevent potential supply problems, some commercial holographers have learned to utilize silver-halide recording materials with characteristics similar to holographic films but which are intended for other industrial uses, such as micro-lithography.)

Currently, the main manufacturers of silver-halide holographic recording materials are Slavich (Russia), HRT (Germany), and Eastman Kodak (USA). The first two companies are actively involved in researching and developing emulsions and substrates suitable for professional applications. In contrast, Kodak's product line has remained unchanged for years and its products are still used mainly by hobbyists and schools. Emulsions from Holdor (China) have been developed and are now available. Royal Holographic Art Gallery (Canada) is also supplying a film called "Red Star," which is reportedly a suitable replacement for Agfa's popular Holotest film. We have included all the information we could gather regarding these holographic recording materials in the following chapter, except for the Kodak specifications, which are available in previous editions of Holography MarketPlace.

Editor's Notes: All the aforementioned manufacturers have, or are developing, distribution channels for their products. Check the Business Directory in this book for an updated listing of suppliers.

Some distributors still have stockpiles of the popular Agfa and Ilford product, though production has been suspended for some time. Please refer to the previous editions of Holography MarketPlace for relevant technical specifications regarding these emulsions.

SELECTING SILVER-HALIDE MATERIALS

The general rules when selecting a silver-halide material are: match the peak sensitivity of the material as closely as possible to the wavelength emission of the laser being used to expose the material; and select an emulsion with the lowest possible graininess characteristics, and highest possible resolution. Let's examine why.

There are five atoms which, because of their atomic similarity, are called the halides. They are chlorine, bromine, iodine, fluorine and astatine. Silver-halide emulsions are made using either silver chloride, silver bromide, or silver iodide. The other two halides are not used because silver fluoride is insoluble in water and astatine is radioactive.

A typical silver-halide emulsion is made by adding a solution of silver nitrate to a solution of potassium bromide and gelatin. Silver bromide crystals form in the emulsion. The emulsion is heated for a certain amount of time, which is called the ripening process.

During the ripening process, the grain size increases and the speed of the emulsion is increased. Some doping agents may be added to the emulsion at this time to foster proper crystal growth. Afterwards, the gelatin is allowed to cool. It is then shredded, and the soluble potassium nitrate is washed out of the emulsion.

The emulsion is heated again, with more gelatin added; then it is cooled and applied to a base. The thickness and hardness of the emulsion is important in holography because emulsions that are too thick tend to deform during development. Emulsions that are too hard can either retard chemical reactions or create vacuoles in the emulsion left by migrating atoms. These vacuoles tend to scatter light.

THE PHOTOCHEMICAL REACTION

Let's assume the emulsion is made and we now want to expose it to light. It sounds surprising, but a perfectly structured crystal of silver bromide does not react to light in any appreciable way. A crystal with defects, however, does react with light. Fortunately, most silver-bromide crystals will have defects which consist of some interstitial (out of order) silver ions displaced in the crystal structure.

The process of the photochemical reaction is not known in exact detail, but it is believed that when light strikes a silver-bromide crystal, enough energy is available to remove an electron from an occasional bromide ion. The electron produced is able to migrate through the crystal until it comes in contact with an interstitial silver ion. The silver ion takes the electron and becomes silver metal. Silver atoms formed by this mechanism apparently act as a nucleus for the formation of aggregates of 10 to 500 silver atoms, known as latent images because they are too small to be seen by the naked eye.

After exposure, the emulsion is developed. The developer goes to the site of any silver-bromide crystal with a latent image and causes all the silver in that particular silver-bromide crystal to be reduced to silver metal and deposited on the already existing latent image of silver metal. This causes a wormlike grain of silver metal to form which is limited in size by the amount of silver available in the silver-bromide crystal. This growth is considerable, amplifying the size of the latent image silver metal by a factor on the order of 10^6.

If the developer is left in contact with the emulsion long enough it eventually attacks all the silver in the emulsion. The speed of development is slow enough, though, that you can use a timer to take the emulsion from the developer just after the latent image, but not the unexposed silver-bromide crystals, have been developed. At this point the developer has converted silver ions to silver metal if, and only if, they belong to a silver-bromide crystal that was exposed to light.

The emulsion is then placed in a fixer solution which attacks all silver-bromide crystals that were not exposed to light. The fixer makes these silver-bromide ions soluble and removes them from the emulsion. The result is an emulsion with black spots where light has struck, and clear spots where no light struck.

THREE MAIN FACTORS TO CONSIDER

An ideal silver-halide emulsion depends somewhat on its use but there are three main factors to consider in any emulsion: thickness of emulsion, grain size of silver-halide crystals, and sensitivity (or density of silver-halide crystals) in the emulsion. We can generally state the following: It is agreed that emulsions of more than 10 mm are neither practical nor theoretically necessary to produce most volume holograms. Thicknesses above this size cause problems in development.

Grain size becomes an important issue in holography because it involves recording fringe patterns that are wavelengths apart. Too large a grain size may create excessive scatter, which may fog or destroy your hologram, and too small a grain size makes the emulsion have no usable sensitivity. It is generally agreed that the ideal grain size is in the range of 01 mm to .035 mm.

The ideal exposure would probably be 100-300 mJ/cm2 to give a useful density (D = 2-3). If exposures are much longer than this, the main attraction of silver-halide emulsion, its speed, comes into question and other emulsions become more attractive.

Since each of the following silver-halide emulsion manufacturers have their own ideas about how to best serve the holography industry, the reader is encouraged to contact them directly in order to find the most suitable product for their particular application.

Presenting an Authoritative Resource for Holographers

Silver-Halide Recording Materials for Holography and Their Processing

H.I. Bjelkhagen, Ph.D.

This book gives a detailed analysis of the theory, characteristics, manufacturing, and processing methods of silver-halide materials used for the recording of holograms. Emphasis is placed on the selection of suitable silver-halide materials for conventional as well as special holographic applications. A detailed account of current developing and bleaching methods used in the production of silver-halide materials is given. In particular, the processing of Russian ultra-high-resolution materials is included, e.g., colloidal development. Various special techniques, such as hypersensitization and latensification, problems due to short exposure time when using Q-switched lasers, reciprocity failure, latent-image fading, etc., are treated. The author also supplies a large number of recipes for different, types of processing baths. The text is complemented by a comprehensive list of references which facilitate further study.

Contents

Introduction. Silver-Halide Materials. Commercial Silver-Halide Materials. Development. Bleaching. Special Techniques. Processing Schemes. After-Treatment. Color, Infrared and Ultraviolet Holography. Recipes and Formulas. References. *Subject Index.*

Springer Series in Optical Sciences, Vol. 66
First Published: 1993/ 440 pp. 64 figs. 46 tabs./Hardcover ISBN 3-540-56576-0.
Second printing: 1995/Paperback ISBN 3-540-58619-9.

The book can be ordered from Springer Verlag or from
INTEGRAF, 745 N. Waukegan Road, Lake Forest, Illinois 60045, USA.
Phone: (847) 234-3756, Fax: (847) 615-0835, Email: tjeong@ao1.com

SLAVICH

SLAVICH HOLOGRAPHY MATERIALS

by Ivan P Anyukhovsky and Yury A. Sazonov (Slavich),
Mikhail Grichine and Stasys Zacharovas (Geola)

The Slavich factory was founded in 1931 under the rather obscure title of the Pereslavl Factory of Cinema Films Nr. 5. The very first Soviet emulsions used for Russian cinematographic films were manufactured at this factory. Initially, Slavich produced only positive films but during the first 10 years of its life the factory broadened its range of products considerably, manufacturing negative cinematographic film, films for standard still photography, X-ray films for medical applications, and even magnetic tapes for audio recording.

After the Second World War this list of manufactured products was further broadened. Technical photographic films were produced for the printing industry, films suitable for micro-recording and the first photo paper was manufactured. Spurred on by the new demand for high-resolution photoplates for microelectronics applications, the Mikron plant was established as an integral part of Slavich in 1976. Konica, a well-known western photographic company, supplied most of the required equipment for this project. Later Mikron began to develop other materials for technical applications and particularly for holography.

EARLY DEVELOPMENTS

The first material specially developed for holography was an ultra-fine-grain photoplate, named PE-2, which was similar to the modem emulsion called PFG-03. Manufacture of these plates constituted the first serial production of photographic materials for holography applications in the U.S.S.R. Shortly thereafter fine grain emulsions bearing the names PFG-Ol and VRP were developed for interferometry applications and for the recording of transmission holograms in respectively the red and green spectral ranges.

For a long time it was unequivocally believed in the U.S.S.R. that the only acceptable material suitable for the recording of high quality reflection holograms was an ultra-fine-grain emulsion (grain size of not more than 10 to 20 nm) . Hence the fine grain emulsions such as the PFG-O 1 and VRP products (having grain sizes between 30 and 40 nm) were thought to be unusable for reflection holography. This conception was in fact so strong that, despite the many potential advantages offered by the fine grain emulsions such as easier manufacturing, radically better sensitivity, longer storage lifetime and a harder emulsion layer, no serious attempts of optimization of these prod-

ucts for reflection holography were attempted for many years. Finally, beginning in 1997, Slavich, with the help of holographers, laser manufacturers and holographic materials distributors, began optimizing their capability to produce a fine-grained emulsion useful for reflection holography. During this same period, Agfa (which was then the major worldwide supplier of holography materials) announced that it would be terminating its holography operation. This news supplied additional financial incentive to proceed along the track of research and development of such fine- grain emulsions.

Research work into the optimization of PFG-01 and VRP also resulted in implications for ultra-fine-grain materials and this led to the parameters of the old PFG-03 being improved. The new version was named PFG-03M which is the version currently produced. The technique of emulsion synthesis was simplified, sensitivity was increased, and the problem of the well-known gray fog on final holograms was eliminated. Prior to 1997 the development of the ultra-fine grain emulsions had also spawned an emulsion suitable for color holograms. This panchromatic version of the PFG-03M emulsion, named PFG-03C, figures today as one of our most advanced products.

We would mention here another achievement of the scientists and technologists at the Slavich company. A material based on dichromated gelatin was developed at the Mikron plant and put into serial production. Slavich is probably the only manufacturer worldwide to have ever produced commercially such a material.

TOWARDS THE PRESENT

Today Slavich is a factory of approximately 4,000 employees of whom about 100 work full-time in the Mikron Holography plant. Since 1997 our holography department has been working hard towards replacing the product vacuum left by Agfa. We have, of course, been aware that such a task could not be instant and would require both significant reorganization of our company together with strong external collaboration.

From our research we knew with certainty that we could produce equivalent products to the Agfa holography materials and, in addition, we realized due to our location in the CIS that we would be able to offer such products on the international market at a much lower price. Our biggest problem to capture the market therefore appeared to rest not so much on the manufacturing side of the business but rather seemed to stem from our lack of an adequate interface for western customers. We further identified specific problem areas in our marketing, sales, customer service, quality checking, packing, and transport sections. To address these problems we embarked on an in-depth process of quality control. In addition we decided to draw on the marketing resources of holographers and existing materials distributors.

Upon our request, in early 1998 one of our customers and collaborators, Geola, formed an international emulsions wholesale office in Vilnius in order to help us to market our holography products. We also asked Geola to look after our promotion on the international arena, to help us with international trade fairs, to properly organize our technical documentation and scientific publications and to generally create an English-speaking western sales interface for the Micron factory. At the date of writing, this office has 16 such distributors covering 21 national territories and a warehouse in Vilnius, from which a second quality check is performed and repacking is effected before final shipment.

In addition to the creation of a network of distributors through UAB Geola, we have maintained a small number of distributors which are presently administered through our Russian international office. This office is also responsible for our Russian clients of both photographic and holographic materials. (Editor's Note: Check the Business Directory section of this book for a more complete listing of companies currently selling Slavich products.)

CURRENT PRODUCTS

Slavich currently produces both silver-halide and dichromated gelatin emulsions for holography applications. Table 1 summarizes all the materials presently available.

The green-sensitive VRP-M emulsions are very close analogues to the old Agfa 8E56 products for both pulsed and CW laser recording. Likewise, the PFG-01 material gives equivalent performance to the Agfa 8E75 product for CW recording. Unfortunately no analogue exists for 8E75 when used with pulsed Ruby radiation.

The PFG-03M emulsion is an ultra-fine-grain red-sensitive emulsion for superior quality imaging. The PFG-03C material is a panchromatic equivalent to the PFG-03M.

Finally, PFG-04 is a long-life dichromated gelatin emulsion for blue and green laser recording.

THE VRP-M AND PFG-01 MATERIALS

Characteristic curves for these emulsions, showing spectral sensitivity versus wavelength, are shown in figure 1. The VRP-M light sensitivity (CW radiation) is seen to peak at approximately 75 micro Joules/cm2 and that of PFG-01 (CW radiation) at approximately 100 microJoules/cm^2.

Figure 2 shows the optical density after exposure by CW radiation and development versus energy.

Grain size characteristics for the VRP-M and PFG-01 emulsions are presented in figure 3.

The diffraction efficiency versus exposure for reflection holograms recorded on PFG-01 (using a CW laser) and on VRP-M (using a pulsed laser) is presented in figure 4. The maximum diffraction efficiency is seen to be >40% for both emulsions.

The VRP-M emulsion may be used equally well with fre-

PRODUCT/DESCRIPTION

VRP-M

Fine-grained green-sensitive holographic plates and film having no antihalation coating and designed for reflection or transmission hologram recording. Average grain size is 35-40 nm, resolving power is more than 3000 lines/mm, 514 nm, 526 nm, 532 nm. The VRP product is identical to VRP-M except for an antihalation coating designed especially for transmission work.

PFG-01

Fine-grained red-sensitive holographic plates and film designed for transmission or reflection hologram recording. Average grain size is 40 nm, resolving power more than 3000 lines/mm, spectral sensitivity range 600-680 nm (including 633 nm, 647 nm).

PFG-03M

Ultra-fine-grained red-sensitive plates and film designed for reflection hologram recording. Average grain size is 8-12 nm, resolving power more than 5000 lines/mm, spectral sensitivity range includes 633 nm, 647 nm.

PFG-03C

Ultra fine-grained panchromatic (full-color holographic plates designed for color reflection hologram recording. Average grain size is 8 nm, resolving power more than 5000 lines/mm, spectral sensitivity range up to 700 nm. (457 nm, 514 nm, 633 nm.)

PFG-04

Dichromated Gelatin holographic plates designed for phase reflection hologram recording. Resolving power greater than 5000 lines/mm, spectral sensitivity range up to 514 nm. (457 nm, 488 nm, 514 nm.)

Table 1. Holographic materials available from Slavich.

---

quency doubled pulsed neodymium lasers as with green CW radiation. The PFG-01 emulsion, however, is not sensitive to pulsed radiation from the ruby laser and as such should be only used with CW sources. Emulsions used with pulsed radiation should be post-sensitized with the technique of latensification (see following section). Material life is more than two years.

Recommended Chemistry for VRP-M

Table 2 shows a summary of various recommended processing schemes for use with pulsed neodymium lasers (526.5 nm, 532 nm). Table 3 describes related formulas. All these processing recipes work equally with CW argon for

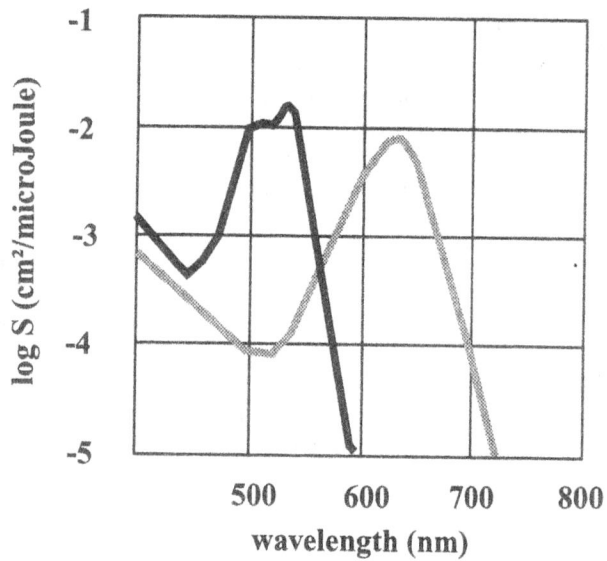

Figure 1. Spectral sensitivity curves for VRP-M (black) and PFG-01 (grey).

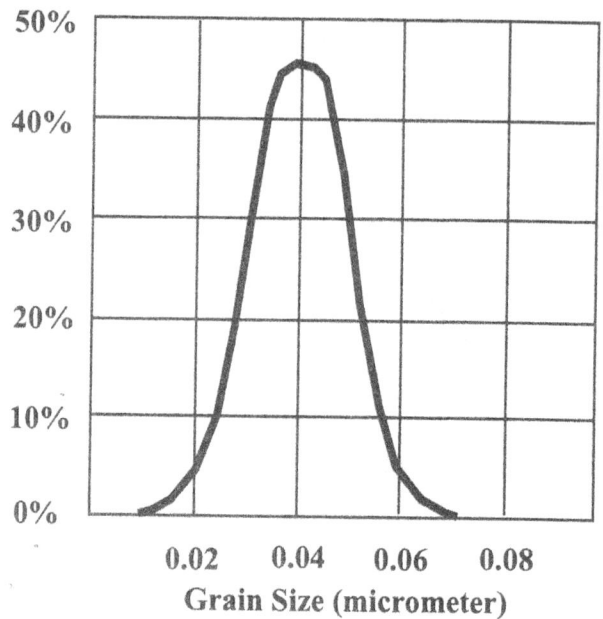

Figure 3. Grain size distribution curve for VRP-M and PFG-01.

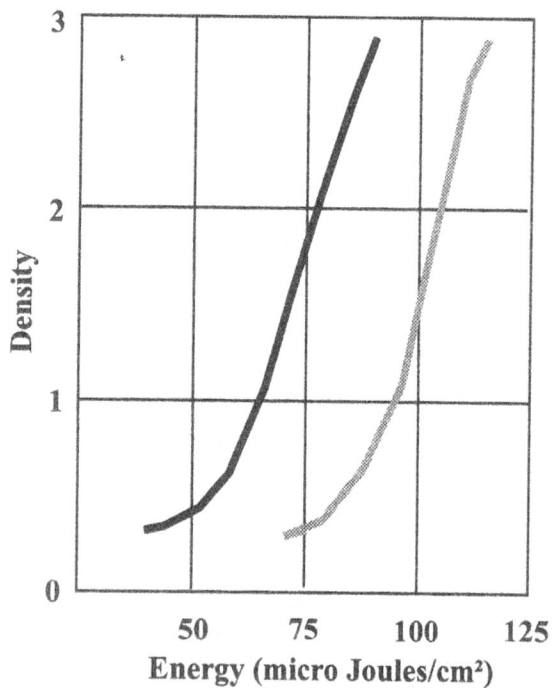

Figure 2. Characteristic curves for VRP-M (left) and PFG-01.

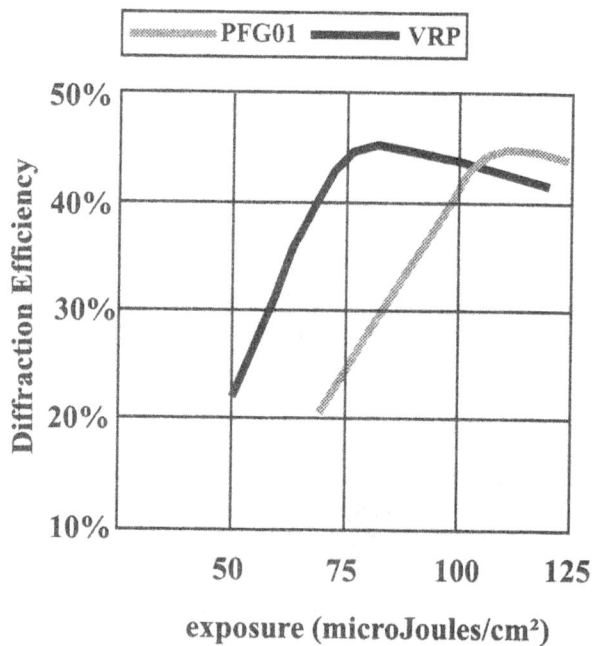

Figure 4. Diffraction efficiency curves for VRP-M and PFG-01.

	Mastering Transmission Holograms	Copying (Reflection Holograms) Final Color: Green	Copying (Reflection Holograms) Final Color: Honey-Green	Copying (Reflection Holograms) Final Color: Orange
Exposure	20-40 microJ/cm2	30-50 microJ/cm2	50-70 microJ/cm2	50-70 microJ/cm2
Latensification	Yes	Yes	Yes	Yes
Development	SM-6, 2-3 min	SM-6, 2-3 min	M-Pyro+, 3-4 min	M-Pyro+, 3-4 min
Wash	Water	Water	Water	Water
Bleach	PBU-Amidol until clear	PBU-Amidol until clear	PBU-Amidol until clear	PBU-Amidol until clear
Wash	10-20 min	10-20 min	10-20 min	10-20 min
Potassium Iodide bath	No	No	No	2 min bath
Washing	No	No	No	1-2 min
Final Wash	Water with wetting agent, 1 min	Water with wetting agent, 1 min	Water with wetting agent,1 min	Water with wetting agent, 1 min
Slow Air Drying	Yes	Yes	Yes	Yes

Table 2. Recommended chemistry for VRP-M with pulsed Nd:YLF/YAG Radiation.

both transmission and reflection holograms; however, in this case latensification is usually not necessary and exposure is a little longer. In addition, for CW, you may obtain better results using the CW-C2 developer depending on your color requirements. Uniform heating in a special oven provides easy color control across the spectrum for pulsed lasers.

Recommended Chemistry for PFG-Ol with CW HeNe and Krypton Radiation

The PFG-OI emulsion can be processed with virtually any chemistry previously applied to the Agfa 8E75 emulsion with the single addition, if necessary, of a post-exposure latensification step before chemical processing. We recommend, in particular, the SM-6 and CW-C2 developers defined in table 4 for both transmission and reflection holograms. Color control is possible with all the normal methods used with the Agfa emulsion including TEA.

Latensification

Slavich PFG-01 and VRP-M emulsions have peak sensitivities to exposures in the millisecond regime. In order to obtain optimal sensitivity to exposures either much longer or much shorter than this time frame, the simple technique of latensification can be used. Practically speaking, this means that some CW laser users and all pulsed laser users will use latensification.

Latensification is usually done directly after the holographic exposure. However, before you can apply the process you must work out a latensification time appropriate for your system. This procedure is as follows:

1) Place a 25 W lamp with a dark filter (green for VRP-M and red for PFG-01, although white light also works) at a distance of 1 m from a test holoplate or film, such that its light uniformly illuminates the emulsion.

2) You will need to try several exposure times ranging from o to around 4 minutes and look at how the emulsion develops. Start with zero exposure time and then (under your normal safelight conditions) develop the plate. The plate will darken a little. This is called the fog level. After development put this control plate into a stop-bath, wash it, and keep it handy.

3) Now you must start to make a series of test exposures with small test plates. Start at about 0.5 mins and go up to around 4 mins. After each exposure, develop your plate and match the darkening of this plate to your control plate. If it is the same, you need more exposure, so go back again and repeat the process. Stop when you get a result that is just marginally darker than the fog level. This is then the correct latensification exposure for your geometry.

Now that you have discovered the proper latensification time, all you must do is after every proper holaplate exposure you must take your plate and illuminate it exactly as described above for the time that you have worked out. Then all processing is as normal. Latensification stabilizes and enhances the latent image formed by the holographic exposure. If required, chemical processing may be done with significant delay after latensification.

PFG-03M

This material is designed for reflection hologram recording using CW radiation in the red spectral range (633 nm -

PBU-AMIDOL BLEACH	
Potassium Persulphate	10.0 g
Citric Acid	50.0 g
Cupric Bromide	1.0 g
Potassium Bromide	20.0 g
Amidol	1.0 g
Water	to 1.0 L
POTASSIUM IODIDE BATH	
Potassium Iodide	18.0 g
Water	to 1.0 L

Table 3. Bleach and bath recommended in table 2.

SM-6 DEVELOPER

Ascorbic Acid	18g
Sodium Hydroxide	12.0g
Phenidone	6.0g
Sodium Phosphate (dibasic)	28.4g
Water	to 1 L

M-PYRO+DEVELOPER

1 part A + 1 part B

Part A

Pyrogallol	20.0g
Phenidone	1.2g
Sodium Metabisulphite	5.0g
Water	to 1.0L

Part B

Sodium Carbonate	130.0g
Water	to 1.0L

CW-C2 DEVELOPER

1 part A + 1 part B

Part A

Catechol	20.0g
Ascorbic Acid	10.0g
Sodium Sulphite (anhydrous)	10.0g
Urea	100.0g
Water	to 1.0L

Part B

Sodium Carbonate	60.0g
Water	to 1.0L

Table 4. Developers recommended for VPR-M and PFG-01.

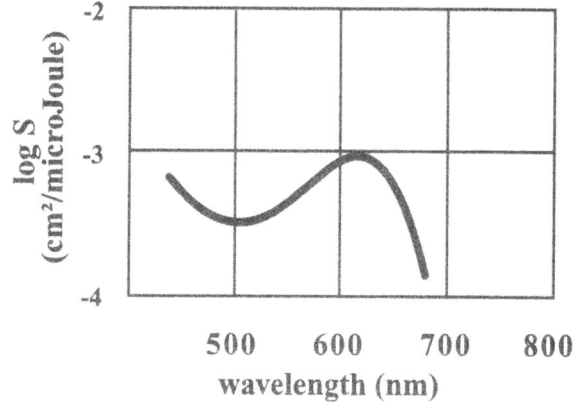

Figure 5. Spectral sensitivity of the PFG-03M material.

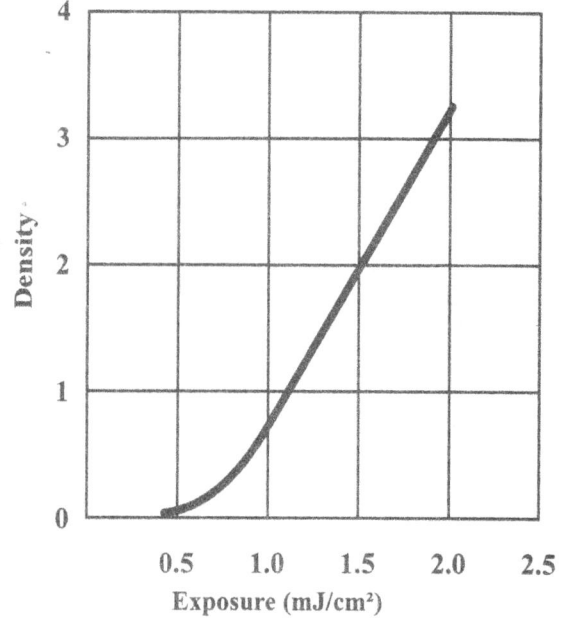

Figure 6. Characteristic curve for PFG-03M.

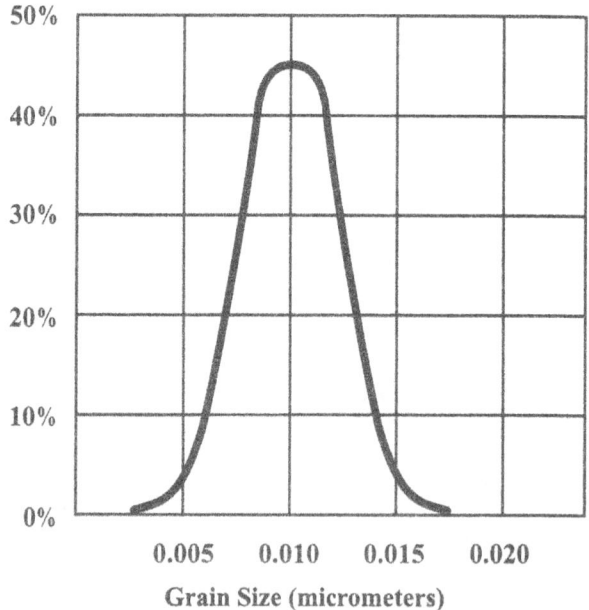

Figure 7. Grain size distribution for the PFG-03M material.

HeNe laser and 647 nm-krypton laser). The spectral sensitivity curve of the material is shown in figure 5. Peak emulsion sensitivity is around 1.5-2 mJ/cm2•

Density versus energy is shown in figure 6 and the grain size distribution curve is shown in figure 7.

Despite a lower sensitivity than VRP-M and PFG-Ol, the PFG-03M material has a higher diffraction efficiency and a very high signal to noise ratio. Holograms recorded on this material have a very clear and powerful object reconstruction and excellent layer transparency. Usually, a physical solution development that acts to create colloidal silver is the preferred processing method. PFG-03M does not need bleaching. Refer to table 5 for processing instructions.

Hardening:
2-3 min

Formalin 37%	10 ml
Potassium Bromide	2 g
Sodium Carbonate	5 g
Water	to 1 L

Washing:
1-2 min

Wash in running (filtered) water.

Development in GP-2:
10-15 min

Concentrated solution:

Methyl Phenidone	0.2 g
Hydroquinone	5 g
Sodium Sulphite (Anhyd)	100 g
Potassium Hydroxide	5 g
Ammonium Thiocyanate	9 g
Water	to 1 L

Working Solution: 40 ml GP-2 + 1 L H20

Washing:
1-2 min

Wash in running (filtered) water.

Fixing:
2 min

Sodium Thiosulphate (cryst.)	160 g
Potassium Metabisulphate	40 g
Water	to 1 L

Washing:
1-2 min

Wash in running (filtered) water.

Drying:
2 min each bath

Ethyl Alcohol	50%,
	75%
	and 96%

Table 5. Recommended processing for PFG-03M (reflection holograms). Note that temperature must be lower than 19°C.

PFG-03C

This material is designed for the production of full-color (bleached) reflection holograms using CW laser radiation in the blue (457 nm-argon laser), green (514 nm-argon laser) and red (633 nm-HeNe laser).

A spectral sensitivity curve of the material is shown in figure 8. The PFG-03C grain size distribution curve has the same shape as for the PFG-03M material.

Diffraction efficiency versus exposure is shown in figure 9. The maximum DE in the blue range is >25 % and in the green and red ranges >45%. Sensitivity values are 2 mJ/cm2 and 3 mJ/cm2 respectively. Refer to table 6 for processing instructions.

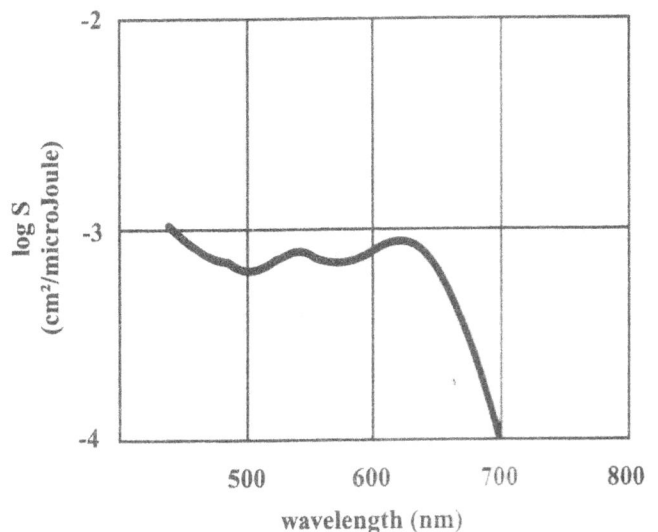

Figure 8. Spectral sensitivity curve for PFG-03C.

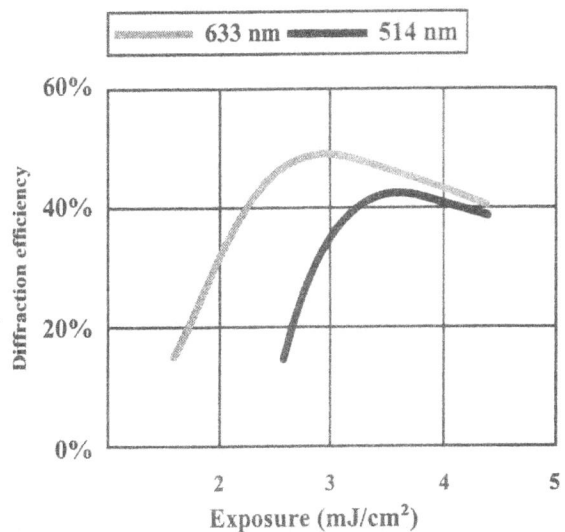

Figure 9. Diffraction efficiency curve for PFG-03C.

Hardening:
6 min

Formalin 37%	10 ml
Potassium Bromide	2 g
Sodium Carbonate	5 g
Water	to 1 L

Washing: 1-2 min

Wash in running (filtered) water.

Development VRP: 4-5 min

Concentrated solution:

Sodium Sulphite (Andydrous)	194 g
Hydoquinone	25 g
Potassium Hydoxide	22 g
Methyl Phenidone	1.5 g
Potassium Bromide	20 g
Potassium Metaborate	140 g
1,2,3-Benzotriazole	0.1 g
Distilled Water	to 1 L

Work Solution: 1 part of VRP developer + 6 parts water.

Washing: 1-2 min

Wash in running (filtered) water.

Bleach in PBU-Amidol:
5 - 8 min

Copper Bromide	1 g
Potassium Persulphate	10 g
Citric Acid	50 g
Potassium Bromide	20 g
Distilled Water	to 1 L
Amidol	1 g

Washing: 1-2 min

Wash in running (filtered) water.

Stop Bath: 2 min

Acetic Acid	20 g
Water	to 1 L

Washing: 1-2min

Wash in running (filtered) water.

Bathing: 2 min

Bathing in distilled water with added wetting agent

Dry: Normally

Table 6. Recommended processing for PFG-03C (Reflection Holograms)

FILM SUBSTRATES

There will be two film substrates introduced in 1999. The original 150 micron TAC film which has been the standard Slavich film product until now will continue to be available and a 180 micron TAC film will be introduced. The thicker film, very similar to the original substrate used by Agfa, will first be introduced for the VRP emulsion. During the course of 1999 we expect to use this substrate in all our current film products (refer to table 7a).

The 180 micron film is suitable for larger format holography and will be available in widths of up to 1200 mm. The 150 micron film is more appropriate for small format applications although may be used for larger format when Lamination techniques are employed. Details of recommended techniques appear in the Slavich technical catalog available from Slavich distributors.

NEW SERVICES FOR 1999

During 1998, we put up our first Web homepage at "www.slavich.com". This is now updated on a regular basis with all the latest information concerning Slavich products. By the end of 1998 the emulsions marketing office will have its own.,Internet server running the Linux operating system and should be connected directly to the Internet via a Swedish landline. This will permit more frequent updates of technical information (see tables 7b and 8) and will also provide an Internet FAQ service. Questions will be submitted by email to the service and answers will be able to be posted by any user subject to verification by the service supervisor. A search facility will allow any user to search the entire database to see if his question has already been asked and answered. Technicians from various testing facilities will also post answers on the service at regular intervals. It is hoped that such a service will be widely used and will provide a focus where a comprehensive knowledge of all technology associated with holography materials will accumulate and be freely available. (Editor s Note : Additional information regarding Slavich products, including processing tips, is available in the Holography MarketPlace 7th Ed., on the 3Deep Web site (www.3deepco.com). and from Integraf Oeong@lfc.edu/physics/holography/).

	1200mmx 10m(Roll)	610mm x 10m(Roll)	304mm x 10m(Roll)	203mm x 20m(Roll)	102mm x 20m(Roll)	76mm x 20m(Roll)	36mm x 20m(Roll)	350mm x 10m(Roll)	203mm x 254mm(x50)	102mm x 127mm(x50)	200mm x 300mm(x 5)
PFG-01	y	n	n	y	y	n	y	y	y	y	y
VRP-M	y	n	n	n	n	n	n	y	y	n	y
PFG-03M	n	n	y	y	y	n	n	n	y	y	y
PFG-03C	n	n	n	n	n	n	n	n	n	n	n
PFG-04	n	n	n	n	n	n	n	n	n	n	n

Table 7a. Currently available film formats.

Although the following Slavich product is a dichromated gelatin coated plate and not a silver-halide emulsion, we've included it in this section for easy reference.

PFG-04

This material is designed for the recording of reflection Denisyuk-type holograms using CW laser radiation (488 nm, 514 nm-argon laser). The material spectral sensitivity curve is shown in figure 10. The sensitivity reaches 100 mJ/cm^2 in the blue spectrum range and 250 mJ/cm^2 in the green. Due to its grainless structure, this material has very high resolving power and a diffraction efficiency of >75% (figure 11).

The recommended processing technique of the exposed hologram consists of the following operations:

1) Thermal hardening after exposure (100 degrees C)- depending on the layer freshness.

2) Cooling to room temperature.

3) Bathing in running filtered water-3 mins.

4) Bathing in 50% Isopropyl Alcohol solution for 2-3 mins.

5) Bathing in 75% Isopropyl Alcohol solution for 2-3 mins.

6) Bathing in 100% Isopropyl Alcohol solution for 2-3 mins.

7) Drying in a desiccator (100 degrees C) for 60 mins.

8) Emulsion layer preservation using optical anhydrous adhesive and protective glass.

Note that the processing solution temperatures must not exceed 20 degrees C for fresh layers. If holograms appear "milky" in color then the processing solution temperature should be decreased or the thermal hardening period should be prolonged. The material shelf life is 12 months (average observed period).

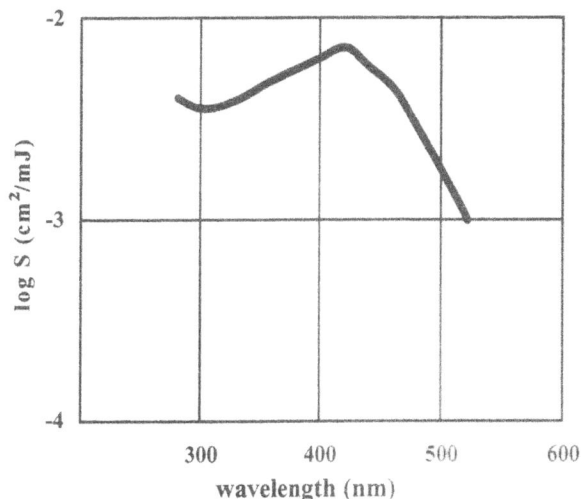

Figure 10: Spectral sensitivity curves for PFG-04

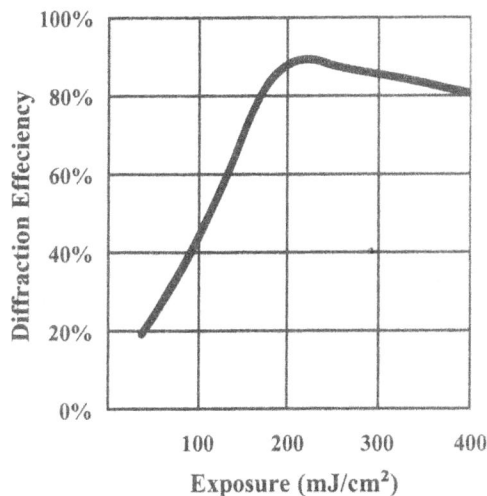

Figure 11: Diffraction efficiency curve for PFG-04.

ACKNOWLEDGEMENTS
We are grateful to Sergey Vorobov who is responsible for the technique of latensification We would also like to acknowledge the work of Hans Bjelkhagen and Nicholas Phillips, on which much of the chemistry listed in this article is based.

About the authors:
Ivan P. Anyukhovshy is the managing director and president of AO Slavich. Yury A. Sazonov is director and manager of the Micron plant and is also a principal director of AO Slavich. Mikhail Grichine is the managing director of UAB Geola. Stasys Zacharovas is sales manager at UAB Geola.

GLASS PLATES	406 x 609 mm	300 x 406 mm	180 x 240 mm	102 x 127 mm	63 x 63 mm
PFG-01	n	y	y	y	y
VRP-M	y	y	y	y	y
PFG-03M	n	y	y	y	y
PFG-03C	n	y	n	y	y
PFG-04	n	y	y	y	y
Plates/box	4	6	6	25	30

Table 7b. Available formats for Slavich glass plates.

SLAVICH HOLOGRAPHIC MATERIALS—TECHNICAL SPECIFICATIONS

Parameter	PFG-01	VRP-M*	PFG-03M	PFG-03C	PFG-04
Holographic Sensitivity @ 457 nm CW microJoules/cm^2	x	x	x	2,000	80,000
Holographic Sensitivity @ 488 nm CW microJoules/cm^2	x	x	x	x	100,000
Holographic Sensitivity @ 514.5 nm CW microJoules/cm^2	x	75	x	3,000	250,000
Holographic Sensitivity @ 526.5 nm 30 ns pulse with latensification -- microJoules/cm^2	x	75	7x	x	x
Holographic Sensitivity @633 nm microJoules/cm^2	100	x	1,500-2,000	3,000	x
Gamma	>6.0	>6.0	>6.0	>6.0	x
Max. Density on Characteristic Curve (D_{max})	<4.0	<4.0	<4.0	<4.0	x
Fog Density (D_0)	<0.01	<0.01	<0.01	<0.01	x
Resolving Power (R), mm^{-1}	3,000	3,000	>5,000	>5,000	grainless
Max. of Spectral Sensitization, nm	633	530	633	457/514/633	415
Swollen Emulsion Layer Strength after Chemical Processing, H (gm force)	900	900	>50	>50	x
Adhesion Between Emulsion Layer and Base After Chemical Processing (Class A-F)	A-C	A-C	A-C	A-C	x
Deformation Temperature of Emulsion Layer in Water (t_{def}),^oC	>90	>90	>35	>35	x
Emulsion Layer Thickness (microns)	7-8	6-7	6-7	9-10	16-17
Normal Diffraction Efficiency for Reflection @ 457 nm, %	x	x	x	>25%	>75%
Normal Diffraction Efficiency for Reflection @ 488 nm, %	x	x	x	x	>75%
Normal Diffraction Efficiency for Reflection @ 514.5 nm, %	x	>40%	x	45%	>75%
Normal Diffraction Efficiency for Reflection @ 526.5 nm, %	x	>40%	x	x	x
Normal Diffraction Efficiency for Reflection @ 633 nm, %	>40%	x	>45%	>45%	x

* VRP and VRP-M plates and films have the same technical specifications.

Table 8. Summary of technical specifications.

HOLOGRAPHIC PLATES

Made in Germany

HRT Holographic Recording Technologies GmbH

We supply a full range of silver halide holographic plates covering all wavelengths from 400 nm up to 700 nm

We are happy to announce that our new line of holographic plates for pulsed ruby work is available now. These plates can be used in identical setups as for Agfa 8E75.

For further information please contact us:

HRT GmbH
Am Steinaubach 19
36396 Steinau
GERMANY

Phone: +49-6663-7668
Fax: +49-6663-7463
e-mail: info@holographic-materials.de
Web: http://www.holographic-materials.de

US-customers please contact:

VinTeq, Ltd.
611 November Lane / Autumn Woods
Willow Springs, NC 27592-7738
USA   *919 556 9924*

Phone: (919) 639-9424
Fax: (919) 639-7523
e-mail: vinson@vinteq.com
Web: http://www.vinteq.com

Silver Halide Holographic Plates:

Wavelengths (A): 6943, 6328, 5320, 5145, 4880, 4416.
Resolution: >5000 1p/mm
Sensitivity: About 2.5 mj/cm²
Emulsion thickness: 7 - 15μ
Storage time: > 1 year (room temp)
Size: 4" x 5", 8" x 10" and so on.
processing: CWC2 Developer or Pyrogallol Developer

Sample package of 4" x 5" Holographic plates (4) is only $20.00

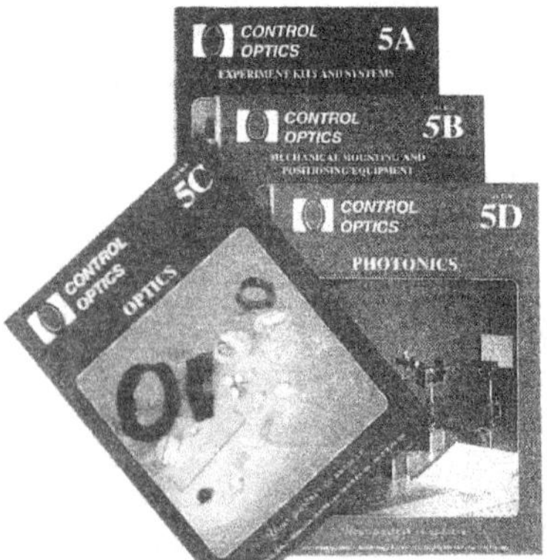

CONTROL OPTICS 5A
EXPERIMENT KITS AND SYSTEMS
CONTROL OPTICS 5B
MECHANICAL MOUNTING AND POSITIONING EQUIPMENT
5C CONTROL OPTICS 5D
OPTICS PHOTONICS

SEND FOR OUR
FREE
SPECTRUM
EYE PROTECTION
WALL CHART
AND/OR OUR
SOURCE CATALOGS
5A-KITS,
5B-MOUNTS,
5C-OPTICS AND
5D-PHOTONICS

CONTROL OPTICS
13111 Brooks Dr., Suite J • Baldwin Park, CA 91706 • U.S.A.
Tel: (626) 813-1991 • Fax: (626) 813-1993
FOR ORDERS ONLY: 888-OPTO-888
E-mail: Liucoc@Sprynet.com

THE BEST
PHOTOPLATES
F O R
HOLOGRAPHY
IN THE WORLD

HOLO.TREND

HOLOGRAPHY
CENTER
AUSTRIA

A - 3042 WUERMLA
Kahlenbergstr 6

Tel +43 2275 8210
Fax +43 2275 8281

HOLOGRAPHIC RECORDING TECHNOLOGIES GmbH

THE BB EMULSION SERIES: CURRENT STANDINGS AND FUTURE DEVELOPMENTS

by Richard Birenheide

As most of those concerned with display holography have experienced, acquiring silver-halide recording material suitable for holography has become more difficult recently. Four years ago we started developing a new silver-halide material co-working with Jeff Blyth. Our goal was to find a way to manufacture an emulsion which had lower scatter, higher diffraction efficiency and a sensitivity similar to existing Western type materials.

The starting point was a special kind of Lippmann emulsion developed by Blyth. These emulsions are characterized by a grain size of approx. 15 nm. This very small grain size (common Western type emulsions have about 35 nm to 40 nm), which would be perfect for holography in principle, has the serious disadvantage of having low sensitivity.

This low sensitivity occurs mainly for three reasons . First of all, the number of photons needed to make a grain developable is fairly independent of its size. Therefore, larger grains are more sensitive because they have a larger cross sectional area to trap the required number of photons.

Second, it is necessary for such a low grain size that the formation of the silver-halide crystals is performed under low concentration of the reaction partners. This leads to a low overall concentration of silver halide in the final emulsion.

Last, but not least: most of the common "tricks" to raise sensitivity are leading to grain growth after the silver-halide particles have formed. Although Blyth's method allowed us to make emulsions of quite high silver-halide concentration, the sensitivity and long-term stability needed improvement.

CHARACTERISTICS OF THE BB EMULSIONS

A way was found at our lab to produce emulsions which are a compromise with respect to grain size and sensitivity. These emulsions have the following characteristics: a grain size between 20 nm and 25 nm and a sensitivity of approx. 100 $\mu J/cm^2$ to 150 $\mu J/cm^2$ (BB-640 exposed at 633 nm and developed to a density of 2.5 with a metoVascorbate developer).

All our emulsions are available coated onto glass with a thickness of approximately 6.5 /μm. The silver-halide density is approx. 3.0 g/m^2. We were able to improve long

term stability such that storage times of one year or more are possible.

BB-640

Figure 1 shows the absorption spectrum of the BB-640 plates. The sensitivity ranges from 580 nm to 660 nm. These plates need approx. three times the exposure of former

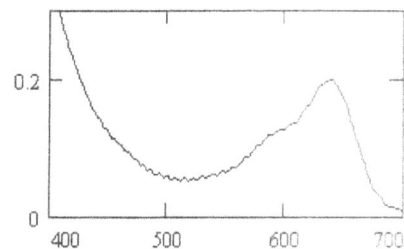

Agfa 8E75HD material (@ 633 nm). This emulsion is not sensitive to ruby wavelength.

Fig. 1. Spectrum of BB-640

BB-520

BB-520 has been designed for work in the blue/green region. The absorption spectrum is shown in figure 2. The sensitivity ranges from 480 nm to 540 nm. The exposure is approx. two times the exposure needed for the former Agfa 8E56HD emulsion. It should be mentioned that scatter is low and grain size small enough to record excel-

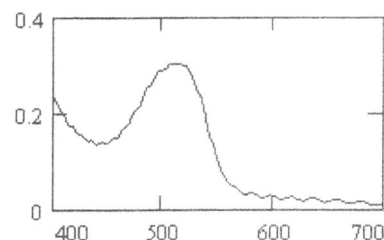

lent reflection holograms with this material, although it is mainly used for transmission mastering.

Fig. 2. Spectrum of BB-520

BB-450

This is a special emulsion for the blue. We designed this emulsion specifically for embossed mastering, e.g., work with HeCd lasers. Figure 3 shows the absorption spectrum. These emulsions have such a low scatter that work with this emulsion is highly recommended if one is concerned with embossed holograms. In contrast to all of our other emulsions this one is not recommended for reflection or Denisyuk holograms. To achieve highest possible sensitivity, dye level is too high to get a suitable fringe contrast throughout the emulsion layer.

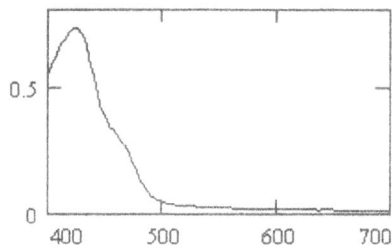

Fig. 3. Spectrum of BB-450

BB-PAN

For those interested in color holography BB-PAN offers the possibility to get into this field. It is a first generation derivative of BB-640 and BB-520 and contains simply the dyes of both. Figure 4 shows the absorption spectrum. Please note the pronounced absorption minimum in the region of 560 nino

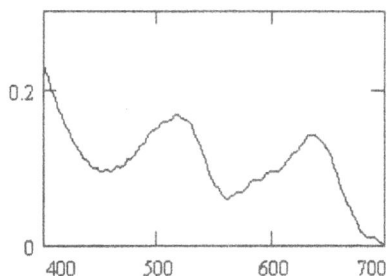

Fig. 4. Spectrum of BB-PAN

BB-700

We now have a preliminary laboratory emulsion sensitized for pulsed ruby available. At the time this document was written, first tests had been carried out at The Holographic Image Studio, London, by Martin Richardson. He used these plates presensitized with 5% TEA under the same lighting conditions as former Agfa 8E75 and yielded excellent results. Please visit our Web site regularly to keep updated about the latest information regarding these plates. At this time, (November 1998) more information is not available.

PROCESSING

To facilitate mastering, the emulsions are hardened to a high extent in order to prevent distortion of the fringes during processing. It is emphasized that nearly any kind of conventional processing works with the BB-plates. For mastering, the following recipe can be recommended:

Developer

- 700 cc water (deionized, if available)
- 70 g sodium carbonate, anhydrous
- 15 g sodium hydroxide
- 4 g metol
- 25 g ascorbic acid
- add water to 1000 cc

Bleach

- 700 cc water (deionized, if available)
- 35 g copper sulfate (pentahydrate)
- 100 g potassium bromide
- 5 g sodium hydrogen sulfate crystals
- add water to 1000 cc

Please note that this combination gives excellent results for mastering but is rather printout sensitive. If this is a problem, a final bath in mild acidified dichromate is recommended.

AVAILABLE FORMATS

The following formats are currently available: 2.5" x 2.5", 4" x 5", 8" x 10", 30 cm x 40 cm, and 50 cm x 60 cm. The first two are on 2 mm plain float glass, all others on 3 mm.

SPECIAL PRODUCTS

Due to the rather small batches run at our facility, a range of special products can be produced. Currently possible are: any format smaller than 50 cm x 60 cm, emulsions with reduced or increased silver content, and emulsions with different dye levels.

SOME RECOMMENDED REFERENCES

H.I. Bjelkhagen's Silver Halide Recording Materials for Holography and Their Processing, Springer, 1988; Graham Saxby's Practical Holography, Prentice Hall, 1988; Graham Saxby's Manual of Practical Holography, Focal Press, 1991.

For further information. contact:
HRT Holographic Recording Technologies GmbH
Am Steinaubach 19
36396 Steinau, Germany
Phone: +49-6663-7668
Fax: +49-6663-7463
Email: hrt.birenheide@t-online.de
Web site: www.holographic-materials.de

ADDITIONAL SILVER HALIDE SUPPLIERS

SILVER-HALIDE PLATES FROM HOLDOR

by Wai-Min Liu (Control Optics) and Chuck Paxton (Photon Cantina)

MANUFACTURER'S SPECIFICATIONS

Wavelengths: 6943, 6470, 6328, 5320, 5145, 4880, 4416

Resolution: 3,000-10,000 lp/mm

Sensitivity: About 2.5 mj/cm^2

Emulsion thickness: 7-15 m

Storage time: > 1 year (room temp.)

Size: 4" x 5",8" x 10" and so on

Processing: CWC2 Developer or Pyrogallol Developer

Applications: Transmission holograms, reflection holograms, embossed holograms, and holographic interferometry

We have been testing these new plates for the last couple of months. The following information should give anyone interested in using these plates a good starting point.

The Emulsion

The emulsion is available in red, green, blue, and panchromatic. We found the panchromatic to be the most sensitive. When compared to Agfa, we found the sensitivity less than that of Agfa 8E75 by a factor of about five. This means the stability of the set-up becomes very important to handle the long exposure times required. One thing we noticed is that the emulsion is very hard and holds up well to processing without the need to harden before processing, which is necessary with the Slavich plates.

Pre-sensitization

We found it important to pre-sensitize prior to exposure with a 2% solution of triethanolomine for two minutes. When using a ReNe laser this will give image playback in the yellow-green region. This also helps to reduce the exposure time by 50%.

Exposure

The laser we used is Spectra-Physics 125 currently producing between 30 and 40 mW Our set-up is single beam reflection type and our exposure time was 10 seconds using the panchromatic plates with a 2 minute pre-sensitizing step.

Processing

We tried several types of developers and bleaches and found the best to be a modified CW-C2 developer and PBU-quinol bleach. The developing time was between three and five minutes, which gave a plate density of about 70% dark. Bleaching time was two to three minutes.

Results

We found the image resolution to be excellent and image brightness to equal that of Agfa 8E75. We feel that this emulsion should be used for single beam or contact copy type holograms. For split-beam holograms, I would recommend the use of a fringe locker to handle the long exposure times.

Processing Steps

1. Pre-sensitize emulsion with a 2% solution of triethanolomine two minutes, then dry plate.

2. Expose plate.

3. Develop plate for three minutes (70% dark).

4. Rinse plate for three minutes.

5. Bleach plate for three minutes.

MODIFIED CWC2 DEVELOPER

Part A	
Distilled Water	400 ml
Catechol	10 g
Ascorbic acid	5 g
Sodium sulfite (anhydrous)	5 g
Add distilled water to make 500 ml	
Part B	
Distilled Water	400 ml
Sodium carbonate (anhydrous)	30 g
Add distilled water to make 500 ml	

Take A:B = 1:1 compounded for use.

PBU QUINOL BLEACH

Distilled Water	750 ml
Cupric Bromide	1 g
Potassium Persulfate	10 g
Citric Acid	50 g
Potassium Bromide	20 g
Add distilled water to make 1 liter	

After the above chemicals have been mixed, add 2 g hydroquinone (quinol) for PBU-quinol bleach. The bleach can be used after six hours.

RED STAR ULTRA FILM NOW AVAILABLE FROM ROYAL HOLOGRAPHIC ART GALLERY

information contributed by Derek Galon

Royal Holographic Art Gallery is now selling a film that it claims is equivalent, if not superior, to the popular Agfa films (which have been discontinued). According to Royal's Derek Galon, the newly available film, called "Red Star Ultra," has received very favorable reviews from the labs which have purchased it.

Galon states the following on the company's web site: "We are pleased to announce the launching of: Red Star Ultra-the new generation film, available both for red and for green lasers. Unlike our standard Red Star film, this one can be developed with formulas normally used for Agfa, and the exposition times are better than HoloTest!"

TECHNICAL DATA

Resolution	over 3000 lines per mm
Light sensitivity	80-100 microJ/1 sq.cm
Diffraction effectiveness	over 40%
Emulsion thickness	7 microns
Acetate base thickness	140 microns

Red Star Ultra R (red-sensitive)

A sensitivity range of 600-680 nm makes it the perfect choice for HeNe lasers and, with just slightly stronger exposure, it works fine with pulsed ruby. For reflection and transmission holograms. Chemicals used for AGFA HoloTest 8E75 work fine on this material (Sodium Hydroxide/Phenidone and Catechol developers are recommended). Fine adjustment of final film color in orange-green range possible with special processing (information available on request). Safe storage time (at 30% humidity, 4-5°C): minimum 16 months.

Red Star Ultra G (green-sensitive)

A sensitivity range of 488-532 nm makes this the perfect choice for argon lasers and pulsed YAG. For reflection and transmission holograms. Chemicals used for AGFA HoloTest 8E56 work fine on this material (Sodium Hydroxide/Phenidone and Catechol developers are recommended). Fine adjustment of final film color in orange-green range possible with special processing (information available on request). Safe storage time (at 30% humidity, 4-5°C): minimum 12 months.

(As a quick comparison, this sensitivity translates to an average exposition of only 1-2 sec. for 5" x 5" film in a reflection setup, using 10 mW HeNe laser!)

For further information, contact:
Royal Holographic Art Gallery
122-560 Johnson St.
Victoria, BC, Canada V8W 3C6
Phone/Fax: (250) 384-0123
Email: office@holograms.bc.ca
Web site: www.hoiograms.bc.ca

POLAROID

Polaroid Corporation announced that 1998 will be the final year that it produces its DMP-128 photopolymer holograms. Polaroid has been at the forefront of three-dimensional imaging since its founder, Dr. Edwin Land, began research on 3D motion pictures in 1938.

Holography research began at Polaroid in the 1960s. The rainbow hologram was developed at Polaroid by Dr. Steven Benton and others during this time. Polaroid chemists were charged with developing a special photopolymer to allow the production of a new type of high performance, white-light hologram. The end result was DMP-128 developed in the mid-1980s. Since that time, photopolymer holograms have been used in a variety of scientific and technical applications ranging from heads-up displays, instrumentation for automotive and aircraft systems, battlefield anti-laser protective goggles, and most recently as a high gain holographic reflector for liquid crystal displays.

Polaroid will continue production of its Light Intensifying Film Technology (LIFT), a white holographic reflector for use in color and black and white transflective and reflective Liquid Crystal Displays (LCDs). In addition, Polaroid will continue its holographic research.

◆Polaroid
Holographic Products

WISHES TO THANK ITS PHOTOPOLYMER CUSTOMERS
WHO HAVE SUPPORTED US OVER THE PAST DECADE
AND INSPIRED US TO MAINTAIN THE HIGHEST LEVEL OF
INNOVATION AND QUALITY

As we phase out production of our DMP-128 photopolymer holograms we also look back at what we've accomplished. Polaroid has been a pioneer in the art and science of holography for more than thirty years. After developing the rainbow hologram, our chemists were charged with creating a special photopolymer which would allow the production of a new type of high performance white-light hologram. The result of their research was DMP-128.

We were the first to mass produce photopolymer holograms, which are recognized as the brightest, sharpest, and clearest holograms available. Polaroid Mirage™ holograms have been used as authentication devices on drivers licenses and Super Bowl tickets, as tax stamps and have been incorporated in more typical products such as trading and phone cards, sunglasses, collectable stickers, and wall art.

Polaroid photopolymer holograms have also been used in a variety of scientific and technical applications including heads-up displays, instrumentation for automotive and aircraft systems, battlefield anti-laser protective goggles, and most recently as a high-gain holographic reflector for liquid crystal displays in our Imagix™ product line.

Our three decades of accomplishments were realized by the many talented and devoted men and women who contributed to holography at Polaroid. They are too numerous to list, but their efforts have been noted many times by the patents and honors which bear their names. Their unwavering work on the development, manufacture, and sale of Polaroid holograms has made our products first in the market and unique in the industry. They are the sole reason for our past successes and deserve our heartfelt thanks and recognition.

Polaroid Corporation
Holographic Products Division
Cambridge and Waltham, MA

PHOTOPOLYMER RECORDING MATERIALS

INTRODUCTION TO PHOTOPOLYMER RECORDING MATERIALS

Photopolymer is a plastic compound that reacts to light. It is formulated so that certain wavelengths of light create specific molecular changes in the material, thereby making it useful as an optical recording medium. These changes in the chemical and physical structure of the film cause changes in the refractive index of the material, which is a useful method for recording a holographic interference pattern.

Chemical processing is not required to produce an image, as is the case with conventional silver-halide materials. Instead, holograms are created "real time" in the photopolymer emulsion during the primary exposure to laser light and "fixed" using a subsequent exposure to ultraviolet light. A simple baking process follows.

Photopolymers have been formulated to produce crisp, bright holographic images. Diffraction efficiencies greater than 95% (for transmission) and to 99% (for reflection) holograms have been measured. This means these holograms can be viewed under the less-than-ideal-lighting conditions that are present in many commercial environments, which is of paramount concern to many potential users. Under proper illumination, photopolymer materials are capable of displaying high quality images with great image depth. Photopolymer compounds can also be used for related industrial applications, such as optical processing and optical storage.

Photopolymer holograms do cost more than embossed holograms to mass-produce due to the fact that the former process utilizes photosensitive materials and optical replication rather than foils and mechanical stamping presses. At present, typical production runs cost 5 to 15 cents/sq. inch for finished product. However, no other holographic technology is more cost-effective for customers that require high impact three-dimensional images that can be seen under "normal" lighting conditions.

ADVANTAGES

There are several advantages of using photopolymer film to record and reproduce display holograms:

• The material is well suited for a variety of commercial applications. Being a flexible plastic film, the photopolymer material is more easily integrated into existing manufacturing, finishing, and distribution systems than other types of reflection holograms. Most notably, photopolymers are less expensive to manufacture than silver-halide emulsions and much easier to handle than dichromates.

The material itself is lightweight, durable and has a long shelf life. It is convenient to store and ship. Finished holograms can easily be cut to shape and attached to other paper or plastic products.

• You can record high quality, fully dimensional deep-image holograms. Photopolymer films that are expressly designed for display holography record "volume phase" reflection holograms (i.e., the image is recorded throughout the entire thickness of the emulsion). This results in a high density of information being recorded, which in turn allows for fully dimensional, highly detailed images to be captured. Images with a very wide viewing angle (up to 180 degrees) and considerable depth (up to several feet) have been recorded and reproduced.

In contrast, the more widely used embossed hologram manufacturing process creates microscopic surface relief patterns on a plastic film or on a sheet of foil. The process results in relatively shallow images measured in inches (or less in most cases). In addition, the techniques used to make most embossed holograms eliminate vertical parallax and limit horizontal viewing angle to approximately 60 degrees.

• The overall color tone, "weight," and "look" of images produced on photopolymer film are much more realistic than those produced by embossing. Photopolymer holograms look more like highly detailed three-dimensional black and white photographs. Most have a green, gold, or orange color tone and a solid black background, instead of the ever-changing prismatic colors and reflective metallic background typical of embossed holograms. The photopolymer emulsion itself is transparent; however, it is usually attached to a dark backing material for higher contrast. The actual color of the holographic image will be similar to that of the laser light used to make the recording.

Typically an argon-ion laser is used to expose the photopolymer film (green 514.5 run) which results in a green tone image. Some hologram reproduction facilities are experimenting with the new diode-pumped solid-state lasers which also emit green (532 run). Methods have been developed to further adjust color. Multicolor and true-color photopolymer holograms should be available in the year ahead.

Until this past year, Polaroid had been actively involved in promoting its in-house production services for its proprietary photopolymer material. As of 1998, over 30 years of research and development has been shelved due to a corporate restructuring. Currently, DuPont is the only supplier available for commercial quantities of this material.

1. Stevenson, et. al. SPIE. Vol. 233, pp. 60-70.
2. Gambogi, et. al. SPIE. Vol. 1555, pp. 256-267.

DUPONT

DuPont Monochromatic Photopolymer Products for Holography

DuPont manufactures photopolymer recording films and replication equipment which it sells directly to authorized replication facilities. Currently there is only a small number of such operations worldwide; however, they are capable of high-volume production (for instance, Doug Miller of Krystal Holographics International reports production capacity at his facility of over 1 million units a day). DuPont continues to provide select holographers with materials for research and development work.

DuPont's line of holographic photopolymer film features OmniDex®706, a blue-green reflection material, currently sold in 500-foot-by-12-inch rolls. R&D sample films are also available: HRF-700 (reflection), HRF-600 (transmission), and HRF-150 (transmission).

In addition, DuPont has developed a full-color photopolymer film which is sensitive to multiple wavelengths, but it is not available for general sale.

One of the attributes of DuPont's materials which makes them so appealing to commercial holographers is the ease of post-exposure processing; i.e., the steps which "fix" an image to make it permanent. Unlike conventional silver-halide emulsions that require complex chemical reactions to occur during the developing process, DuPont's photopolymer materials are "developed" using ultraviolet light and heat.

Exposure to ultraviolet light fixes the image, then heat processing (in a forced air convection oven) increases the brightness of the image. This "dry" developing-technique lends itself to automated mass production. Holograms made on silver-halide materials are much more cumbersome to replicate, which has severely impeded their widespread application.

OmniDex® 706-A Detailed Look

OmniDex®706 Holographic Recording Film is a photopolymer film that records volume-phase reflection holograms. It can record HI holograms directly, or it can record H2 holograms by contact-copying from a master hologram. Masters produced on silver-halide materials, dichromate, or even on another piece of photopolymer have been successfully copied.

The 706 film is comprised of three layers: a Mylar poly ester film base, a middle layer of photosensitive photopolymer and a polyvinyl chloride (PVC) cover sheet. (See table I and figure 1.) The middle layer is a mix of poly-meric binders, monomers, an initiator system, plasticizers, and dyes sensitive to visible light. Photopolymerization and diffusion of the monomers is responsible for image formation.

The product is sold with a matching amount of OmniDex® CTF-75, a color tuning film which is laminated to the recording material during the processing procedure to enhance image color and brightness. The color tuning film is also comprised of three layers.

Manufacturer's Specs

Film performance is characterized by the sensitivity of the unexposed film (absorption and film speed) and the brightness of the resulting holographic image (reflection efficiency, bandwidth, and integrated efficiency). The brightness of the image is dependent on the magnitude of the variation of refractive index of the periodic micro features recorded in the film (i.e., index modulation).

Photosensitivity

As shown in figure 2, OmniDex! 706 film is sensitive to blue-green lines of an argon-ion laser (458-528 run). Upon processing the dye absorption diminishes as shown by the dotted line.

Film Speed

Film speed is defined as the minimum amount of laser exposure necessary to produce a saturated image in terms of its brightness. Experimentally, film speed is determined by making a series of holographic exposures at a fixed laser intensity with various exposure times. The exposure energy (mJ/cm^2) is simply calculated by multiplying laser intensity (mW/cm^2) by exposure time (sec.). Film speed curves are shown in figure 3. Brightness of the hologram initially increases with exposure energy of about 25 mJ/cm^2.

Mass Production Process

High-volume reproduction utilizes a master hologram, a hologram copy setup, a roll of OmniDex®706, a roll of OmniDex®CTF-75, an ultraviolet source, an oven, lamination machinery and transport equipment. Here is the basic procedure used to mass-produce these photopolymer holograms:

1. Contact copies are recorded from a master hologram in a step-and-repeat process. DuPont's replicating system employs a web transport mechanism in conjunction with a laser and optics. The master hologram is loaded into a frame and OmniDex®706 film is contacted to it (this procedure calls for the direct lamination of the film onto a glass plate or another piece of photopolymer. Contact DuPont for lamination instructions).

2. A flood or scanning laser beam copies the image. Laser exposure of 40-50 mJ/cm^2 at 514.5 nmis recommended.

3. The film is advanced another frame and the process is repeated. Production rates of 1-5 sq. feet/min. are common.

4. Ultraviolet exposure is done in line (l 00 mJ/cm2 at 300-366 nm).

5. A roll of exposed film and a roll of color tuning film are loaded on a DuPont OmniDex® laminator. The machine removes the cover sheets from the film and the color tuning film while laminating them together at 6 ft.lmin. (It is very important that the two films laminated together are free of tension mismatch and wrinkles. Also, removing the cover sheet generates a great deal of static electricity. This causes any suspended particulate in the air to be attracted to the film. This is one of the main causes of defective product. To avoid excessive wastage, manufacturing must be done in a dust-free "clean" room.)

6. The roll is fed through a DuPont OmniDex® scroll oven at 140 degrees C for eight minutes. Both the temperature and the time used in the baking processing influences the color and the brightness of the hologram. If color tuning film is not used, heating the hologram is still recommended to increase brightness. However, it is advisable to cover the bare photopolymer with a sheet of protective material such as Mylar (not PVC) so the film is not damaged in the oven (as it is softened by the heat) or exposed to air (which would cause a blue shift).

7. The finished roll still has polyester base sheets attached (they were laminated face to face) that can also be removed. Different adhesives materials or protective coatings can be applied depending on the requirements of the user, followed by die-cutting or trimming.

Those interested in replication should contact DuPont. Some replication facilities are mass-producing photopolymer with custom-built equipment, although the basic process used is the same as the one DuPont recommends. Company spokespersons are available for consultation.

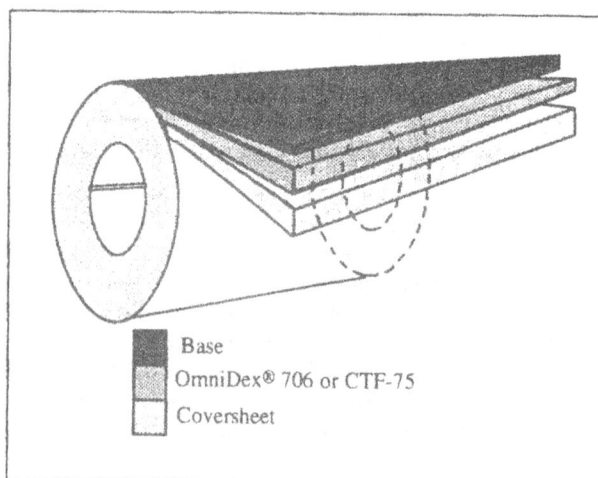

Figure 1. OmniDex®706 , CTF-75 film structure

OMNIDEX® 706 FILM STRUCTURE		
Layer	Material	Thickness
Cover sheet	PVC	60.9 μm
Photopolymer	compound	20.0μm
Base Sheet	200D Mylar®	50.8μm
OMNIDEX® CTF-75 FILM STRUCTURE		
Layer	Material	Thickness
Cover Sheet	Polypropylene	17.8μm
Photopolymer		24.6 μm
Base Sheet	200D Mylar®	50.8μm

Table 1. OmniDex®706, CTF-75 film structure.

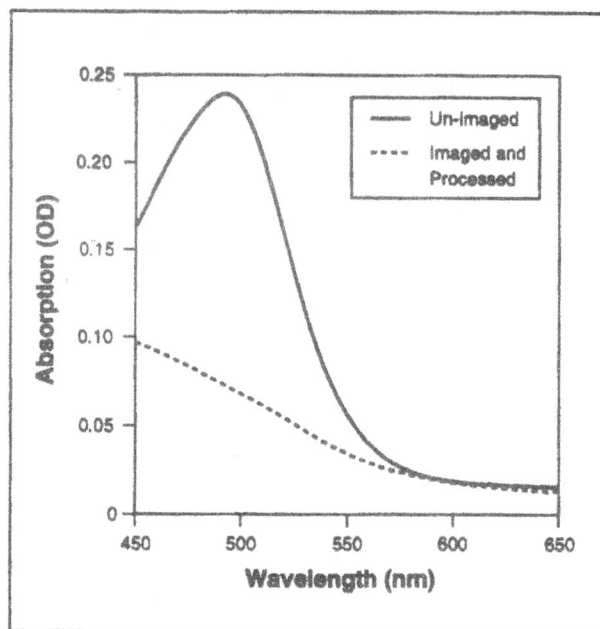

Fig. 2. Absorption spectra of OmniDex® 706 film (dotted line UV cured and baked).

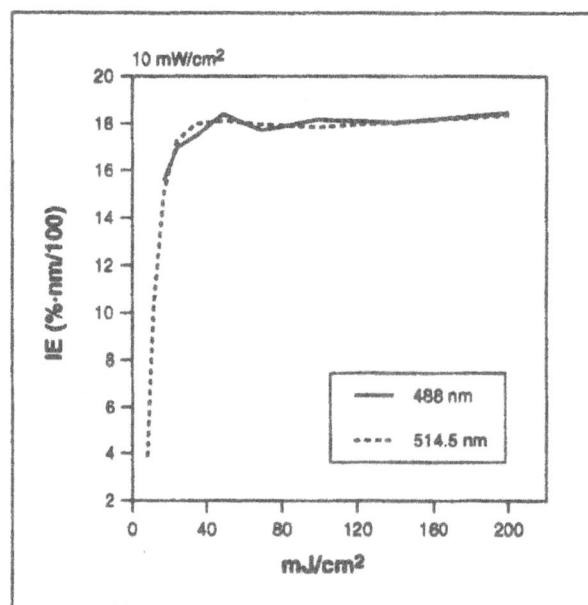

Fig. 3. Film speed curves of OmniDex®706 film at 514.5 nm and 488 nm.

Most commercial holographic applications utilize embossed holograms. A crucial step in the production of embossed holograms is transforming the microscopic optical recording into a useful production tool. To accomplish this, holographers record their images onto a high-resolution photosensitive material called "photoresist". Once an image is recorded onto a photoresist emulsion, it can be then be processed in a manner suitable for mechanical mass replication.

Although most holographers shooting holograms on photoresist materials prefer to use pre-coated, ready-made recording plates, some holographers might want to (or need to) coat and process their own plates. Shipley Chemical Company is the main U.S. manufacturer of these emulsions and sells its product in liquid form mainly to high volume users. Although it is mostly used by the microelectronic and semiconductor industries, their photosensitive formula has proven suitable for holographic recordings. (Editor's note: The process for coating plates from scratch is described in Holography MarketPlace 7th Ed.)

Many holographers buy pre-coated plates from Towne Technologies - a major U.S. manufacturer and distributor of photoresist recording plates. Towne is not the world's only supplier of these plates; however, it has extensive experience serving the holographic industry and its products have been used successfully by many embossed hologram production facilities.

TOWNE'S PRE-COATED PLATES

The materials that are used to produce the large iron-oxide coated holographic plates are purchased to specification requirements of the microelectronic, semiconductor, and printed circuit board industries for which the product was originally designed.

For example, an optical grade polished (both sides) soda lime, float glass substrate 24" x 32" x 0.190" (609.6 x 812.8 x 4.83 mm) has a flatness tolerance of ISO x 10.6 inches per linear inch (flat to within ISO micro inches per linear inch). Before it is acceptable for Fe02 coating, each piece of glass is cleaned and surface-polished to ensure that the slightest surface imperfections and even micro-dust particles are removed. The pure iron-pentacarbonyl used has a controlled specific gravity of 1.44-1.47 @ 20°C and its deposition is carried out in Class 100 clean room conditions. After the Fe02 deposition, 100 Angstroms thick, the plate is inspected for integrity of coating. Pinholes are marked, and when the 24-inch-by-32-inch plate is cut into final working plates, the pinholes are avoided.

The plates are dried in a thermostatically controlled Class 100 environment, then cleaned and inspected again. From that time to the deposition of the photosensitive coating, nothing is permitted to contact the surface of the plate. The Microposit-S-1800 highly sensitive photoresist is used for coating because it is specifically formulated to be striation-free. On plates up to 15 inch by 15 inch, (381 mm square) the photoresist is applied by a spinning process to a final standard thickness of I.S +/- 10% micrometers subsequent to a 0.2 micrometer filtration process.

The success of the iron-oxide coating is owed primarily to two inherent characteristics of FeO, coating. E.g., the iron-oxide coating effectively absorbs any laser light that may be transmitted through the photosensitive coating. This virtually eliminates light back scatter and the possibility of damage to the primary image. Second, and possibly more important, the iron-oxide coating greatly increases the adhesive quality of the photosensitive coating, thus ensuring the integrity of the imaging and electroplating processes to follow.

Towne has not needed to make any specific modifications to its plates to accommodate holographers, except for the need to install two "oversize" spinners. Plates for the electronic industry are usually dip coated when the size is above 7 inch by 7 inch. Holographers require the smoother finish of spin coated plates. Towne can spin up to 15-inch square or 18-inch octagonal plates.

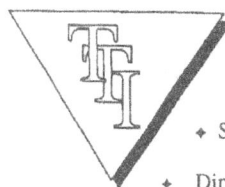

DICHROMATED GELATIN (DCG)

Dichromated Gelatin (DCG) has the highest index of refraction of any emulsion used in holography. Therefore, holograms recorded on this emulsion create the brightest and most easily viewable images under a variety of lighting conditions. This makes it an ideal material to use for commercial displays, art, and giftware. In addition, DCG produces little scatter in blue light, making it valuable material to use when manufacturing precision optical components, such as HOEs.

MAKING DCG PLATES

This emulsion consists of ammonium dichromate or potassium dichromate, gelatin, and water. Dichromates are available through chemical supply houses. Gelatin is easily obtained from gelatin manufacturers, by the barrel. Recording plates can be made by coating a piece of glass with a uniform thickness of the liquid emulsion using standard application methods, such as spin coating. To obtain the best results, this should take place in a dust-free environment.

DCG is one of the easiest emulsions to work with and high quality holograms can be consistently produced once manufacturing variables are identified and controlled. The major variables to be aware of are the concentration of dichromate used in the emulsion, the "hardness" of the gelatin used in the emulsion, and the temperature and amount of humidity that the emulsion experiences once it is coated onto the recording plate.

PRE-COATED DCG PLATES

Russian emulsion maker Slavich lists a dichromated gelatin coated plate PFG-04 in its latest product catalog. The company literature states that these plates "are designed to record holograms in contrary beams by the Denisyuk method using continuous laser emission in the blue and green spectrum (for instance using helium-cadmium, argon, or neodymium lasers) More detailed information about PFG-04 can be found in the section about Slavich products on the preceding pages.

MASTERING AND PROCESSING

Mastering set-ups depend on the size and complexity of the particular job. A single beam "Denisyuk" type of set-up is most commonly employed when recording small objects, such as miniature models. Due to the ease of preparing a shot, mastering charges for dichromate production runs are often comparable to, or even lower than, mastering charges for other types of holograms.

The laser power necessary to expose DCG emulsions varies according to the formula used, but a rule of thumb is that the more dichromate in the emulsion, the shorter the exposure. DCG is blue-green sensitive-the shorter the wavelength, the more sensitive the emulsion becomes.

Most holographers working with DCG use 5 watt argon lasers, but small holograms can be made with smaller argon (40 mW) or helium-cadmium lasers.

Color control is somewhat limited in DCG since exposures are made using the shorter wavelengths of light. Some holographers shoot wide-band, which results in a white or silver tone image. Others shoot narrowband, which often produces a gold or bronze colored image. However, some laboratories have perfected multicolor production techniques by selecting certain wavelengths and painting the subject matter to match.

Processing is quite simple - starting with a fixer, then a water rinse, then drying with isopropyl alcohol. After processing, one must isolate the emulsion from atmospheric moisture and direct contact with water. This is done by laminating another piece of glass over the emulsion and sealing the sandwich with an appropriate glue. Otherwise, the emulsion will dissolve and the holographic image will disappear.

MASS PRODUCTION

It is very cost-effective to produce short runs of DCGs. Large runs are more expensive due to the amount of time consuming hand labor used throughout the process, especially in the sealing and finishing stages. Many production facilities mass-produce DCG holograms by repeatedly "remastering" each shot using the original model or subject matter. Higher quality results can usually be obtained by copying from a master hologram.

The most popular finished product is glass discs (they can be more easily sealed) which are used as watch faces and as jewelry items. Other companies produce ready-to-frame plates for wall decor and executive gifts. A few artists have integrated DCG holograms with glass sculpture, creating elegant and unusual artworks.

OTHER CONSIDERATIONS

The major drawback to using DCGs for a wider range of commercial applications is the fact that they are thick and fragile, since they are usually sealed in glass. This usually makes them impractical to use in product packaging and publishing. In the past, manufacturers have attempted to mass-produce DCGs by laminating them in plastic, but holograms sealed in this way tend to fade more quickly than those sealed in glass.

Editor's Request: We are unaware of any holographers using pre-coated DCG plates in their work. If you have successfully used these products, please notify us.

ONE METHOD FOR MAKING DICHROMATED GELATIN (DCG) HOLOGRAMS

by CD. Leonard and B.D. Guenther

The following article explains how to make and use DCG. Although it is not the most efficient method to employ in order to produce DCG holograms in commercial quantities, the information is relevant for use in the laboratory and the classroom. (For instance, high-volume manufacturers typically employ mechanical spin-coating techniques instead of the process described below.)

WARNING: Improper handling of DCG is dangerous. Avoid inhaling fumes, ingesting and unnecessary contact. Always use proper laboratory techniques and appropriate disposal methods. Use safe eyewear. Work in a well-ventilated area.

INTRODUCTION

Gelatin sensitized by ammonium dichromate produces a near optimum medium for holography. Holograms produced in this medium have very low scattering noise and very high diffraction efficiencies. This report summarizes techniques found to be useful in producing high quality holograms.

The processing described is largely based on the procedure of Colburn and Chang that was developed at the En-
vironmental Research Institute of Michigan. Credit should be given to Zech[2] of Radiation, Inc., for the generalized procedure of incorporating the preprocessing techniques of Chang[3] and the post-processing techniques of Lin[4]. Their procedures have been modified to reduce the overall processing time.

Note: The procedures outlined in this report will result in high quality holograms, but none of the procedures may be optimal. A more detailed evaluation of DCG and its chemistry has been described[5].

Five areas will be covered: general preparation, plate preparation, exposure, development, and assembly.

GENERAL PREPARATION

For processing of DCG, it is imperative that the relative humidity of the laboratory be kept under 40% from gelatin sensitization to final assembly of the hologram. The relative humidity can vary markedly even within a single room; therefore, it is recommended that several hydrometer gauges be used to monitor the laboratory environment. It should be noted that hydrometers, even as received from the manufacturer, can be out of calibration. It is recommended that a wet-dry bulb thermometer be used for periodic calibration. Several small room dehumidifiers will usually reduce the relative humidity below 40%. Distilled water is used in making all solutions; washes were done using temperature-controlled tap water.

PLATE PREPARATION

For convenience, it is usually desirable to initially purchase commercially available photographic films or plates. The silver halides and dyes can be removed and the remaining gelatin sensitized with ammonium dichromate. The controls used by the film manufacturer in the production of silver-halide holographic films and plates do not appear to be coincident with the controls needed to produce good DCG holograms. Thus, good holographic films may not be good DCG films. We have had satisfactory results using Kodak 649F and 131 emulsions. The performance of these emulsions will vary according to batch and/or age. Exposure and processing tests should be performed before every critical exposure, at least until experience is gained in working with DCG.

Having selected one or more emulsions for testing, the plates are preprocessed and sensitized according to the following procedure:

a) Preprocessing–For as many plates as can be simultaneously processed:

P1 - Fix 10 minutes in Kodak Rapid Fixer without hardener.

P2 - (Kodak 649F emulsion only) Wash 5 minutes at 90°F, initially raising (and then lowering) the temperature gradually (5 to 10 min.) from room temperature (~ 70°F).

P3 - (Kodak 649F emulsion only) Wash 5 minutes in Kodak Rapid Fixer with hardener.

P4 - Wash 10 to 12 minutes at 70°F, running water.

P5 - Rinse 1 minute in Photoflo (one drop Kodak Photoflo 200 in 500 ml of water).

P6 - Dry by wiping with wiper blade.

b) Sensitizing–For plates to be used within 8 hours:

S1 - Sensitize 5 minutes in X% solution of ammonium dichromate (X gm of ammonium Bichromate in 100 ml of water with O.X ml of Photoflo).

S2 - Wipe with wiper blade.

S3 - Dry on hotplate for 3 minutes at 160°F.

S4 - Allow to cool to room temperature for 1 hour.

The concentration of ammonium bichromate used in the sensitization of the gelatin controls the exposure time and the Bragg angle shift. The amount of hardener used in the fixer for both preprocessing and development controls the diffraction efficiency and scattering noise. These two facts again point out the need for making test exposures.

Two methods can be used for drying the sensitized plates. Softer emulsions may be dried in a horizontal position at room temperature. A second, faster technique useful with harder emulsions is to wipe the surface of the gelatin with a wiper blade (windshield wiper blades designed for cars with flat windshields are excellent). After wiping, the plate is placed on a preheated hotplate (~ 160°F–you should be able to touch the hotplate but not leave your finger on it). The actual temperature of the hotplate is not very important; we use one that dries the plates in approximately 3 minutes. If crystalization occurs on the gelatin surface beyond a narrow edge region, then the plate should be placed into the sensitizing solution again. A high room temperature or a high substrate temperature can result in rapid crystalization before the wiping process is complete (this would occur when using very high concentrations) and can be solved by simply lowering the temperature of the plates. After drying, the plates should appear clean with no streaks, smudges, or crystals.

The rapid drying procedure appears to affect the dichromate concentration required for a given holographic performance. When the wiping process is used, three times the dichromate concentration used in the nonwiping process is needed to produce a similar effect. For example, to obtain no Bragg angle shift, holograms made using the nonwiping technique required concentrations of 6 to 8%. When the wiping technique was used, concentrations of 20 to 25% were required with the lower concentration corresponding to the requirements of reflection holograms, while the higher values correspond to transmission holograms.

The sensitizing solution and the sensitized plates should be exposed only to red light (for example, Kodak Safelight Filter 1A). The color of the plates and solution changes from a yellow to a brown with exposure to light or through aging. Low concentrations can be stored by refrigeration. At high concentrations, the solution lasts approximately 8 hours and the plates approximately 4 hours. The plates can usually be washed and reused if some darkening has occurred. The reader should note that the hardness of the gelatin is modified as the plates darken.

EXPOSURE

Once the plates have been sensitized, the relative humidity should be kept below 40%. It has been found that the sensitized emulsion can be covered with a glass plate using xylene as a liquid gate. By doing this, the relative humidity of the exposure laboratory need not be held below 40%. The gate can drain and evaporate during long exposures. Operating times of over an hour have been obtained by sealing the gate with aluminum tape (the type used for making lantern slides). Exposure times can range from 50 to 1000 mJ/cm^2 at 5145A depending on the dichromate concentration and the percent modulation required. A good starting point would be an exposure of 250 mJ/cm^2 at a dichromate concentration of 5%.

DEVELOPMENT

For processing of plates immediately after exposure:

D1 - Agitate 5 minutes in 0.5% ammonium dichromate solution (can be made from the sensitizing solution).

D2 - Agitate 5 minutes in Rapid fixer without hardener (with hardener for 649F). USE ADEQUATE VENTILATION!

D3 - Wash 10 minutes in running water (70°F for 649F and 90°F for other emulsions).

D4 -Agitate 3 minutes in 50:50 solution of isopropyl alcohot and water; then wipe off back and edges. USE ADEQUATE VENTILATION!

D5- Vigorously agitate 3 minutes in 100% isopropyl alcohol. USE ADEQUATE VENTILATION!

D6 - Dry standing on edge (649F), or wipe with blade and heat on hotplate.

Except for the 50:50 solution of isopropyl alcohol in water, all solutions should be discarded after one use. To reduce the quantity of chemicals used, tray development is used with only enough liquid to barely cover the plates.

The gelatin emulsions have different degrees of hardness depending on their age and on the manufacturing process. For example, the gelatin on thin plates has been found to be harder than the same emulsion type on microflats. It is useful to have stock fixer solutions with and without hardener to allow the correct percentage of hardener to be selected for Step D2 during the initial test exposures. This type of processing control need be applied only for critical control of the DCG performance.

The most important developing step is the 100% isopropyl alcohol bath. In this step, rapid and complete dehydration of the gelatin is required. Excess liquid is wiped from the back and edges of the plate (not the emulsion). The plate is placed, emulsion side up, in a tray and the alcohol is poured directly onto the plate. Rapid agitation should begin immediately and should continue for the entire 3-minute wash. The agitation cannot be too rapid. The 100% isopropyl alcohol can be retained for later use in making up 50% alcohol/water solutions.

Drying may be accomplished in two ways. Softer emulsions may be placed on end to drain and dry. The results using this drying process will be variable. Absolute control of the relative humidity is required for consistent results.

The second drying technique involves first removing the excess liquid from the emulsion with a wiper blade and drying on a hotplate. The hotplate should provide uniform heating of the DCG. If processing non-uniformities are observed after drying, the plate may be returned to the 50:50 alcohol/water solution and reprocessed.

After 1 minute at Step D2, room lights may be used. Steps D3 through D6 may be repeated if the processed plates are accidentally exposed to moisture. These steps may also be repeated with the wash temperature at D3 increased. This reprocessing can often increase the diffraction efficiency to recover what would normally be an unusable hologram.6 There does not appear to be a limit to the number of times these final processing steps may be repeated.

ASSEMBLY

Moisture will degrade the holograms . To make DCG rugged, a cover plate is cemented over the gelatin. The same type of protection is obtained when multiple holograms are assembled emulsion-to-emulsion. Before assembly, the plates are baked at 160°F in a vacuum for approximately 2 hours to remove water from the gelatin. American Handicraft Company's "Clear Cast" liquid casting plastic or

Summers "Lens Bond" optical cement (available from Edmund Scientific Company) were used as the cement. Five cubic centimeters of the cement are mixed with one drop of catalyst and placed in a vacuum of 25 to 26 in. until all of the bubbles were removed. If too high a vacuum is used, the solvent evaporates, making the cement unusable. After the bubbles have been removed from the cement, it is poured over the gelatin surface of the hologram in an "X". The cover plate or the gelatin surface of another hologram is then slowly rolled onto the X allowing the cement X to spread to the plate edges. Care should be taken that no air is trapped as the X spreads. The sandwich of glass and cement is then heated on a hotplate at 140 to 160°F until the plastic becomes firm, but tacky. A razor blade can then be used to trim away excess from the edges of the plates, and acetone can be used to remove the cement from the plate faces.

The following is a listing of the previously described steps of the assembly procedure:

A1 - Bake holograms at 160°F for 2 hours under vacuum.

A2 - Mix 5 cc of cement and one drop catalyst.

A3 - Place cement in vacuum to remove air bubbles.

A4 - Cement glass plates together.

A5 - Bake cemented plate at 140 to 160°F until cement is tacky.

If the need arises, the plates can later be separated by dissolving the cement in acetone. The holograms are not damaged by exposure to acetone, but must be reprocessed starting at Step D3.

The most important step in this final assembly is Step A1, the vacuum bake-out. If this step is carried out, the covered hologram will survive exposure to boiling water.

A dry box is a convenient accessory for storing processed plates until one is able to cover the gelatin with a protective plate.

ADDITIONAL PROCESSING TECHNIQUES

The procedures used here are derived from those of BJ. Chang. One procedure is used for 649F emulsions and a second is used for all other emulsions. Exposure and processing are done in rooms where the relative humidity is held below 40%. Unless otherwise noted, all washes are temperature controlled and last for 10 minutes. All plates are dried by first wiping off all excess liquid with a rubber wiper blade and then drying them in a horizontal position on a hotplate set at 160 of. A thin photographic plate is removed after 3 minutes and a thick (0.25 in.) photographic plate is removed after 4 minutes.

Preprocessing

For 649F plates, the following procedure is used:

P1 - Fix 10 minutes in Kodak Rapid Fixer without hardener.

P2 - Wash 10 minutes in 90°F water.

P3 - Wash minutes in Kodak Rapid Fixer with hardener.

P4 - Wash 10 minutes.

P5 - Rinse 1 minute in Photofio.

P6 - Wipe with wiper blade.

All other plates are preprocessed without Steps P2 and P3.

Sensitizing

The following procedure is the same for all emulsions. It is conducted in red light (Kodak Safelight Filter 1A):

S1 - 5 minutes in X% solution of ammonium dichromate with O.x% of Kodak Photofio.

S2 - Wipe with wiper blade.

S3 - Dry on hotplate for 3 minutes at - 160°F.

S4 - Remove crystallized ammonium dichromate from the back of the plate.

S5 - Allow to cool for 1 hour.

At the end of the sensitizing process, the plates should appear clean with no streaks, smudges, or crystals. If crystals of ammonium dichromate form on the gelatin, the sensitizing process must be repeated.

Exposure

It is recommended that the sensitizing plates be used the same day. If the sensitizing solution was 10% or less, the plates can be stored overnight in a refrigerator, provided the plates are sealed from moisture.

In previous tests, the plates required from 50 to 1000 mJl cm^2 at 5145 A to produce a hologram. The actual exposure energy is determined by the percent modulation required and the concentration of ammonium dichromate used. Test exposures should be run to ensure optimum performance.

Development

The following developing procedure was used with modifications required for Kodak 649F plates shown (Steps Dl and D2 are performed in red light):

D1 - Agitate 5 minutes in a 0.5% ammonium dichromate solution.

D2 - Agitate 5 minutes in fixer without hardener (use fixer with hardener for 649F).

D3 - Wash at 90°F (70°F for 649F).

D4 - Rinse 3 minutes in a 50:50 solution of distilled water and isopropyl alcohol. Then wipe off back and edges of plate.

D5 - Rinse 3 minutes in 100% isopropyl alcohol with rapid agitation.

D6 - Dry (do not use a hotplate for 649F plates).

The 100% isopropyl alcohol used in Step D5 can only be used to process one plate. The index of refraction modulation can be changed by repeating Steps D3 through D6. To increase the modulation, a higher wash temperature can be used [up fo 110°F (90°F for 649F)].

Assembly

The assembly procedure is as follows:

A1 - Vacuum bake DCG at 160°F for 1 hour.

A2 - Mix 5 cc cement and one drop catalyst.

A3 - Place cement in vacuum to remove air bubbles.

A4 - Cement glass plates together.

A5 - Heat cemented plates at 140-160°F until cement is tacky.

REFERENCES
1. Colburn, W.S. and BJ. Chang, Holographic Combiner for Head-up Displays, Report No. AFAL-TR-77-110, 1977, pp. 44-46 .
2. Zech, KG., Data Storage in Volume Holograms, Ph.D. Thesis, University of Michigan, 1974, pp. 217-222.
3. Chang, M., "Dichromated Gelatin of Improved Optical Quality," Applied Optics, Vol. 10, 1971, p. 2550.
4. Lin, L.H., "Hologram Formation in Hardened Dichromated Gelatin Films," Applied Optics, 1969, pp. 963-966.
5. Chang, BJ. and Leonard, C. D., "Dichromated Gelatin for the Fabrication of Holographic Optical Elements," Applied Optics, May 1978.
6. Chang, BJ., "Post-Processing of Developed Dichromated Gelatin Holograms," Optical Communications, Vol. 17, 1976, pp. 270-272.

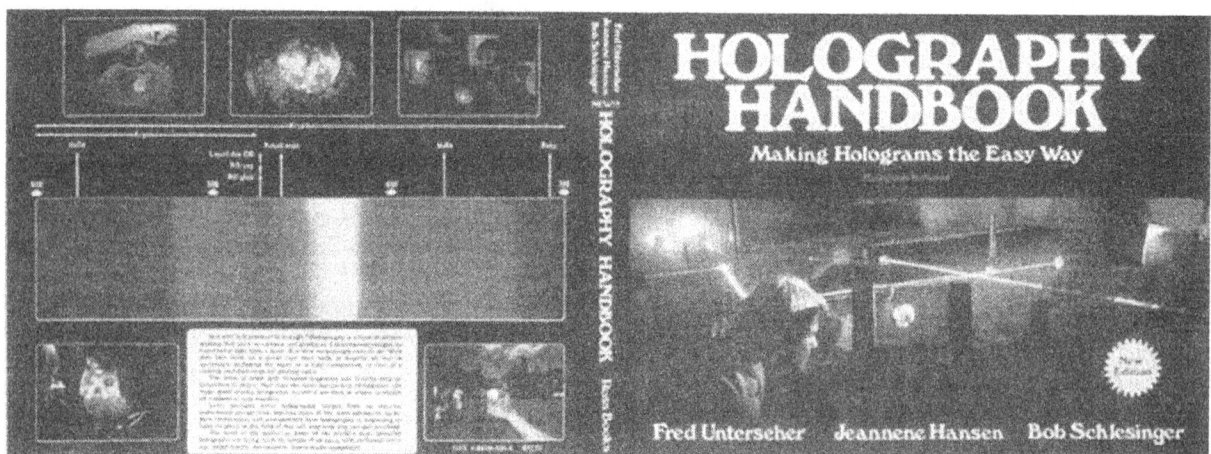

SECTION 4
Industry Directories

Chapter 13 - Business Listings

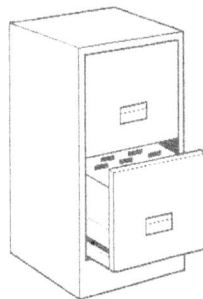

Chapter 14 - Cross-Index Tables

"Alan Rhody, editor of the trade journal Holography Marketplace, predicts: 'As laser technology progresses, optical barriers are crossed, and venture capitalists become actively involved in the hologram industry, theatrical holographic presentations will become a reality.'"

NEWSWEEK Magazine, Winter 1997

ABOUT ROSS BOOKS

Founded in 1977, Ross Books is a general trade publisher of books in print and electronic format. Our catalog includes books about a variety of topics including health, cooking, music, sports, science, and more! Many of our titles are instructional and educational. We also publish computer software. We are widely known for our books on holography which we have been publishing since 1982.

In addition to our publications, we also provide affordable consulting and research assistance for companies seeking either general or very specific information about the holography industry. We have the world's most comprehensive and extensive listing of companies and individuals involved with commercial applications, research, artistic endeavors, etc. We also have thousands of related and pertinent publications in our database. We can provide reports, database searches and manuscripts custom-tailored to your needs. Call us today for further details about the services we provide.

Franz H. Ross is the President of Ross Books (email: franz@rossbooks.com). He founded Ross Books in April of 1977. The company has published a wide variety of books, as well as computer software. The holography community has especially benefited from Mr. Ross's efforts. In 1982, he published the world's best-selling holography instruction manual The *Holography Handbook - Making Holograms the Easy Way*. In 1989 he published the first edition of the *Holography MarketPlace*. The seventh edition of HMP was released in 1998. Mr. Ross holds a BA degree in physics from the University of California, Berkeley.

Alan E. Rhody is the Managing Editor of Ross Books (email: rhody@rossbooks.com). Mr. Rhody first studied holography as part of his undergraduate curriculum at Clark University in 1978. He has worked in the holography industry consistently since 1985, mainly in the sale and distribution of holographic art and giftware. He also has experience in the design and production of holograms for various commercial applications. In addition, his own holographic limited-edition fine art has been displayed in galleries throughout the United States. Currently, Mr. Rhody writes, edits and publishes holography related articles for Ross Books, including major parts of the *Holography MarketPlace Editions 5, 6, 7and 8*. His work has also appeared in *Newsweek Magazine, Laser Focus World, International Designer's Network, Signs of the Times, Inside Finishing, Video Toaster User, Holography News* and other publications, both printed and electronic.

13

Business Listings

Here is an alphabetical listing of businesses that are involved in the holography industry, followed by a listing of individuals named as the primary contacts for these businesses. Please notify us of relevant updates to this directory as telephone prefixes, email addresses, and Internet Website addresses are changing with increasing frequency.

International Business Directory A-Z

21 st Century Finishing Inc.
215 Pennsylvania Avenue
City: Paterson
State: NI
Postal Code: 07503
Country: United States of America
Voice Phone: (1) 973 279 2100
Fax Phone: (1) 973 279 5659
Contact #1: Anthony Olmo
Description: Multi-faceted converting specialists. 12 years experience working with trade and corporate clients. Full range of web or sheet finishing services offered including: cutting, hot-stamping, labels, lamination, overprinting, etc. Specialists in applying holograms to a variety of substrates.
****

3-D Hologrammen
1012 GA Amsterdam
City: Holland
Country: Netherlands
Voice Phone: (31) 20 6247225
Fax Phone: (31) 20 6247225
Email: j.kraak@inter.nl.net
Contact # 1: Erik Swetter
Description: Wholesale & Distribution - Retail - Art. Gallery/shop since 1987.
****

3-D Systems
Linel of Address: P.O. Box 145
City: Pt. Arena
State: CA
Postal Code: 95468
Country: United States of America
Voice Phone: (I) 707 882 1910
Email: 19cross@intercoastal.com
Contact #1: Lloyd Cros

Description: Currently produces "virtual optics laboratory" software for holographers and optics research. Pioneered integral holograms and other holographic techniques.
****

3D Images Ltd.
31 The Chine
Grange Park
City: London
State: England
Postal Code: N2 1 2EA
Country: United Kingdom
Voice Phone: (44) 181 364 0022
Fax Phone: , (44) 181 364 1828
Email: burder3d@aol.com
Contact #1: David Burder
Description: Manufacturer and distributor of 3D images. Also supplies lenticular 3D products. Producers of "Virtual Video". 3D supplies and glasses.
****

3D Ltd.
Gewerbestrasse 19
City: Unteraegeri,
Postal Code: CH-6314
Country: Switzerland
Voice Phone: (41) 41 750 4972
Fax Phone: (41) 41 750 3236
Email: 3d-ltd@3d-ltd~ com
Web: http: //www.3d-ltd.coml
Contact #1: Glen Lloyd
Description: In house design, mastering, electroforming, recombination and shim production facilities, 3D Ltd. can offer full production services. Specialist origination capabilities include a range of holographic security products. 3D Ltd. is a leading supplier of diffraction and prismatic patterns. Currently available range of over 300 patterns is available to any bonafide foil producer under a standard licensing and royalty contract.
SEE OUR ADVERTISEMENT
****

3D Optical Illusions
P.O. Box 765
City: Buddina

State: Queensland
Postal Code: 3153
Country: Australia
Voice Phone: (61) 39 729 6337
Voice Phone: (61) 414 776 226
Fax Phone: (61) 39 729 6020
Contact #1: Trevor McGaw
Description : Specializing in lenticular and holographic movement illusions with up to 20 different motion images.
****

3D Technologies & Arts
Scopolijeva 19
City: Ljubljana
Country: Slovenia
Voice Phone: (386) 61 558463
Fax Phone: (386) 61 1330189
Email: holography&tripod.net
Web: http://members. tripod.coml-holography
Contact #1: Nikola Ielic
****

3D Vision
Hologramme-Laserprodukte
Ostertorsteinweg 1-2
City: Bremen
Postal Code: D-28203
Country: Germany
Voice Phone: (49) (0)421 767 97
Fax Phone: (49) (0)421 767 97
Contact #1: Uwe Reichert
Description: Holograms-Holographic projects. General commercial holograms for sale and distribution.
****

3D-4D Holographics
97 St John Street
City: London
State: England
Postal Code: EC1M 4AS
Country: United Kingdom
Voice Phone: (44) 171 250 3545
Fax Phone: (44) 171 2503566
Email: graham@hologram.demon.co.uk
Web: hologram.demon.co.uk/home/index.html

Contact #1: Graham Tunnadine
Description: Production house for all types of 3D visual displays. Large format, silver halide mastering facility on premises.
****

3Deep Company
540 Massachusetts Ave.
City: Boston
State: MA
Postal Code: 02118
Country: United States of America
Voice Phone: (1) 617 912-1040
Fax Phone: (1) 617 912-1040
Email: acheimets@3deepco.com
Web: www.3deepco.com
Contact #1: Alex Cheimets
Description: Supplies Russian silver halide and DCG emulsions on plates and films. Green sensitive, red sensitive, and new full color silver halide emulsions available. Can also recommend appropriate developing and processing procedures.
SEE OUR ADVERTISEMENT
****

3DIMAGE
G.P.O. Box 95
City: Sydney
Postal Code: 2001
Country: Australia
Voice Phone: (61) 148 804 905
Voice Phone: (61) 88 383 7255
Fax Phone: (61) 88 383 7244
Email: info@3dimage.com.au
Web: http://www.mcm.com.au
Contact # 1: Simon Edhouse
Description: General Holographic design consultancy based on seven years experience in manufacturing. Specializing in design of large scale thematic exhibitions and alternative 3D effects.
****

3M - Safety and Security Systems
3M Center, Bldg. 225-4N-14
City: St. Paul
State: MN
Postal Code: 55144-1000
Country: United States of America
Voice Phone: (1) 800 328 7098
Voice Phone: (1) 651 733-3957
Fax Phone: (1) 800 223 5563
Email: info@mmm.com
Web: http://www.mmm.com
Contact # 1: Maureen Tholen
Description: Supplier of authenticating labels, including some holographic applications. Full range of origination services offered.
****

AB Riick Holoart
Ropers Weide 26
City: Hamburg
Postal Code: D-22605
Country: Germany
Voice Phone: (49) (0)40 8807151
Contact #1: A. B. Rueck
Description: Holograms for conventional presentations-refection and transmission.
****

Abrams, Claudette
22 Bayview Avenue
City, Province: Toronto, Wards Island
Postal Code: M5l IZI
Country: Canada
Voice Phone: (1) 416 203 7243
Fax Phone: (1) 416 203 7243
Contact #1: Claudette Abrams
Description: Holographic artist and technician.
****

Abtei Brauweiler
Landschaftsverband Rheinland
Ehrenfriedstrasse 19

City: Puhlheim
Postal Code: D-50259
Country: Germany
Description: Former Matthias Lauk Collection Museum for holography
****

Academy of Media Arts Cologne
Peter-Welter Platz 2
City: Cologne
Postal Code: D-50676
Country: Germany
Voice Phone: (49) (0)221 201 89 115
Fax Phone: (49) (0)221 201 89 124
Contact #1: Dieter lung
Description: International Academy for Media Arts. Extensive holography lab and teaching.
****

Accuwave Corp.
1651 19th St.
City: Santa Monica
State: CA
Postal Code: 90404
Country: United States of America
Voice Phone: (1) 310 449 5540
Fax Phone: (1) 310 449 5539
Email: staff@accuwave.com
Contact #1: Nevin Karlovac
Description: Manufacturer of Holographic Optical Elements for wavelength measurement applications for fiber optic industry.
****

Acme Holography
12 Sunset Road
City: West Somerville
State: MA
Postal Code: 02144
Country: United States of America
Voice Phone: (1) 617 623 0578
Fax Phone: (1) 617 625 1409
Email: bconn@media.mit.edu
Contact #1: Betsy Connors
Description: Acme Holography is Boston's first private holography lab. We offer full service in reflection, transmission and computer generated holography, including design consultation and large-scale environmental holography.
****

ACR GmbH Laser- und Medientechnik
Auf dem Hahnenberg 4
City: Miilheim - Kiirlich
Postal Code: D-56218
Country: Germany
Voice Phone: (49) (0)263049725
Fax Phone: (49) (0)263049827
Email: acr@acr-group.com
Description: Distributor of high power laser equipment.
****

Action Tapes
Unit 5
Boundary Road
City: Brackley
State: England
Postal Code: NN13 7ES
Country: United Kingdom
Voice Phone: (44) 1280 700 591
Fax Phone: (44) 1208 700 590
Email: action.tapes@dial.pipex.com
Web: www.actiontapes.com
Contact #1: Alan 1. Phillips
Description: Adhesive tapes for industry: close tolerance custom slit rolls; varying length stroke diameter; high speed die cut; large and small volume runs. Fast and experienced service. Experienced staff to provide professional
and technical advise.
****

AD 2000, Inc.

780 State Street
City: New Haven
State: CT
Postal Code: 06511
Country: United States of America
Voice Phone: (1) 203 624 6405
Voice Phone: (1) 800 334 4633
Fax Phone: (1) 203 624 1780
Email: info@ad2000.com
Web: www.ad2000.com
Contact # 1: Peter Scheir
Description: Fully custom & customized stock image, embossed and photopolymer holograms. Our HOLO-BANK contains the world's largest selection of stock image embossed holograms - available plain, as labels, foil , magnets, pins, roll stock, etc . (visit: www.holo-bank.com). LOW RUN SECURITY OUR SPECIALITY!
****

AD HOC Public Relations GmbH
Thesings Allee 10
City: Giitersloh
Postal Code: D-33332
Country: Germany
Voice Phone: (49) (0)5241 500990
Fax Phone: (49) (0)5241 500999
Contact # 1: Wolfpeter Hocke
Description: Press agency-specialized on holography-public relation concepts-promotionbrochures.
****

Advanced Deposition Technologies, Inc.
(A.D. Tech)
580 Myles Standish Blvd.
Myles Standish Industrial Park
City: Taunton
State: MA
Postal Code: 02780
Country: United States of America
Voice Phone: (1) 508 823 0707
Voice Phone: (1) 800 767-5432
Fax Phone: (1) 508 823 4434
Email: staff@adv-dep.com
Web: http ://www.adv-dep.com
Contact #1: Glenn 1. Walters
Description: Advanced Deposition Technologies, Inc. (A.D. TECH) was founded in 1983. The company has devoted itself to the field of vapor depositions on thin polymer films frequently utilizing specialized patterns. Specialists in holographic security solutions using the company's patented manufacturing techniques and products. (NASDAQ: ADTC)
****

Advanced Optics, Inc.
110 N.E. Trilein Drive
Suite #4
City: Ankeny
State: IA
Postal Code: 50021
Country: United States of America
Voice Phone: (1) 515 964 5050
Fax Phone: (1) 515 964 5050
Email: sales@advancedoptics.com
Web: http ://www.advancedoptics.com
Contact # 1: Wendy Heil
Description: Manufactures of custom and precision optics including first surface mirrors for use in holographic equipment.
****

Advanced Photonics
83, 3rd fl. , Bhandup Industrial Estate L.B.S.
City, Province: Marg, Bhundup (W)
State: Mumbai
Postal Code: 400 078
Country: India
Voice Phone: (91) 22 591 0539
Fax Phone: (91) 22 577 8954
****

Advanced Precision Technology, Inc.
4669 Hillside Drive
City: Castro Valley
State: CA
Postal Code: 94546
Country: United States of America
Voice Phone: (1) 415 397 1545
Fax Phone: (1) 510 889 8195
Contact #1: Bruce Pastorius
Description: Makes holographic optical elements used in company's fingerprint recognition system and other biometric devices.
****
Advanced Technology Program
Bldg. 101- Room A402
City: Gaithersburg
State: MD
Postal Code: 20899
Country: United States of America
Voice Phone: (1) 800 287 3863
Fax Phone: (1) 301 926 9524
Web: http: //www.atp.nist.gov
Description: United States Government program providing funding for R&D projects.
****
Aerospatiale
Ets D' Aquitaine
Saint-Medard-En-Jalles
City: Bordeaux
Postal Code: F-33165
Country: France
Voice Phone: (33) 56 57 34 80
Fax Phone: (33) 56 57 30 70
Description: Scientific and industrial research, NDT testing.
****
AF Elektronik
Implerstrasse 62
City: Miinchen
Postal Code: D-81371
Country: Germany
Voice Phone: (49) (0)89 7213481
Fax Phone: (49) (0)89 72579667
Email: afendel@t-online.de
Contact # 1: Achim Fendel
Description: Distributor of optical/mechanical and chemical holographic equipment.
****
Ag Electro-Optics Ltd.
Tarporley Business Centre
City, Province: Tarporley, Cheshire
State: England
Postal Code: CW6 9UY
Country: United Kingdom
Voice Phone: (44) 1829 733 305
Fax Phone: (44) 1829 733 679
Email: sales@ageo.co.uk
Web: http: //www.ageo .co.uk/html/
ag_electro_home.html
Contact #1: J.A. Gibson
Description: Distributor of lasers, optics, lab equipment and fiber optics.
****
Agfa - Gevaert N. V
Septestraat 27
City, Province: Mortsel, Antwerp
Postal Code: B-2640
Country: Belgium
Voice Phone: (32) 3 444 8251
Fax Phone: (32) 3 444 8243
Email: bebay2uk@ibmmail.com
Contact # 1: Frank Mortier
Description: Headquarters. Manufacturer of silver halide recording materials.
****
Ahead Optoelectronics, Inco
No. 13, Chin-ho Road
City, Province: Chung-ho, Taipei Hsien, 235

Country: Taiwan
Voice Phone: (886) 2 2226 5554
Fax Phone: (886) 2 2226 5536
Email: sales@ahead.com.tw
Web: http://www.ahead.com.tw
Contact #1: Julie Lee
Description: A manufacturer specializing in holographic products, holographic systems, optomechanical systems. Can produce custom masters (conventional, 3D dot matrix, Lippmann) and holographic technology (recombining, hot-stamping, embossing, tamperevident). Can offer two different dot matrix systems (one 200-400 dpi, one high security at 150-1300dpi and over 10 dots/sec).
SEE OUR ADVERTISEMENT
****
AHT 3D-Medien
Association for Hologram Techniques & 3D
Media (AH
Niederesch 28
City: Bad Rothenfelde
Postal Code: D-49214
Country: Germany
Voice Phone: (49) (0)542 5365
Fax Phone: (49) (0)542 5359
Contact #1: Gunther Deutschmann
Description: Association of German hologram manufacturers and 3D media-engaged in the commercialization of holographic products. Realization of projects and campaigns in cooperation between the member companies. From consultation, design, mastering, and hologram production to overprinting, converting, and integration into the finished product.
****
AKS Holographie-Galerie GmbH
Pots darner Strasse 10
City: Essen
Postal Code: D-45145
Country: Germany
Voice Phone: (49) (0)201 756455
Fax Phone: (49) (0)201 753582
Email : akshol@aol.com
Web: http://members.aol.com/akshol/
hhome_d.htm
Contact #1: Detlev Abendroth
Description: Producing all kinds of holograms incl. computer-generated up to a size of 80 x 100 cm. One Df the world's greatest film-hologram- edition. Internationally represented by distributing partners. Owns great holographic gallery.
****
Alabama A&M University
Center for Applied Optical Sciences
P.O. Box 1268
City: Normal
State: AL
Postal Code: 35762
Country: United States of America
Voice Phone: (1) 256 858 8118
Fax Phone: (1) 256 851 5622
Email: kukhtar@caos.aamu.edu
Web: http://www.caos.aoamu.edu
Contact #1 : Nicholai Kukhtarev
Description: R&D in all applications of holography including dynamic (real time) holography and photorefractive materials.
****
Alfred Dirksen + Sohn Modellwerkstatten
Gruner Weg 8-10
City: Wesseling
Postal Code: D-50389
Country: Germany
Voice Phone: (49) (0)2236 42827
Contact # 1: Alfred Dirksen
Description: Consulting and design for holographic projects. Building of models for holography- hologram displays.
****

Amagie Holographies, Ineo
1652 Deere Ave.
City: Irvine
State: CA
Postal Code: 92606
Ad-Am
Country: United States of America
Voice Phone: (1) 949 474 3978
Voice Phone: (1) 800 262 4421
Fax Phone: (1) 949 474 3979
Email: amagicl68@aol.com
Description: Fully integrated manufacturer of holographic custom images, stock images, diffraction/digitized patterns as polyester, PVC/ vinyl and polypropylene. Produces hot stamp foil, pressure sensitive labels, and film to 39.37 inches.
SEE OUR ADVERTISEMENT
****
American Bank Note Holographies, Ineo
399 Executive Blvd.
City: Elmsford
State: NY
Postal Code: 10523
Country: United States of America
Voice Phone: (1) 914 592 2355
Voice Phone: (1) 800 966 2264
Fax Phone: (1) 914 592-3248
Email: abnh@abnh.com
Web: www.abnh.com
Contact #1: Bill Wheeler
Description: World leader in development of embossed holography for security and commercial applications. Produces embossed holograms in a range of formats: foil, pressuresensitive & tamper-evident labels, clear & wide-web laminates for packaging.
SEE OUR ADVERTISEMENT
****
American Holographic hic.
60 I River Street
City: Fitchburg
State: MA
Postal Code: 01420
Country: United States of America
Voice Phone: (1) 978 343 0096
Fax Phone: (1) 978 348 1864
Email: amerholo@tiac.net
Contact #1: Thomas Mikes
Description: Design, develop and manufacture optical components and instruments for use in industrial and medical measuring devices. We are using holographic diffraction grating design and manufacture capability to produce components for unique measurement instruments.
****
American Laser Corporation
1832 South 3850 West
City: Salt Lake City
State: UT
Postal Code: 84104
Country: United States of America
Voice Phone: (1) 801 972 1311
Fax Phone: (1) 801 972 5251
Email: sales@amlaser.com
Web: http://www.amlaser.com
Contact # 1: David Cox
Description: Established 1970. Manufacrure: of Argon, Krypton and mixed gas laser systems and subsystems from 3 mw to lOw. ill air or water-cooled configuration.
****
American Paper Optics Inc.
3080 Barlett Corporate Drive
City: Barlett
State: TN
Postal Code: 38133
Country: United States of Amenca
Voice Phone: (1) 901 381 0 1515
Voice Phone: (1) 800 767 8427

INTERNATIONAL BUSINESS DIRECTORY Am - Ar

Fax Phone: (I) 901 381 1517
Email: optics3d@lunaweb.net
Contact #1: John Jerit
Description: American Paper Optics is the leading manufacturer of paper 3D glasses in the world. The products include Diffraction glasses, Polarized and ChromaDepth glasses. The latest holographic 3D glass is called HoloSpex and creates an image of a logo when viewing a direct point of light.

American Society for Nondestructive Testing
1711 Arlingate Lane
P.O. Box 28518
City: Columbus
State: OH
Postal Code: 43228-0518
Country: United States of America
Voice Phone: (1) 800 222 2768
Voice Phone: (1) 614 274 6003
Fax Phone: (I) 614 274 6899
Email: webmaster@asnt.org
Web: http: //www.asnt.org/
Contact # 1: Robert K. Windsor
Description: ASNT is a non-profit organization representing the field of non-destructive testing. They offer certification programs, Information services (Patent searches), and a Magazine/Journal called "Materials Evaluation".

Ana MacArthur
P.O. Box 15234
City: Santa Fe
State: NM
Postal Code: 87506
Country: United States of America
Voice Phone: (1) 505 438 8739
Fax Phone: (I) 505 438 8224
Email: aurorean@nets .com
Contact #1: Ana MacArthur
Description: Holographic installation artist - unique installations. Also produces limited edition dichromatr holograms and holographic sculptures. Experimental use of realtime interferometry.

Anaspec Ltd., Holography and Image Systems
International Marketing & Technical Services
PO Box 25
City, Province: Newbury, Berkshire
State: England
Postal Code: RG20 8BQ
Country: United Kingdom
Voice Phone: (44) 1635 248080
Fax Phone: (44) 1635 248745
Email: spectrolab@compuserve.com
Web: http://www.spectrolab.co.ukl
Contact #1: Bill Vince
Description: Design and manufacture of complete holography studios. Manufacturer of vibration isolation tables. Supplier of lasers and film. Design and production of large format color holograms. Holographic constancy.

Anderson Lasers, Inc.
2883 W. Royalton Rd.
City: Broadview Hts.
State: OH
Postal Code: 44147
Country: United States of America
Voice Phone: (1) 440 237 6629
Fax Phone: (1) 440 237 0656
Email : anderr@netcom.com
Web: http://www.andersonlasers.com!
Contact # 1: Dick Anderson
Description: We have many different types of lasers including: Pulsed Ruby Lasers, Argon Lasers, Scientific PulsedlCW YAGs, made-to-order Holographic Systems, New Rods (YAG/ Ruby/Glass,) Mirrors, and Cavities. We also carry electronic test equipment including: High-Speed Oscilloscopes, Fiberoptic Test Equipment, and Spectrum Analyzers. We have Newport optical mounts and tables, laser power meters, and components. Our inventory changes weekly. Also, we offer Argon, Krypton, and HeNe laser refurbishment. Please call us for your needs.
SEE OUR ADVERTISEMENT

Andreas Wappelt - Photonics Direct
Weisestrasse II
City: Berlin
Postal Code: D-12049
Country: Germany
Voice Phone: (49) (0)30 62709372
Voice Phone: (49) to)30 62709370
Fax Phone: (49) (0)30 62709371
Email: wappelt@t-online.de
Web: http ://home.t-online.de/home/wappelt/
Contact #1: Andreas Wappelt
Description: Distribution of holograms (fine art & entertainment), He-ne lasers, laser pointer, laser diodes, diode pumped blue and green solid-state lasers, holographic equipment.

Ann Arbor Optical Co.
see Data Optics, Inc.
115 Holmes Road
City: Ypsilanti
State: MI
Postal Code: 48198-3020
Country: United States of America
Voice Phone: (1) 734 483 8228
Voice Phone: (1) 800 321 9026
Fax Phone: (1) 734 483-9879
Email: webmaster@dataoptics .com
Web: www.dataoptics.com
Contact #1: David Shindell
Description: Manufacturer of the Model C and Model E Optical Testers. Ronchi-type testers, they can be used to visualize and measure distortion and focal length variation in lenses, mirrors, and optical ""'"windows.'"'"' Typical applications include distortion testing of telescope mirrors, fighter pilot visors, laser protective goggles for the military and eyeglass lens blanks. Part of Data Optics, Inc.

Another Dimension Inc. (Spectore/ADI)
637 NW 12th Ave.
City: Deerfield Beach
State: FL
Postal Code: 33442
Country: United States of America
Voice Phone: (1) 800 422 0220
Voice Phone: (1) 954 481-8422
Fax Phone: (1) 954 421 2391
Email: info@spectore.com
Web: http ://www.spectore.com
Contact #1 : Carol Stoller
Description: Another Dimension, Inc. merged with Spectore Corp. in January 1997. Spectore/ADI distributes and manufacturers holograms and 3DFX (Stero optics/Lenticular) both domestically and internationally. Primary focus is retailers.

API Foils
3841 Greenway Circle
City: Lawrence
State: KS
Postal Code: 66046-5444
Country: United States of America
Voice Phone: (1) 800 255 4605
Voice Phone: (1) 785 842 7674
Fax Phone: (1) 785 842 9748
Email: astoruniversal@worldnet.att.net

Contact #1: John Thoma
Contact #2: Sheryl Hicklin
Description: Worldwide designers, originators and manufacturers of holographic pattern and image foils and films. (Ed -Formerly listed as Astor Universal, Advanced Holographic Laboratories, Dri-Print Foils and Whiley Foils.)
SEE OUR ADVERTISEMENT

API Foils
Eccles New Road
City: Salford
State: England
Postal Code: M5 2DA
Country: United Kingdom
Voice Phone: (44) (0)161 7898131
Voice Phone: (44) (0)181 987 9463
Fax Phone: (44) (0) 161 787 8348
Contact #1: Francis Tuffy
Contact # 2: Laurence Holden
Description: Worldwide designers, originators and manufacturers of holographic pattern and image foils and film. (Ed.- Formerly listed as Astor Universal, Advanced Holographic Laboratories, Dri-Print Foils and Whiley Foils.) F. Tuffy can also be reached at our London office: (44) (0)181 9879463.
SEE OUR ADVERTISEMENT

Applied Holographics, Plc.
40 Phoenix Road, Crowther
Washington 3
City: Tywe & Wear
State: England
Postal Code: NE38 OAD
Country: United Kingdom
Voice Phone: (44) 191 417 5434
Fax Phone: (44) 191 416 3292
Email: sales@appplied-holographics.com
Web: http ://www.applied-holographics .com
Contact # 1: David Tidmarsh
Description: Holographic designers and manufacturers of secure optically variable devices in the form of hot stamping foils, tamperevident labels, semi-transparent laminates and secure transfer films for the protection of security printed documents and branded products from counterfeiting.

Applied Optics
2662 Valley Drive
City: Ann Arbor
State: MI
Postal Code: 48103-2748
Country: United States of America
Voice Phone: (1) 734 998 0425
Fax Phone: (1) 734 998 0425
Email: upatnks@applopt.com
Contact #1: Juris Upatnieks
Description: Applied Optics provides consulting services and laboratory breadboard testing of optical systems in coherent optics, holography, diffractive optical elements, and light control. System analysis using the ZEMAX optical design program.

Arbeitskreis Holografie B.V
Gelderstrasse 22
City: Geldern
Postal Code: D-47608
Country: Germany
Voice Phone: (49) (0)2831 94750
Contact #1: Herman-Josef Bianchi
Description: Artistic holography

Ardee International (div. of Ardee Lighting Japan)
3F., Shichi Bldg.
2-10-15 Kasuga, Bunkyo-ku

City: Tokyo
Postal Code: 11 2-0003
Country: Japan
Voice Phone: «81» 3 5800-1 721
Fax Phone: «8 1» 3 5800- 1723
Email: kodera@jp.interramp.com
Contact # 1: Mitsuo Kodera
Description: Producer and coordinator in Holographic Art, Light Art, Computer Art and other Art and Science field for art exhibition, public art in public space, monumental objects for commercial facilities. Importer and exporter. Produced the Hologram Exhibition of "Alice in the Light World" in Shinjuku, Tokyo; "International Exhibition World of Holography" ; Nagoya, "Art Hologram of Ludie Berkhout", Fukuoka and Tokyo; "Hologram Now", "Hologram Annual In Ginza", Ginza, Tokyo, etc.
****

Armin Klix Holographie
Postfach 260218
City: Dusseldorf
Postal Code: D-40095
Country: Germany
Voice Phone: (49) (0)211 317775
Fax Phone: (49) (0)211 317 749
Contact # 1: Armin Klix
Description: Anfertigung von Displayhologrammen im Auftrag von Werbung und Industrie und Einzel- und Grosshandel. Katalog von Lieferbaren Hologrammen vorhanden. Einzelstucke und Grosserien.
****

Art Agentur Kiiln
Venloer Strasse 461
City: Koeln
Postal Code: D-50825
Country: Germany
Voice Phone: (49) (0)221 5441 00
Voice Phone: (49) (0)221 54 1400
Contact #1: Liesel Dr. Hollmann-Langecker
Description: Mediation agency for modern art-focused on holography.
****

Art Foil Graphic Machinery
Unit 21
Modema Business Park
City, Province: Mytholmroyd, Halifax
State: England
Postal Code: HX7 5QQ
Country: United Kingdom
Voice Phone: (44) 01422 886 550
Fax Phone: (44) 01422 886 614
Contact #1: S.D. Smith
Description: Large sheet hot foil stamping machines for graphic enhancement and hologram application. Maximum sheet size 29 x 41 (73cm x 104cm).
****

Art Ig Wohndesign - Holografie
Muchener Strasse 29
City: Dachau
Postal Code: D-85221
Country: Germany
Voice Phone: (49) (0)8131 352749
Contact #1: Mr. Fiedler
Description: Distributor of all types of holograms.
****

Art Illusions Ltd.
Uziel2
City: Bat-yam
Country: Israel
Voice Phone: (972) 6 264959
Fax Phone: (972) 3 507 2079
****

Art Institute Of Chicago (The School of the)
Holography Department
112 South Michigan Ave.

City: Chicago
State: IL
Postal Code: 60603
Country: United States of America
Voice Phone: (1) 312 345 3998
Fax Phone: (1) 312 345 3565
Web: http: //www.artic.edu
Contact # 1: E Kac
Description: The School of the Art Institute of Chicago offers an MFA degree with a concentration in holography, and is equipped with three tables, with one containing a stereogram printer for recording computer generated imagery or live subjects.
****

Art Lab
1000 Richmond Terrace
City: Staten Island
State: NY
Postal Code: 10301
Country: United States of America
Voice Phone: (1) 718 447 8667
Fax Phone: (1) 718 447 8668
Contact # 1: Carol Summer
Description: The Art Lab is an art school which periodically offers classes and workshops in the art of holography. Free brochure available.
****

Art, Science & Technology Institute (ASTI)
1400 South Joy St.
Suite 226
City: Arlington
State: VA
Postal Code: 22202
Country: United States of America
Voice Phone: (1) 703 892-2226
Fax Phone: (1) 703 892-2161
Contact #1: Laurent Bussaut
Description: Research and educational corporation dedicated to the advancement of art, science and technology in holography, laser, optics and photonics industries. The museum features a permanent collection of masterpiece holograms including one of the largest holograms in the.. world. The museum is visited by guided tour only, and reservations are required.
****

Artbridge Light Studios
Madam Weg 77
City: Braunschweig
Postal Code: D-38120
Country: Germany
Voice Phone: (49) (0)531 352 816
Fax Phone: (49) (0)53 1 352 816
Email: holodietma@aol.com
Web: www.holonet.khm.de/holodietma/od/index.htm
Contact #1: Odile Meulien
Description: Design & Production of Holograms and Light Show - Organization of special happenings in coordination with multidisciplinary artists-engineers and scientists.
****

Artplay Holographika Studio
Ady Endre ut. 8
City: Budapest
Postal Code: H-1191
Country: Hungary
Voice Phone: (36) 1 281-9114
Fax Phone: (36) I 282-4921
Email: baloghho@caesar.elte.hu
Contact #1: Tibor Balogh
Description: A.P.Holografika provides mastering, whole process for custom embossed holograms (mainly in security applications), optical design using HOEs. Wholesaler of holographic novelties and diffraction foils.

Asahi Glass Co.
BD Division,
2-1-2 Marunouchi Chiyoda-Ku
City: Tokyo
Postal Code: 100-0005
Country: Japan
Voice Phone: «81» 3 3218 5473
Fax Phone: «81» 3 3211 1327
Contact # 1: Fumihiko Koizumi
Description: R&D on holography and single copy holograms.
****

Atelier Holographique De Paris
13 pass Courtois
City: Paris
Postal Code: F-75011
Country: France
Voice Phone: (33) 1 43 79 69 18
Fax Phone: (33) I 40 09 05 20
Description: Artistic holography; Buying & Selling; Consulting
****

Automated Holographic Systems (AHS)
24 Mainline Dr.
City: Westfield
State: MA
Postal Code: 01085
Country: United States of America
Voice Phone: (1) 413 562 9992
Fax Phone: (1) 413 562 9993
Email: ahologram@earthlink.net
Contact #1: Jim Gibb
Description: R&D of holographic coating, mastering, combining and production machinery. Consulting services of technological development and procedures. Production of embossed hologram replication, combining, DCG mastering, photo polymer, silver halide, photoresist, and HOE product development.
****

Avant-Garde Studio
34 North Rochdale Ave.
P.O. Box 296
City: Roosevelt
State: NJ
Postal Code: 08555-0296
Country: United States of America
Voice Phone: (1) 609 448 6433
Fax Phone: (1) 609 448 6433
Email: avantgardestudiio@home.com
Contact # 1: Amy E. Medford
Description: Model makers. Sculptors, excelling in relief (forced perspective, texture, form), for holographic models. We will design and/or work from, photographs, drawings or other material.
****

b+g Banse und Grohrnann GmbH
Iisenburger Strasse 40
City: Wernigerode
Postal Code: D-38855
Country: Germany
Voice Phone : (49) (0)3943 5440-33
Fax Phone: (49) (0)3943 5440-30
Contact # 1: Mr. Grohmann
Description: Distribute holographic material
****

Baier Praegepressen
Maschinenfabrik Gebr. Baier KG
Lindenthaler Str. 78
City: Rudersberg
Postal Code: D-73635
Country: Germany
Voice Phone: (49) (0)7183 532
Fax Phone: (49) (0)7183 3481
Description: Manufacturer of machines for production of embossed holograms.
****

Batelle Pacific Northwest National Lab
P.O. Box 999, K8-09
City: Richland
State: WA
Postal Code: 99352
Country: United States of America
Voice Phone: (1) 509 375 212 1
Email: ma_lind@pnl.gov
Web: http ://www.pnl.gov
Contact # 1: Michael Lind
Description : Theoretical and experimental research and development in applying holographic techniques to active and passive optical, radar, acoustic, ultrasound, EEG and EKG imaging. R&D for Dept. of Energy.
****

Bauer-Josef
Augustenstrasse 121
City: Munich
Postal Code: D-80798
Country: Germany
Voice Phone: (49) (0)89 2710248
Contact #1: Josef Bauer
Description: Educational courses in holography.
****

Beijing Dongfang Laser Printing Tech. Co.
Yuqun Hu-Tong Jia- # 18
Dong-Cheng-Qu
City: Beijing
Postal Code: 100010
Country: China
Voice Phone: (86) 10 64033223
Fax Phone: (86) 10 64034996
Contact # 1: Liu Haidong
****

Beijing Fantastic Hologram Products
Xue-Yuan-Lu #5
Hai-Dian-Qu
City: Beijing
Postal Code: 100083
Country: China
Voice Phone: (86) 10 62083373
Fax Phone: (86) 1,62083810
Contact #1: Wang Yao
Description: Sales of holograms.
****

Beijing Hologram Printing Tech. Co
PO Box 9622-#2
City: Beijing
Postal Code: 100086
Country: China
Voice Phone: (86) 10 68379223
Fax Phone: (86) 10 68378302
Contact #1: Fu Ziping
Description: Holography sales and distribution.
****

Beijing Sanyou Laser Images Co.
Bei-San-Huan-Zhong-Lu #40
City: Beijing
Postal Code: 100088
Country: China
Contact #1: Li Qiang
Description: hologram distribution.
****

Beijing University of Posts & Telecommunications
P.O. Box 163
City: Beijing
Postal Code: 100088
Country: China
Voice Phone: (86) 1062281535
Fax Phone: (86) 10 62281774
Email: xudx@bupt.edu.cn
Contact #1: Hsu Da-hsiung
Description: Holography class.
****

Beijing Xi Ji Wo Computer Graphic Wkg Co.
Dong-Huang-Cheng-Gen-Bei-Jie Jia-#20

City: Beijing
Postal Code: 100010
Country: China
Voice Phone: (86) 10 64255163
Fax Phone: (86) 1064255163
Contact # 1: Chen Hui
Description: Holography classes.
****

Bellini, Victor
850 Howard Ave. # 5J
City: Staten Island
State: NY
Postal Code: 10301
Country: United States of America
Voice Phone: (1) 7 18 442 4726
Email: victor3D@webtv.net
Contact # 1: Victor Bellini
Description: Extensive collection and archive of holographic, lenticular, and other 3D collectibles. Appraisals for trading cards.
****

Benyon, Margaret - Holography Studio
40 Springdale Avenue
City, Province: Broadstone, Dorset
State: England
Postal Code: BHI8 9EU
Country: United Kingdom
Voice Phone: (44) 1202 698 067
Fax Phone: (44) 1202 698 067
Email: benyon@holography.demon.co.uk
Contact # 1: Margaret Benyon .
Description: Fine art holography, established in 1968. Limited edition and unique works available. Included in a large number of private art collections world-wide.
****

Berkhout, Rudie
223 West 21 st Street
City: New York
State: NY
Postal Code: 10011
Country: United States of America
Voice Phone: (1) 212 255 7569
Fax Phone: (1) 212 727 0532
Email: rudieberkhout@mindspring.com
Web: http://
www.rudieberkhout.home.mindspring.com
Contact # 1: Rudie Berkhout
Description: Holographic Fine Artist who has had work exhibited worldwide, including at the Whitney Museum of American Art (New York). Also teaches holography at the School of Visual Arts NY.
****

Bernhard Halle Nachf. GmbH & Co.
Hubertusstr. 10-11
City: Berlin
Postal Code: D-13589
Country: Germany
Voice Phone: (49) (0)30 7 91 60 77
Fax Phone: (49) (0)30 7 91 85 27
Contact #1: A. Frank
Description: Precision optics-polarization optics-catalog and custom made optics.
****

BIAS - Bremer Institute Applied Beam
Klagenfurter S tr 2
City: Bremen
Postal Code: D-28359
Country: Germany
Voice Phone: (49) (0)421 218 5002
Fax Phone: (49) (0)421 218 5059
Contact #1: Werner Prof. Juptner
Description: Industrial research; holographic non-destructive testing.
****

BfFO
Berliner Institut fur Optik GmbH
Rudower Chaussee 6

City: Berlin
Postal Code: D-12484
Country: Germany
Voice Phone: (49) (0)30 6392-3450
Fax Phone: (49) (0)30 6392-3452
Description: Manufacturer and designer of HOEs.
****

Bjelkhagen, Hans
Center for Modern Optics
De Monfort University
City: Leicester
Postal Code: LEI 9BH
Country: United Kingdom
Voice Phone: 44 116 250 6145
Fax Phone: 44 116 250 6144
hansholo@aol.com
Description: R&D in many types of holography. Specializing in color holography. Author. Currently pursuing work with Lippmann photography.
****

Blue Ridge Holographics, Inc.
511 Stewart St.
City: Charlottesville
State: VA
Postal Code: 22902
Country: United States of America
Voice Phone: (1) 804 296 1110
Fax Phone: (1) 804 296 1182
Email: steve@blueridgeholo.com
Web: http://www.blueridgeholo.com
Contact # 1: Steve Provence
Description: Mastering facility for the production of embossed holograms. Consulting and design services offered. Extensive client list.
****

Bobst Group Inc.
146 Harrison Avenue
City: Roseland
State: NJ
Postal Code: 07068
Country: United States of America
Voice Phone: (1) 973 226 8000
Fax Phone: (1) 973 226-8625
Email: sales@bobstgroup.com
Web: http ://www.bobstgroup.com
Contact #1: Doug Herr
Description: One of the world's largest manufacturers of hologram hot stamp machinery.
****

Booth, Roberta
5326 Sunset Blvd.
City: Los Angeles
State: CA
Postal Code: 90027
Country: United States of America
Voice Phone: (1) 213 466 5767
Fax Phone: (1) 213 465 5767
Contact #1: Roberta Booth
Description: I am a holographic artist working in transmission and reflection holography. I also work as a consultant for holographic projects and curate holography shows.
****

Boyd, Patrick
Archway House, Church Street
Easton On The Hill
City: Stamford
State: England
Postal Code: PE9 3LL
Country: United Kingdom
Voice Phone: (44) 976 298 578
Voice Phone: (44) 1780 755 647
Fax Phone: (44) 1780 756 002
Email: boyd@atlas.co.uk
Contact #1: Patrick Boyd
Description: Holographic Fine Artist. Extensive portfolio.
****

Brainet Corporation - International Division
4F. Asset Bldg.
3-31-5 Honkomagome Bunkyo-ku
City: Tokyo
Postal Code: 113-0021
Country: Japan
Voice Phone: (81) 3 5395 7030
Fax Phone: (81) 3 5395 7029
Email: brainet@bnn-net.or.jp
Contact # 1: Yutaka Inoue
Description: Distributor and producer of framed holograms and other processed holographic products from a variety of manufacturers. Sole agent distributor of LightrixlUSA and Laza/UK for Japan and Korea. Also provide consulting for custom work and exhibition services for galleries and museums .
****

Brandtjen & Kluge, Inc.,
539 Blanding Woods Road
City: St. Croix Falls
State: WI
Postal Code: 55024
Country: United States of America
Voice Phone: (1) 715 483 3265
Voice Phone: (1) 800 826 7320
Fax Phone: (1) 715 483 1640
Contact # 1: John Edgar
Description: With over 75 years experience in manufacturing what are acknowledged as the industry's most reliable presses, Brandtjen & Kluge is uniquely qualified to deliver presses for efficient, trouble free application of hologram foils .
****

Bridgestone Technologies, Inc.
375 Howard Ave.
City: Bridgeport
State: CT
Postal Code: 06605
Country: United States of America
Voice Phone: (1) 203 366 1595
Fax Phone: (1) 203 366 1667
Email: bgtechno@pcnet.com
Web: www.bridgestonetech.com
Contact #1: Richard Zucker
Description: Product authentication systems. Anti-counterfeiting technology. Fully integrated provider of security printed products and services including holography, microtracers, biocoding and field investigation.
SEE OUR ADVERTISEMENT
****

British Aerospace Pic.
Sowerby Research Centre
Fpc: 267 PO Box 5
City, Province: Filton, Bristol
State: England
Postal Code: BSI2 7QW
Country: United Kingdom
Voice Phone: (44) 1179 366 842
Fax Phone: (44) I 179 363 733
Web: www.bae.co.uk
Contact #1: Dr. Phillip Salter
Description: Sheer Holography and NDT interferometry as it applies to material stress. Will work on projects that are of mutual benefit to British Aerospace and client.
****

Broadbent Consulting
1070 Commerce Street - Suite A
City: San Marcos
State: CA
Postal Code: 92069
Country: United States of America
Voice Phone: (1) 760 752 1039
Fax Phone: (1) 760 752 1009
Email: hologram@fda.net
Contact #1: Donald Broadbent
Description: An independent, privately owned, holographic facility producing HOEs and display holograms in various recording materials. Donald Broadbent has 35 years experience in the field of holography.
****

Bruck, Richard Holography
175 South Lake Street
City: Grayslake
State: IL
Postal Code: 60030
Country: United States of America
Voice Phone: (1) 847 543 4385
Fax Phone: (1) 847 543 4385
Email: rabhol@juno.com
Contact #1: Richard Bruck
Description: Specialists in large format holography. Extensive experience with live models and commercial work. We are accustomed to the advertising world, and know the importance of quality and service.
****

Bundesdruckerei
Oranienstrasse 9 I
City: Berlin
Postal Code: D-I0958
Country: Germany
Voice Phone: (49) (0)30 2598 -1146
Fax Phone: (49) (0)30 2598-1012
Email: 101640.1466@compuserve.com
Contact #1: Dr. R. Paugstadt
Description: mass production of security holograms for German / European banknotes.
****

Buntstift
Oststrasse 24
City: Harsum
Postal Code: D-3I 177
Country: Germany
Voice Phone: (49) (0)5127 93 I II 3
Fax Phone: (49) (0)5127 931 I 16
Email: buntstift@t-online.de
Contact # 1: Guido Hoppe
Description: Holography lab.
****

Burgmer Brigitte
Grabengasse 27
City: Cologne
Postal Code: D-50679
Country: Germany
Voice Phone: (49) (0)221 884374
Fax Phone: (49) (0)221 814091
Contact #1: Brigitte Burgrner
Description: Artist-initiator of nUmerous holography projects.
****

Burleigh Instruments, Inc.
Burleigh Park
City: Fishers
State: NY
Postal Code: 14453
Country: United States of America
Voice Phone: (1) 716 924 9355
Fax Phone: (1) 716 944 9072
Web: http://www.burleigh.com
Contact #1: Tim Klimasewski
Description: Burleigh Instruments, Inc. is a manufacturer of electro-optical equipment including wavelength meters, laser spectrum analyzers, interferometers, and nanopositioning devices .
****

C. Roth GrnH + Co.
Schoemperlenstrasse 1-5
City: Karlsruhe
Postal Code: D-76185
Country: Germany
Voice Phone: (49) (0)721 5606-0
Fax Phone: (49) (0)72 1 5606-149
Email: carl.roth@t-online.de
Description: Distribute chemical lab supplies.

California Institute of the Arts
School of Critical Studies
24700 McBean Parkway
City: Valencia
State: Ca
Postal Code: 91355
Country: United States of America
Voice Phone: (1) 805 255 1050 x2406
Voice Phone: (1) 415 337-6475
Fax Phone: (1) 805 255 01 77
Email: walschulr@hotrnail.com
Description: We teach introductory holography and Lippmann photography.
****

Cambridge Laser Labs
853 Brown Road
City: Fremont
State: CA
Postal Code: 94539
Country: United States of America
Voice Phone: (1) 510 651 0110
Fax Phone: (1) 510 651 1690
Email: camlaser@cambridgelaser.com
Web: http: //www.cambridgelaser.com
Contact # 1: Brian Bohan
Description: World renowned specialist in ion laser repair. Rental systems and used laser system sales. Price guide furnished on request.
****

Canon Inc. R&D Headquarters
3-30-2 Shimomaruko,
Oota-ku,
City: Tokyo
Postal Code: 146
Country: Japan
Voice Phone: ((81» 3 3757-6268
Contact #1: Tetsuro Kuwayama
Description: Research. Gourses in holography.
****

Capilano College
Physics Department - Holography Research
Lab
2055 Purcell Way
City, Province: N. Vancouver, British Columbia
Postal Code: V7J 3H5
Country: Canada
Voice Phone: (1) 604 983 7571
Fax Phone: (1) 604 983 7520
Email: bsimson@capcollege.bc.ca
Web: http: //merlin.capcollege. bc.ca/physics/bsimsonl
Contact # 1: Milessa Crenshaw
Description: Digital holography research. Synthesizing holograms from video (integrals).
****

Capitol Converting Equipment, Inc.
500 North Redfield Court
City: Park Ridge
State: IL
Postal Code: 60068
Country: United States of America
Voice Phone: (1) 847 825 789 I
Fax Phone: (1) 847 825 8661
Contact # 1: Kenneth Singer
Description: Manufacturers of Hot stamping presses. Press Sizes: 29"x41", 33"x47", 39"x55".
****

Carl M. Rodia And Associates
13 Locust St.
City: Trumbull
State: CT
Postal Code: 06611
Country: United States of America
Voice Phone: (1) 203 261 1365
Fax Phone: (1) 203 268 1619
Email: carlrodia@aol.com
Contact #1: Carl M. Rodia

INTERNATIONAL BUSINESS DIRECTORY Ca - Ch

Description: Comprehensive engineering consultation services in precision hologram manufacturing. Plant design and engineering, process engineering, troubleshooting and seminar training of manufacturing personnel.
****

Carls Zeiss GmbH
Bereich PQG
Forstweg 3
City: Jena
Postal Code: D-07745
Country: Germany
Voice Phone: (49) (0)3461 64 -2435
Fax Phone: (49) (0)3461 64-2734
Contact #1 : Rainer Hultzsch
Description: Manufacturer and designer of HOEs.
****

Casdin-Silver Holography
99 Pond Avenue Suite D403
City: Brookline
State: MA
Postal Code: 02445
Country: United States of America
Voice Phone: (1) 617 739 6869
Voice Phone: (1) 617 423 4717
Fax Phone: (1) 617 739 6869
Contact #1 : Harriet Casdin-Silver
Description: I have been creating holographic art and interactive holographic installations since 1968. Our company specializes in original holograms for advertising, architectural and theater settings, and expositions. We are also consultants and exhibition organizers/designers.
****

Cavomit
22 Pipinou Steet
City: Athens
Postal Code: GR-11257
Country: Greece
Voice Phone: (30) I 823 2355
Fax Phone: (30) I 231 4499
Contact #1: Alkis Lembessus
Description: Hot-stamping equipment (cylinders - platen), hologram registration systems, foils and consumables. Local distributor for Astor Universal, Kluge, Light Impressions, Applied Holographics, Revere Graphic Products.
****

Center for Applied Research in Art & Tech.
University of Gent
72 Lange Boongaardstraat
City: Gent
Postal Code: B-9000
Country: Belgium
Voice Phone: (32) 91 626384
Fax Phone: (32) 91 237326
Contact #1: Prof. Pierre Boone
Description: research.
****

Center for the Holographic Arts
45-10 Court Square
City: Long Island City
State: NY
Postal Code: 11101
Country: United States of America
Voice Phone: (1) 718 784 5065
Fax Phone: (I) 718 784 5065
Email: holocenter@mindspring.com
Web: http://www.holocenter.com
Contact #1: Ana Nicholson
Description: The Center offers an Artist In Residence program, small exhibitions of interesting, current holographic work, larger traveling exhibitions, a series of talks, and will maintain a web page with news and developments. The Center is equipped with a pulse laser camera, a c.w. table and Argon and Helium-Neon lasers. Under the guidance of the two directors, artists will be encouraged to explore the medium as part of their artistic vocabulary for a maximum of four weeks. The Artist In Residence Program is at the heart of the Center's activities. Funded in part by The Shearwater Foundation.
****

Central Glass Co., Ltd.
Kowa-Hitosubashi Bldg
7-1 Kanda-Nishikicho 3-Chom
City: Tokyo (Chiyoda-Ku)
Postal Code: 101
Country: Japan
Voice Phone: ((81» 3 3259 7354
Contact #1: Chikara Hashimoto
Description: Research - Heads Up Display.
****

Centre d'Art Holographique et Photonique
College de Maisonneuve (c-2200)
3800 Sherbrooke est
City, Province: Montreal, Quebec
Postal Code: HIX 2A2
Country: Canada
Voice Phone: (1) 514 254 7131 ex4509
Fax Phone: (I) 514 253 8909
Email: holo@cmaisonneuve.qc.ca
Web: www.cmaisonneuve .qc .ca/holostar.
Contact # 1: Eric Bosco
Description: We are situated in a CEGEP (post-secondary school). We have 2 fully equipped tables with HeNe's and an argon (with eta Ions). We do all types of holograms up to 2 feet by 3 feet. We give courses, workshops, do production. Lab space rental
****

Centro de Investigaciones en Optica, A.C.
Lorna del Bosque liS
Col. Lomas del Campestre
City, Province: Leon, Gto.
Country: Mexico
Voice Phone: (52) 47 731017
Voice Phone: (52) 47731018
Fax Phone: (52) 47 175000
Email: dfa@riscl.cio.mx
Contact #1: Fernando Mendoza, Ph.D.
Description: Our Optical Research Center offers: optical design and construction of a wide variety of optical components (lenses, mirrors, prisms, etc .); high standards in R&D in optical NDT for industrial applications, optical fiber sensors, rare earth doped fibers and optical shop testing.
****

CFC Applied Holographics
500 State St.
City: Chicago Heights
State: IL
Postal Code: 60411
Country: United States of America
Voice Phone: (1) 800 438 4656
Voice Phone: (1) 708 891 3456
Fax Phone: (1) 708 758 5989
Email: patswineah@aol.com
Web: http://www.cfcintl.com
Contact #1: Dave Beeching
Contact #2: Jim Vaughn
Description: For security, decoration and packaging applications; we offer a full range of tamper evident and authentication labels, hot stamping foils, and release and size coat combinations. Print treatments and custom metallizing available.
SEE OUR ADVERTISEMENT
****

Cfc Northern Bank Note Co.
5400 East Ave
City: Countryside
State: IL
Postal Code: 60525
Country: United States of America
Voice Phone: (1) 708 482 3900
Fax Phone: (1) 708 482 3332
Email: sales@nbnco.com
Web: http: //www.cfcintl.com
Contact #1: Mark Lamb
Description: Part of cfc Applied Holographics. For security, decoration and packaging applications; we offer a full range of tamper evident and authentication labels, hot stamping foils, and release and size coat combinations. Print treatments and custom metallizing available.
SEE OUR ADVERTISEMENT
****

Checkpoint Security Services Limited
Export Dept
115 Chatham Street
City, Province: Reading, Berkshire
State: England
Postal Code: RGi 7JX
Country: United Kingdom
Voice Phone: (44) 118 925 8250
Voice Phone: (44) 118 925 8251
Fax Phone: (44) 118 958 3749
Email: icarmichael@checkpoint.co.uk
Web: http ://www.checkpoint.co.uk
Contact # 1: Lisa Carmichael
Description: Point of issue hologram applicators for continuous and sheet documents, to authenticate or protect high value, negotiable documents. Available with metallized, and demetallized foil in various die block styles.
****

Chemnitzer Werstoffmechanik GmbH
Postfach 344
City: Chemnitz
Postal Code: D-09003
Country: Germany
Voice Phone: (49) (0)371 5397 -478
Fax Phone: (49) (0)371 5397-463
Description: Industrial holographic interferometry.

Chengdu Xinxing Institute for Development of Tech.
Bin-Jiang-Lu-Wai Dong-Xia-He-Ba # 60
City, Province: Chengdu, Sichuan
Postal Code: 610061
Country: China
Contact #1: Zeng Xuzhang
Description: Holography class.
****

Cherry Optical Holography
2047 Blucher Valley Road
City: Sebastopol
State: CA
Postal Code: 95472
Country: United States of America
Voice Phone: (1) 707 823 7171
Fax Phone: (1) 707 823 8073
Contact # 1: Greg Cherry
Description: Highest quality display holography available. Stock and custom reflection! transmission holograms on glass plates or film up to 40"" x 72"" in size. Open and Limited edition fine art holograms. Custom mastering services offered for silver halide and photopolymer.
****

Chiba University
Engineering Department
School Of Science & Engineering
City: Chiba
Postal Code: 263-0022
Country: Japan
Voice Phone: ((81» 43 290 3469
Fax Phone: ((8 1» 3 290 3490
Contact #1: Miss Prof. Toshio Honda

Description: Research holographic display, electro holography.

****

Chiba University
#702, 7-27-14 Matsue
Edogawa-ku
City: Tokyo
Postal Code: 132-0025
Country: Japan
Voice Phone: ((81» 3 3656-1215
Fax Phone: ((81» 3 3656-1429
Contact #1: Jumpei Tsujiuchi.
Description: Scientific; holographic research.

****

China Ann Arbor Holographical Institute
Rm. 403 Bldg. 22
Zhong Lou Xing Chen
City, Province: Jiangsu, Suzhou
Postal Code: 215006
Country: China
Voice Phone: (86) 512 227 461
Contact #1: Yaguang Jiang
Description: Holography sales, distribution and classes.

****

CHIRON Technolas GmbH
Max-Planck Strasse 6
City: Dornach
Postal Code: D-85609
Country: Germany
Voice Phone: (49) (0)89 945514 0
Fax Phone: (49) (0)89 945514 70
Contact #1: Hr. Junger
Description: Ophthalmologic Systems.

****

Chongqing Yinhe Laser Products Ltd.
Shi-Ma-He Xia-Hua-Yuan #6,
Jiang-Bei-Qu
City: Chongqing
Postal Code: 630021
Country: China
Voice Phone: (86) 811 5312071
Fax Phone: (86) 811 5312050
Contact #1: Zhang Zheng
Description: Laser distributor.

****

Chromagem Inc.
573 South Schenley
City: Youngstown
State: OH
Postal Code: 44509
Country: United States of America
Voice Phone: (1) 330 793 3515
Fax Phone: (1) 330 793 3515
Email: chromagem@aol.com
Contact #1: Thomas J. Cvetkovich
Description: Established in 1981. Hologram mastering facility specializing in shooting photoresist and photopolymer masters for use in commercial mass-production of embossed holograms: 2D, 3D, stereogram, large format rainbow holograms (up to 40 x 40 inches) and dot matrix. Extensive experience working with major corporate accounts . Design and consultation services.

****

Chronomotion
424 Ninth St.
City: Santa Monica
State: CA
Postal Code: 90402
Country: United States of America
Voice Phone: (1) 310 393 9859
Email: mburney@ix.netcom.com
Contact #1: Michael Burney
Description: Developed and patented a general process for producing "electronic holograms" with a real image projected into the room with the viewer. Also works with liquid crystal displays.

****

Cifelli, Dan
712 Bancroft Road # 332
City: Walnut Creek
State: CA
Postal Code: 94598
Country: United States of America
Voice Phone: (1) 925 930 8033
Fax Phone: (1) 925 930 8033
Email: c3ddan@aol.com
Contact # 1: Dan Cifelli
Description: Holography consulting and brokering since 1979 for stock and custom products. Evaluate, match, and develop your product for holography mastering and production techniques; sales potential and market positioning; manufacturing and converting processes/materials; cost analysis; and patent/licensing potential. Experience with bio-metrics.

****

City Chemical
100 Hoboken Ave.
City: Jersey City
State: NJ
Postal Code: 07310
Country: United States of America
Voice Phone: (1) 201 653 6900
Voice Phone: (1) 800 248 2436
Fax Phone: (1) 201 653 4468
Description: Photo-chemicals and chemical supplies.

****

Coburn Corporation
1650 Corporate Road West
City: Lakewood
State: NJ
Postal Code: 08701
Country: United States of America
Voice Phone: (1) 732 367 5511
Fax Phone: (1) 732 367 2908
Email: coburncorp@aol.com
Web: http ://www.coburn.com
Contact #1: John White
Description: Offers a wide range of stock repeating geometric holographic designs in pressure sensitive film; conventionally printable substrate; v,arious traditional and designer oriented colors available. Introductory sample kits available.

****

Coburn Europe GmbH
Sc hiiferstrasse 5
City: Berka (Werra)
Postal Code: D-99837
Country: Germany
Email: coburncorp@aol.com
Contact # 1: Mike Wahl
Description: Manufacturer and designer of HOEs in form of adhentsive films.

****

Coherent Deutschland GmbH
Dieselstrasse 5b
City: Dieburg
Postal Code: D-64807
Country: Germany
Voice Phone: (49) (0)6071 968-0
Fax Phone: (49) (0)6071 968-499
Contact #1: Mrs. Beck
Description: Manufacturer of lasers.

****

Coherent Lubeck GmbH
See land Strasse 9
City: Lubeck
Postal Code: D-23569
Country: Germany
Voice Phone: (49) (0)451 3909 300
Fax Phone: (49) (0)451 3909 399
Contact # 1: Marina Schmidt
Description: Established 1986. manufacturer of diode laser-pumped solid state lasers which operate in CW and pulsed mode with wavelengths in IR-visible and UV Branch office: 636 Great Road-Stow-MA 01775-USA Owned by Coherent-Santa Clara -CA-USA

****

Coherent, Inc. - Laser Group
5100 Patrick Henry Drive
City: Santa Clara
State: CA
Postal Code: 95054
Country: United States of America
Voice Phone: (1) 408 764 4000
Voice Phone: (1) 800 527 3786
Fax Phone: (1) 408 764 4800
Email: tech_sales@cohr.com
Web: http ://www.cohr.com
Contact #1: Paul Ginouves
Description: Coherent is the world leader in high-power ion and diode-pumped solid-slale (DPSS) lasers. Our products for professional holography include argon ion lasers (up to W at 488.0 nm, krypton lasers (up to 3.5 W al 647.1 nm), and DPSS lasers (up to 5 W at 532 nm).
SEE OUR ADVERTISEMENT ON THE BACK COVER

****

Collector's Castle
P. O. Box 262
City: Factoryville
State: PA.
Postal Code: 18419
Country: United States of America
Voice Phone: (1) 717 945-9326
Voice Phone: (1) 800 246-0694
Fax Phone: (1) 717 9,45-7231
Email: hologram@epix.net
Web: www.Cardmall.com/castie
Contact #1: Ronald Evans
Description: This site offers a selection of over 900 different holograms from spinning disks, non-sport, post cards, comic books, sport cards, jewelry, cereal boxes, stickers special offers, and commercial art prints.

****

Collimage International Co., Ltd.
2F-4 No.72 Sec.4 Hsing-Lung Road
City: Taipei
Postal Code: 116
Country: Taiwan R.O.C.
Voice Phone: (886) 2 2234 3073
Fax Phone: (886) 2 2234 3075
Email: c888@msl.hinet.net

****

Colour Holographics
Unit 7a, 1-2 Domingo Street
City: London
State: England
Postal Code: ECI YOTA
Country: United Kingdom
Voice Phone: (44) 171 251 0511
Fax Phone: (44) 171 7364710
Email: colourholographics.@bt.co.uk
Contact #1: Michael Medora
Description: We produce large format full color stereo grams 50 x 60 cm.

****

Competitive Edge, Inc.
3500 109th St.
City: Irvingdale
State: lA
Postal Code: 50322
Country: United States of America
Voice Phone: (1) 800458 3343
Fax Phone: (1) 515 288 3343
Description: Full line of hologram merchandise including keychains, buttons, T-shirts, calculators. Custom or stock designs.

****

INTERNATIONAL BUSINESS DIRECTORY Co - Da

Concordia University
Communications Studies
7141 Sherbrooke St. W
City, Province: Montreal, Quebec
Postal Code: PQ H4B lR6
Country: Canada
Voice Phone: (1) 514 848 2539
Voice Phone: (1) 514 848 2424
Fax Phone: (1) 514 848 3492
Email: hal@vax2.concordia.ca
Web: http://www.concordia.ca/
Description: 3 Dimension Research Center.
****

Continental Optical
15 Power Drive
City: Hauppauge
State: NY
Postal Code: 11788
Country: United States of America
Voice Phone: (1) 516 582 3388
Fax Phone: (1) 516 582 1054
Contact # 1: Mark Grindel
Description: Optics and custom orders.
****

Control Module Inc.
227 Brainard Road
City: Enfield
State: CT
Postal Code: 06082
Country: United States of America
Voice Phone: (1) 860 745 2433
Voice Phone: (1) 800 722 6654
Fax Phone: (1) 860 741 6064
Email: knorman@controlmod.com
Web: http ://www.controlmod.com
Contact #1: Kenneth Norman
Description: Experience in the design and manufacture
of automatic data colle cti on equipment and systems,
CMI offers exciting innovations in 1996, including
Holonetics TM, a machine-readable hologram, offer-
ing the highest security protection available for access
control and labor management.
****

Control Optics
13111 Brooks Drive, Unit J
City: Baldwin Park
State: CA
Postal Code: 91706
Country: United States of America
Voice Phone: (1) 626 813 1991
Fax Phone: (1) 626 813 1993
Email: liucoc@interserv.com
Web: http ://www.controloptics.coml
Contact #1: Wai-Min Liu
Description: Provides full-service optical engineering
supporting industry and education. Offers full range of
holographic table top optics, positioning devices and
mounts. New products include holography and fiber-
optic experimenter's kits.
SEE OUR ADVERTISEMENT
****

Corion Corp.
8 East Forge Parkway
City: Franklin
State: MA
Postal Code: 02038
Country: United States of America
Voice Phone: (1) 508 528 4411
Fax Phone: (1) 508 520 7583
Email: sales@corion.com
Web: http://www.Corion.com
Contact #1: Don McLeod
Description: Corion Corp. manufactures volume and
one-of-a-kind custom and stock optical components
including coatings, filters, optics and optical assem-
blies for use in the UV-Visible-IR spectrum. Mostly
biomedical applications.

****
Corning Incorporated
City: Corning
State: NY
Postal Code: 14831
Country: United States of America
Voice Phone: (1) 800222-7740
Voice Phone: (1) 800 492 1110
Email: info@corning.com
Web: http ://www.coming.comlindex.html
Description: On April 24, 1997 Corning announced the
acquisition of Optical Corporation
of America who produces precision, large aperture (to
36 inch diameter) aspheric mirrors for holographic pro-
duction systems.
****

Corporacion Mexicana De Impresion S.A. de
C.Y.
General Victorian·o Zepeda 22, Col.
Observatorio
Delegacion Miguel Hidalgo, c.P.
City: Mexico D.E
Postal Code: 11840
Country: Mexico
Voice Phone: (52) 5 273-5583
Fax Phone : (52) 5 272-2916
Email: bety@fenix.ifisicocu.unam.mx
Contact #1: Salvador Nava-Calvillo
Description: Comisa is Mexico City government's
printing company. We have installed a holographic
production line in order to use holograms for high
securi ty purposes in all out official documents artd
certificates.
****

Creative Holography Index, The
The International Catalog for Holography
46 Crosby Road
City, Province: West Bridgford, Nottingham
State: England
Postal Code: NG2 5GH
Country: United Kingdom
Voice Phone: (44) 7050 133 624
Fax Phone: (44) 7050 133 625
Email: pepper@monand.demon.co.uk
Web: http ://www.holo.comlpeper/search.html
Contact #1: Andrew Pepper
Description: The Creative Holography Index is an in-
ternational catalogue, in colour, and includes an artist
produced hologram. Available as the complete collec-
tion. It features artists working with holography as a
creative medium and includes critical essays, biogra-
phies, statements and a hologram. Cost 89 .95 Stlg.
****

Creative Label
2450 Estes Drive
City: Elk Grove Village
State: IL
Postal Code: 60007
Country: United States of America
Voice Phone: (1) 847 956 6960
Fax Phone: (1) 847 956 8755
Contact #1: Jerry Koril
Description: Full range decorative graphic fmishers.
Large volume capability. Bindery application of ho-
lograms to paper, cardboard, and plastics. Kluge (2
stream) and Bobst (4 stream) machines. Call for more
information.
****

Crown Roll Leaf, Inc.
91 Illinois Ave.
City: Paterson
State: NJ
Postal Code: 07503
Country: United States of America
Voice Phone: (1) 973 742 4000
Voice Phone: (1) 800 631 3831
Fax Phone: (1) 973 742 0219

Email: sales@crownrollleaf.com
Web: http://www.crownrollleaf.com
Contact #1: Kathy Kassover
Description: Crown Roll Leaf is a major manufacturer
of hot stamp foils suited for holographic applications. In
addition, our in house production facilities are capable
of full origination through finish ing and converting.
SEE OUR ADVERTISEMENT
****

Customer Service Instrumentation
7 Meadowfield Park South
City, Province: Stocksfield, Northumberland
State: England
Postal Code: NE43 7QA
Country: United Kingdom
Voice Phone: (44) 1661 842 741
Fax Phone: (44) 1661 842 288
Email: ghscott@netcom.co.uk
Contact #1: G.H. Scott
Description: Manufacture front surface mirrors and
optics for holography.
****

CVI Laser Corporation
200 Dorado Place
City: Albuquerque
State: NM
Postal Code: 87 123
Country: United States of America
Voice Phone: (1) 505 296 9541
Fax Phone: (1) 505 298 9908
Email: cvi@cvilaser.com
Web: http ://www.cvilaser.com
Contact # 1: Bob Soales
Description: Manufactures holographic quality single
and multiple element lenses, mirrors, windows, and
beamsplitters for all standard holographic laser sourc-
es. Free 104-page catalog available.
****

D. Brooker & Associates
Rt. 1, Box 12A
City: Derby
State: Iowa
Postal Code: 50068
Country: United States of America
Voice Phone: (1) 5155332103
Fax Phone: (I) 515 533 2104
Email: dbrooker@netins.net
Contact # 1: Dennis Brooker
Description: NEW kit - Enables user with inkjet or
laser printer to apply images and text to holographic
vinyls. Kit includes materials, instructions and patterns
- make bus & greeting cards, labels, ornaments and
much more! Call for details!
****

Dai Nippon Printing Co., Ltd.
Central Research Institute
250-1 Wakashiba
City: Kashiwa-City, Chiba
Postal Code: 277-0871
Country: Japan
Voice Phone: ((8 1» 471 34-0512
Fax Phone: ((8 1» 471 33-2540
Emai l: Hotta-T@mail.dnp.co.jp
Contact # 1: Tsuyoshi Hotta
Description: Volume type holograms such as color ho-
lograms and holographic optical elements. Embossed
holograms for security application.
SEE OUR ADVERTISEMENT AND
THE HOLOGRAM ON FRONT COVER.
****

Daimler Benz Aerospace
Dornier Medizintechnik GmbH
Industriestrasse 15
City: Germering
Postal Code: D-82 110
Country: Germany
Voice Phone: (49) (0)89 84108 0

Fax Phone: (49) (0)89 84108 575
Contact #1: Ms. Thiemon
Description: Industrial Research; holographic
non-destructive testing. HOE research .
****
Dan Han Optics
188-261 An Nyeong-Ri Tean-Eup
City, Province: Hwasong-Gun, Kyung Ki-Do
Country: Korea
Voice Phone: (82) 0331 351 030
Fax Phone: (82) 0331 351 031
Contact # 1: Chung Song
Description: General optical supplies.
****
Data Optics, Inc.
115 Holmes Road
City: Ypsilanti
State: MI
Postal Code: 48198-3020
Country: United States of America
Voice Phone: (1) 800 321-9026
Voice Phone: (1) 734 483-8228
Fax Phone: (1) 734 483-9879
Email: webmaster@dataoptics.com
Web: www.dataoptics.com
Contact # 1: David Shindell
Description: For over 30 years, Data Optics has manu-
factured precision optical research equipment for de-
manding applications at competitive prices. We lead
the industry in quality with active fringe control sys-
tems, environmental isolation enclosures, spatial fil-
ters, hologram and x-y plate holders, tablerails, optical
benches, carriers, liquid gates and film drives. Best of
all, we work closely with our customers to satisfY their
special needs, using custom, semi-custom and standard
components. Let us work with you.
SEE OUR ADVERTISEMENT
****
Datacard Corporation
11111 Bren Road West
City: Minneapolis
State: MN
Postal Code: 55343
Country: United States of America
Voice Phone: (1) 612 933 1223
Voice Phone: (1) 800 621 6972
Fax Phone: (1) 612 931 0418
Email: info@datacard.com
Web: http: //www.datacard.coml
Contact #1: Mark Iverson
Description: Security and authentication applications
utilizing holographic technologies. Capable of high
volume runs for government and commercial users.
****
Datasights Ltd.
Alma Road
Ponders End
City, Province: Enfield, Middlesex
State: England
Postal Code: EN3 7BB
Country: United Kingdom
Voice Phone: (44) 181 8054151
Fax Phone: (44) 181 805 8084
Email: dsights@netcom.uk.co
Contact #1: Frank Sharpe
Description: Manufacture mirrors for use in hologra-
phy. Beamsplitters and gratings also available.
****
David Dann Modelmaking Studios
PO Box 396, 4 East Hill Rd.
City: White Sulphur Springs
State: NY
Postal Code: 12787
Country: United States of America
Voice Phone: (1) 914 292 1679
Fax Phone: (1) 914 292 1679
Email: davidann@catskills.net

Web: www.fastwww.comldaviddanstudio
Contact #1: David Dann
Description: A maker of holographic models for more
than a decade, David Dann's work has been seen on
the covers of National Geographic, Omni, Marvel
and Malibu comics, and many, many other places.
Clients have included American Bank Note, Polaroid,
HoloGrafx, Blue Ridge Holography, and Bridgestone
Technologies. Brochure and samples avail.
****
De La Rue Holographics
Stroudley Road
Daneshill Industrial Estate
City, Province: Basingstoke, Hampshire
State: England
Postal Code: RG24 8FW
Country: United Kingdom
Voice Phone: (44) 1256 463 000
Fax Phone: (44) 1256 460 800
Contact #1: Annette Kiely
Description: De La Rue Holographics, a division of De
La Rue International Ltd., offers customers high tech-
nology protection against product counterfeit and tam-
pering though security optical microstructures. It oper-
ates in two main markets: security products and brand
protection.
****
Decolux GmbH
Verdistrasse 7
City: Munich
Postal Code: D-81247
Country: Germany
Voice Phone: (49) (0)89 8112044
Voice Phone: (49) (0)89 8112045
Fax Phone: (49) (0)89 8118582
Contact # 1: Horst Mairiedl
Description: 28 years self adhesive foils-16 years
effect-foils-12 years diffraction foils-6 years hologra-
phy-5 years holographic stamping foils . Contacts to
master hologram manufacturers and all important com-
panies in this field.
****
Deem, Rebecca
709 112 West Glen Oaks Blvd
City: Glengale
State: CA
Postal Code: 91202
Country: United States of America
Voice Phone: (1) 818 549 0534
Fax Phone: (1) 818 549 0534
Email: 100142.1543@compuserve.com
Contact 411: Rebecca Deem
Description: Holographic artist. Originates masters for
mass production holograms in embossed, DCG and
photopolymer materials. Both pulsed and CW lasers.
****
Deep Space Holograph~s
1337 Rockland Ave., Suite 10
City, Province: Victoria, British Colombia
Postal Code: V8S IV4
Country: Canada
Voice Phone: (1) 250 384 3927
Email: eyetrek@islandnet.com
Contact # 1: Marc de Roos
Description: Exotic fine artIcommercial sculpture/ani-
mation, conceptual/industrial design, display merchan-
dising, exhibits and special effects. Since 1980 secured
worldwide distribution of our DCG designs via Holo-
crafts, including Star Trek holograms design.
****
DeFreitas Holography Studio
815 Allen Street
City: Allentown
Da-De
State: PA
Postal Code: 18102
Country: United States of America

Voice Phone: (1) 800 458 3525
Fax Phone: (1) 800 458 3525
Email: director@holoworld.com
Web: http: //www.holoworld.com
Contact #1 : Frank DeFreitas
Description: A full service holography studio family
owned and operated since 1983.
****
Dell Optics Company, Inc.
25 Bergen Blvd.
City: Fairview
State: NJ
Postal Code: 07022
Country: United States of America
Voice Phone: (I) 201 941 1010
Fax Phone: (I) 201 941 9524
Contact # 1: Belle Steinfeld
Description: Custom working of precision optical
components. Established 1950. 15 Employees at this
address.
****
Denisyuk, Yuri N.
A.F.loffe Physicotechnical Institute
Politechnicheskaya 26
City: St. Petersburg
Postal Code: 194021
Country: Russia
Voice Phone: (7) 812 247 9384
Contact # 1: Yuri N. Denisyuk
Description: Holography teacher. One of the founders
of holography.
****
Design + Kunst e.V Chernnitz
Theaterstrasse 27
City: Chernnitz
Postal Code: D-09111
Country: Germany
Voice Phone: (49) (0)371 642723
Email: i.bruhn@abo.freiepresse.de
Contact #1: Prof. Ines Bruhn
Description: artistic society
****
Deutsche Gesellschaft fur Holografie e. V.
Marienstrasse 28
City: Halle
Postal Code: D-06108
Country: Germany
Voice Phone: (49) (0)345 2026751
Fax Phone: (49) (0)345 2026752
Email: nimoe@burg-halle.de
Web: http://www.burg-halle.de/-nimoe/
Contact # 1: Niklas Moller
Description: The society was founded to promote
awareness of holography-and its members are mainly
holographers and artists. To this end-the group intends
to organize exhibitions. Interferenzen is a periodical
published by this organization.
SEE OUR ADVERTISEMENT
****
Deutscher Drucker Verlagsgesellschaft mbH
& Co. KG
Senefelderstrasse 12
City: OstfildemlRuit
Postal Code: D-73760
Country: Germany
Voice Phone: (49) (0)711 442096
Voice Phone: (49) (0)711 442098
Fax Phone: (49) (0)711 442099
Contact #1: Theodor J. Anton
Description: Publisher of worldwide journals
for the printing industry. Deutscher Drucker
42 issues/year-distributed in more than 50
countries. World-Wide Printer 6 issues/yeardistributed
in 158 countries. Supplement EI Arte Tipografico in 34
countries. Some holography related articles with real
holograms.

****

Diamond Images, Inc.
P.O. Box 170 I
City: Miami
State: FL
Postal Code: 33133
Country: United States of America
Voice Phone: (1) 305 854 4656
Voice Phone: (1) 305 323 8406
Fax Phone: (1) 305 854 6965
Email: mark@Diamondlmages.com
Web: http://www.Diamondlmages.com
Contact #1: Mark Diamond
Description: Holographer Mark Diamond brings 24 years experience to full color stereograms. Work is featured in museums and collections in 15 countries. Founding member of Museum of Holography New York. Specializing in portraiture and animated digital compositing.

****

Diastases Epee
Not Bistro I
City: Iranian
Postal Code: 45444
Country: Greece
Voice Phone: (306) 517 0658
Fax Phone: (306) 517 4034
Email: stylos@otenet.gr

****

Diavy sri
Via Vivaldi 108
City: soli era (Modena)
Postal Code: 1-41019
Country: Italy
Voice Phone: (39) 59 565758
Fax Phone: (39) 59 566074
Description: Subsidiary of Diaures; producer of holographic metallic paper as part of a venture with Scharr Industries, USA.

****

Die Dritte Dimension
Frankfurter Strasse , 132-134
City: Neu Isenburg
Postal Code: D-63263
Country: Germany
Voice Phone: (49) (0)610 33367
Fax Phone: (49) (0)610 326709
Contact #1: Elke Hein
Description: Always over 1000 different holograms in stock. Very comprehensive fine art section. Branch office: Nordwest Zentrum, Tituscorso, 60439 Frankfurt/M. Germany.

****

Dietmar Oehlmann
Mergesstrasse 16
City: Braunschweig
Postal Code: D-38108
Country: Germany
Voice Phone: (49) (0)531 352 816
Fax Phone: (49) (0)531 352 816
Email: holodietma@aol.com
Web: www.holonet.khm.de/holodietma/od/index.htm
Contact # 1: Dietmar Oehlmann
Description: Master of Arts in Holography from the Royal College of Arts-with own light creation lab to design and produce special effects in holography for artworks-performances and stage decoration.

****

Diffraction Ltd.
P.O. Box 909
Route 100
City: Waitsfield
State: VT
Postal Code: 05673
Country: United States of America
Voice Phone: (1) 802 496 6642

Fax Phone: (1) 802 496 6644
154 Holography MarketPlace - 8'" Edition
Email: hologram@madriver.com
Web: http ://www.diffraction.com
Contact # 1: Bill Parker
Description: Products and services relating to diffractive optics and holographic optical elements (HOEs) including micro fabrication and photomask production.

****

Digital Matrix Corp.
67 Whitson Street
City: Hempstead
State: NY
Postal Code: 11550
Country: United States of America
Voice Phone: (1) 516 481 7990
Fax Phone: (1) 516481 7320
Email: digmat@galvanics.com
Web: http ://www.ga~vanics.com
Contact #1: Alex Greenspan
Description: Manufacturers of high precision computerized electroforming systems for the production of nickel embossing shims for the holography industry. Turnkey systems, training and consultation.
SEE OUR ADVERTISEMENT

****

Dilas Diodenlaser GmbH
Galileo Galilei Strasse 10
City: Mainz
Postal Code: D-55129
Country: Germany
Voice Phone: (49) (0)61 31 92,26-0
Fax Phone: (49) (0)6131 9226-55
Contact # 1: B. De Odorico
Description: Manufacturer of laser diodes.

****

Dimension 3
3380 Francis-Hughes St.
City, Province: Laval, Quebec
Postal Code: H7L 5A 7
Country: Canada
Voice Phone: (1) 514 662 0610
Fax Phone: (1) 514 662 0047
Email: pierre@dimension3 .net
Web: http://www.dimension3.net
Contact #1: Pierre Gougeon
Description: We offer creative solutions to holographic projects. We are a full holographic production house (DCG, foil , transmission, photopolymer), large format and micro embossed with animation and colour control Photograms (TM) (lenticular photography/ printing).

****

Dimensional Arts
40 I Carver Road
City: Las Cruces
State: NM
Postal Code: 88005
Country: United States of America
Voice Phone: (1) 505 527 9183
Fax Phone: (1) 505 527 9927
Email: arts@holo.com
Web: http ://www.holo.com
Contact #1: Ken Harris
Description: Exclusive manufacturer of the Light Machine, a patent protected digital origination system. Custom stock dot matrix patterns available. Capable of 2D, 3D and full color stereogram work. Can transfer technology worldwide.
SEE OUR ADVERTISEMENT

****

Dimensional Foods Co.
8 Faneuil Hall Market Place
City: Boston
State: MA
Postal Code: 02109
Country: United States of America
Voice Phone: (1) 617 973 6465

Fax Phone: (1) 617 973 6406
Email: holo@lightvision.com
Web: http://lightvision.coml
Contact #1: Erich Begleiter
Description: Scientific and artistic research. Licensing a proprietary "micro relief" process to food manufacturers for producing chocolate and hard candy holograms.

****

Dimensions
Taj Pura
City: Sialkot
Country: Pakistan
Voice Phone: (92) 432 85197
Voice Phone: (92) 432 66006
Fax Phone: (92) 432 558336
Contact #1: Mr. Shahjahan
Description: International agents and importers of holograms, diffraction foils and other holographic products.

****

Dimuken (GB)
33 Stapledon Rd
Orton Southgate
City: Peterborough
State: England
Postal Code: PE2 6TD
Country: United Kingdom
Voice Phone: (44) 1733 230 044
Fax Phone: (44) 1733230012
Email: sales@dimuken.demon.co.uk
Web : www.dimuken.demon.co.uk
Contact # 1: John Bentley
Description: Manufactures holographic hot stamping machinery which can do hot stamping or blind embossing by switching machinery components.

****

Directa GmbH
Hammer Strasse 40
City: Munster
Postal Code: D-48153
Country: Germany
Voice Phone: (49) (0)251 521551
Voice Phone: (49) (0)251 521411
Contact # 1: Ute Schulze
Description: Shop (holography and gifts)wholesale of holograms and accessories.

****

DLR e.V
Energietechnik und Solarchemie
Linder Hohe
City: Cologne
Postal Code: D-51147
Country: Germany
Voice Phone: (49) (0)2203 6012868
Fax Phone: (49) (0)2203 66900
Email: luepfert@dlr.de
Contact # 1: Eckhard Lupfert
Description: Using HOEs for high efficiency solarpanels.

****

Doris Vila Holographics
445 Grand Street
City: Brooklyn
State: NY
Postal Code: 11211
Country: United States of America
Voice Phone: (1) 718 388 6533
Fax Phone: (1) 718 388 6533
Email: dorvila@earthlink.net
Contact #1: Doris Vila
Description: Custom holography in state-ofthe- art inhouse lab, silver halide limited editions, architectural-scale & fine-art originals, mastering and transfers, consultations and classes available by appointment.

****

Dornier Medizintechnik GmbH
Industriestrasse 15

City: Germering
Postal Code: 0-82110
Country: Germany
Voice Phone: (49) (0)89 84108 0
Fax Phone: (49) (0)89 84108 575
Contact # 1: W. Langer
Description: Industrial Research; holographic non-destructive testing. HOE research. Medical systems in Lithotripsy-Surgery-Orthopaedics.
****

Dorra, Bodo
Zieblandstrasse 31II
City: Munich
Country: Germany
Voice Phone: (49) (0)89 2711720
Fax Phone: (49) (0)89 2711720
Description: Author of articles about applied holography.
****

Dri-Print Foils
(see API Foils)
****

Dt. Gesellschaft fiir Elektronenmikroskopie
eV
c/o MPI fur Kohlenforschung Abtl. EM
Kaier Wilhelm Platz I
City: Miihlheirnl Ruhr
Country: Germany
Voice Phone: (49) (0)208 306-2130
Fax Phone: (49) (0)208 306-2980
Contact # 1: Dr. Bernd Tesche
Description: Association.
****

Dt. Gesellschaft fur Angewandte Optik - DGaO
c/o Carl Zeiss Jena GmbH
Tatzenpromenade I a
City: Jena
Postal Code: 0-07740
Country: Germany
Voice Phone: (49) (0)3641 64-2304
Email: dgao@zeiss.de
Web: www.iof.fhg.de
Contact #1: Prof. Dr. Theo Tschudi
Description: Association.
****

Dt. Gesellschaft fur Stereografie - DGS
Kurt Schumacher Ring 50
City: Bruchk6bel
Postal Code: D-63486
Country: Germany
Voice Phone: (49) (0)6181 740904
Fax Phone: (49) (0)6181 740904
Email: dgs@t-online.de
Contact #1: Jiirgen Horn
Description: Organizer of 3D workshops, meetings.
****

Dt. Physikalische Gesellschaft e.V
Magnus Haus
Am Kupfergraben 7
City: Berlin
Postal Code: D-I0117
Country: Germany
Voice Phone: (49) (0)30 201748-0
Description: Association.
****

DuPont (E.I. DuPont De Nemours & Co.)
Holographic Materials Division
p. O. Box 80352
City: Wilmington
State: DE
Postal Code: 19880-0352
Country: United States of America
Voice Phone: (1) 302 695 4893
Fax Phone: (1) 302 695 9631
Email: paulafreeman.
bobeck@usa.dupont.com
Web: http ://www.dupont.com
Contact #1: Paula Bobeck

Contact #2: Andrew Weber
Description: Manufacturer of photopolymer emulsions for sale to holography businesses.
SEE OUR ADVERTISEMENT
****

Dutch Holographic Laboratory BV
Kanaaldijk Noord 61
City: Eindhoven
Postal Code: NL-5642JA
Country: Netherlands
Voice Phone: (31) 40 281 7250
Fax Phone: (31) 40 281 4865
Email: Spierings@holoprint.com
Web: http ://www.holoprint.com
Contact # 1: Walter Spierings
Description: Manufacturer of Holoprinter and Holo-track equipment. Production of holograms on silver halide, photoresist and photopolymer. Computer-generated holograms and multiple photo-generated holograms (MPGH). Also traditional recording techniques.
SEE OUR ADVERTISEMENT
****

Eastman Kodak Company
343 State St.
City: Rochester
State: NY
Postal Code: 14650-0811
Country: United States of America
Voice Phone: (1) 800 242 2424
Voice Phone: (1) 800 823 4474
Fax Phone: (1) 800 755-6993
Description: Manufacturer of silver halide recording materials. Glass plates & film.
****

Edmund Scientific Company
101 East Gloucester Pike
City: Barrington
State: NJ
Postal Code: 08007
Country: United States of America
Voice Phone: (1) 609 573 6250
Fax Phone: (1) 609 573 6295
Email: scientifics@edsci.com
Web: http ://www.edsci.com
Contact # 1: John Stack
Description: Mail-order catalogue, wholesale, and retail. We offer one of the largest selections of precision optics and optical components and accessories for the optical lab.
****

El Don Engineering
4629 Platt Rd.
City: Ann Arbor
State: MI
Postal Code: 48408
Country: United States of America
Voice Phone: (1) 313 973 0330
Email: eldonlaser@aol.com
Contact #1: Don Gillespie
Description: Surplus and refurbished lasers for the holographer. Full warranty. Technical request calls welcomed.
SEE OUR ADVERTISEMENT
****

Electro Optical Industries, Inc.
859 Ward Drive
City: Santa Barbara
State: CA
Postal Code: 93111
Country: United States of America
Voice Phone: (1) 805 964 670 I
Fax Phone: (1) 805 967 8590
Contact # 1: Steve White
Description: Manufacturer of infrared test and calibration instrumentation including: collimators, choppers, blackbody sources, differential temperature sources, FLIR test equipment, radiometers, LLL-TV target simulators.
****

Electro Optics Developments Ltd.
Howards Chase
Pipps Hill Industrial Estate
City, Province: Basildon, Essex
State: England
Postal Code: SSI4 3BE
Country: United Kingdom
Voice Phone: (44) 1268 531 344
Fax Phone: (44) 1268 531 342
Contact #1 : Chris Varney
Do-En
Description: Custom optics, HOEs, gratings.
****

Electro-Optics Lab, NECTEC
King Mongkut's Institute of Technology
Chalongkrung Road, Ladkrabang
City: Bangkok
Postal Code: 10520
Country: Thailand
Voice Phone: (66) 2 326 9045
Fax Phone: (66) 2 326 9045
Email: fkh@nwg.nectec.or.th
Description: A national lab that is also Thailand's first hologram manufacturer. Produces embossed holograms and photopolymer holograms. Provides service in training, consulting and hologram mastering. Also conducts academic research in holography, photonics, optoelectronics.
****

Embossing Technology Ltd
Steepmarsh, Nr Petersfield
City: Hants
State: England
Postal Code: GU32 2BN
Country: United Kingdom
Voice Phone: (44) 1730895 390
Fax Phone: (44) 1730 894 383
Description: Wide web embossing by contract. Also stock images. Also for sale is complete system for originating embossed holograms including laser.
****

Empaques y Envolturas Holograficas, S.A.
de C.V
Pino 343 locales 42-42
Col. Atlampa
City: 06450 Mexico City
State: D.F.
Country: Mexico
Voice Phone: (52) 5 5472033
Voice Phone: (52) 5 5471983
Fax Phone: (52) 5 5410719
Email: eehdir@holomex.com.mx
Web: http :/ www.eeh.com.IDX
Contact #1: Ramon Bautista
Description: Established in 1990, the company is a producer of continuous forms with or without holographic applications such as hot-stamping, automatic labeling, bar coding, etc. The company also has a labels flexographic division for overprinting on holographic or non-holographic substrates. The company specializes in government forms, Bank Checks, Credit Cards, etc.
****

Engineering Animation, Inc.
2321 N. Loop Drive
City: Ames
State: IA
Postal Code: 50010
Country: United States of America
Voice Phone: (1) 515 296 9908
Fax Phone: (1) 515 296 7025
Contact # 1: Brad Shafer
Description: EAI develops, produces and sells 3D animation products that address visualization, animation and graphics needs of its customers. Products include: 3D interactive software titles on CD ROM; animation software (UNIX); and custom 3D computer animations.
****

INTERNATIONAL BUSINESS DIRECTORY Ep - Fi

EPA - Elektro-Physik Aachen GmbH
Jiilicher Strasse 336-340
City: Aachen
Postal Code: D-52070
Country: Germany
Voice Phone: (49) (0)241 531778
Fax Phone: (49) (0)241 1822100
Contact #1: Eva Schulze Brockhausen
Description: EPA delivers HOEs-dichromated holograms for advertisement-processing to jewelry.
****

es - Lasersysteme D. Baur
Neue Rottenburger Strasse 37
City: Hechingen
Postal Code: D-72379
Country: Germany
Voice Phone: (49) (0)7471 9166-1
Fax Phone: (49) (0)7471 9166-6
Email: ml@es-laser.de
Description: Distributor of optical and mechanical holographic equipment.
****

ETA-Optik Gmbh
Niethausener Strasse 15
City: Heinsberg
Postal Code: D-52525
Country: Germany
Voice Phone: (49) (0)245 66654
Fax Phone: (49) (0)245 64433
Contact # 1: Wilbert Dr. Windeln
Description: DCG pendants-diffraction gratings and custom HOEs.
****

ETH - Eidgenossische Technische Hochschule
Laboratorium fur physikalische Chemie
Universitiitstrasse 22
City: Ziirich
Postal Code: CH- 8092
Country: Switzerland
Voice Phone: (41) (0)1 632-4381
Voice Phone: (41) (0)1 632-4381
Fax Phone: (41) (0)1 632-1021
Email: wild@phys.chem.ethz.ch
Contact # 1: Urs. P. Wild
****

Ettemeyer GmbH & Co
Memminger Strasse 72 / 207
City: Neu - Ulm
Postal Code: D-89231
Country: Germany
Voice Phone: (49) (0)731 850-31
Fax Phone: (49) (0)731 850-33
Email: sales@ettemeyer.de
Description: Industrial holographic interferometry.
****

Evergreen Laser Corp.
9G Commerce Circle
City: Durham
State: CT
Postal Code: 06422
Country: United States of America
Voice Phone: (I) 860 349 1797
Fax Phone: (I) 860 349 3873
Contact # 1: Cheryl Smith
Description: Repair service & systems sales of Argon, Krypton, Mixed gas, Helium Neon, Xenon & Carbon Dioxide lasers. Full technical support for all systems, including those no longer supported by the original manufacturer.
****

Excalibur Engineering
1260 North 200 East #2
City: Logan
State: UT
Postal Code: 84341
Country: United States of America
Voice Phone: (1) 435 755 9221
Fax Phone: (1) 4357559321

Description: Fringe stabilizers and HOEs.
SEE OUR ADVERTISEMENT
****

Excitek Inc.
320 Mt. Pleasant Ave.
City: Newark
State: NJ
Postal Code: 07104
Country: United States of America
Voice Phone: (1) 973 483-2415
Fax Phone: (1) 973 482-2274
Email: excitek.inc@gte .net
Contact #1: George Cubberly
Description: Supplier of re-manufactured argon and krypton ion laser tubes, and used laser systems. Established in 1984. 5 Employees at this address.
****

Expanded Optics Limited
Moon Lane
City, Province: Barnet, Hertfordshire
State: England
Postal Code: EN5 5ST
Country: United Kingdom
Voice Phone: (44) 18] 441 2283
Fax Phone: (44) 18] 4496143
Email: info@expandedopticsltd.com
Web: www.expandedopticsltd.com
Contact # 1: T.R. Hollinsworth
Description: Manufacturer of medical and industrial endoscopes; HOEs used in microprecision optics for medical viewing.
****

F & E Labor
Dieding 7
City: Ebersberg
Postal Code: D-85560
Country: Germany
Voice Phone: (49) (0)8092 83541
Contact # 1: Frank Knocke
Description: Constructional engineering in optics and mechanics.
****

Fachhochschule St. Polten
Herzogenburgerstrasse 68
City: St. Pol ten
Postal Code: ACountry:
Austria
Voice Phone: (43) (0)2742 313228
Email: direktion@fh-stpoelten.ac.at
Contact #1: Prof. Dr. Werner Sobotka
Description: Courses in holography.
****

Fachhochschule Ulm a IITA
Schwambergerstrasse 35
City: Ulm
Postal Code: D-89073
Country: Germany
Voice Phone: (49) (0)731 502-8187
Fax Phone: (49) (0)731 502-8258
Email: vapetrov@bild.lab.fh-ulm.de
Contact # 1: Prf. Dr. B. Lau
Description: R&D for new holographic photochemical processes.
****

Fantastic Holograms
P.O. BOX 492026
8400 Pena Blvd. (DIA Terminal - Level 5)
City: Denver
State: CO
Postal Code: 80249
Country: United States of America
Voice Phone: (1) 303 342 3440
Fax Phone: (1) 303 342 3440
Contact #1: RB Osada
Description: Well stocked holography store selling a variety of unique giftware including holographic pictures, jewelry, executive gifts, books, and optical novelties.

****
Far East Holographics
12/F Hang Wai Commercial Bldg
231-233 Queen's Road East
City, Province: Wanchai, Hong Kong
Country: China
Voice Phone: (852) 2 893 9773
Fax Phone: (852) 2 893 0640
Contact # 1: Adrian 1. Halkes
Description: Finisher and distributor of holograms and holographic products.
****

Feinmechanische Optische Betriebs GmbH
Karolingerstrasse 45
City: Salzburg
Postal Code: A- 5020
Country: Austria
Voice Phone: (43) (0)662 832320
Description: Manufacturer of optical and mechanical holographic equipment.
****

Feofaniya Ltd.
P.O. Box 164
City: Kiev
Postal Code: 252191
Country: Ukraine
Voice Phone: (380) 44 261 4343
Voice Phone: (380) 044 261 4343
Fax Phone: (380) 044 261 4343
Email: eeic@gluk.apc.org
Contact # 1: Sergey Komienko
Description: Non destructive testing, pulse holography, embossing & shim making, production of holography stickers. Holography portraits studio.
****

Feroe Holographic Consulting
1420 45th Street #33
City: Emeryville
State: CA
Postal Code: 94608
Country: United States of America
Voice Phone: (1) 510 658 9787
Fax Phone: (1) 510 658 9788
Email : jferoe@dnai.com
Web: www.dnai.coml-jferoe/
Contact # 1: James F eroe
Description: Consultant with 20 years handson experience in holography: silver-halide reflection and transmission, photoresist and embossed. Will travel to your site.
****

FhG Fraunhofer Gesellschaft
fur graphische Datenverarbeitung
Wilhelminenstrasse 7
City: Darmstadt
Postal Code: D-64283
Country: Germany
Voice Phone: (49) (0)6151 55-299
Email: ferri@igd.fhg.de
Contact #1: Lucilla Croce Ferri
Description: R&D in CGH.
****

FhG Fraunhofer Gesellschaft
ILT Institut fur Lasertechnik
S teinbachstrasse 15
City: Aachen
Postal Code: D-52074
Country: Germany
Voice Phone: (49) (0)241 8906-207
Fax Phone: (49) (0)241 8906-121
Email: puetz@ilt.fhg.de
Contact # 1: Dr. Reinhard Noll
Description: R&D in holographic interferometry.
****

Fiber Engineering
106 Stratford Way
City: Signal Mountain
State: TN

Postal Code: 37377-2521
Country: United States of America
Voice Phone: (1) 423 886 3783
Fax Phone: (1) 423 886 7865
Email: mmakansi@aol.com
Contact #1: Munzer Makansi
Description: Patent pending process for embossing dynamic rainbow and hologram images directly on fabrics without lamination. These holographic fabrics retain breathability and other fabric attributes. The holographic colors are not adversely affected by handling, washing and dyeing and they remain visible when wet. This technology can replace or supplement traditional dye and print coloring. Looking for funding and/or partners to assist commercialization.
****

Fielmann-Verwaltung KG
Weidestrasse 118a
City: Hamburg
Postal Code: D-22083
Country: Germany
Voice Phone: (49) (0)40 27076 0
Fax Phone: (49) (0)40 27076 410
Contact #1: Mrs. Nussbaum
Description: Optician collector of holograms.
****

FilmoTec GmbH
Areal A a Rontgenstrasse Geb. 415
City: Wolfen
Postal Code: D-06766
Country: Germany
Voice Phone: (49) (0)3494 369 -683
Fax Phone: (49) (0)3494 369 -682
Email: pshcg@stud.com.urz.uni-halle.de
Contact # 1: Dr. Rainer Redmann
Description: former East Germany Agfa company ãORWOÒ. Producer of silver halide holographic material.
****

Fink Feinoptik
Hammermiihlweg 18
City: Rehau
Postal Code: D-95111
Country: Germany
Voice Phone: (49) (0)9283 2762
Fax Phone: (49) (0)9283 2762
Contact #1: Detlef Fink
Description: handmade optical and mechanical elements.
****

Fisher Scientific
Educational Materials Division
485 South Frontage Road
City: Burr Ridge
State: IL
Postal Code: 60521
Country: United States of America
Voice Phone: (1) 800 955 II 77
Voice Phone: (1) 630 655 4410
Fax Phone: (1) 630 655 4335
Web: http://fisheredu.com
Description: Supplies science lab equipment including holography kits, lab manuals, lasers and laser related equipment.
****

FLEXcon
I FLEXcon Industrial Park
City: Spencer
State: MA
Postal Code: 01562-2642
Country: United States of America
Voice Phone: (1) 508 885 8200
Fax Phone: (1) 508 885 8400
Email: staff@flexcon.com
Web: http://www.flexcon.com
Contact #1: John Pannace
Description: Manufacturer of holographic and prismatic materials used for authentication and decoration. Holograms can be combined with overt and covert security features to provide unique solutions to graphic films, packaging and security applications. Wide web embossing in excess of 60 inch width.
****

FlexSystems USA
8517 Earhart Road, Suite 160
City: Oakland
State: CA
Postal Code: 94621
Country: United States of America
Voice Phone: (1) 510 635 0545
Fax Phone: (1) 510 635 0565
Email: dianec@flexsystems.com
Web: www.flexsystems.com
Contact #1: Diane Chapman
Description: Business takes photopolymer holograms and attaches them to rubber backing. Final product can be used as tags which you can use for zipper pulls, hat straps, sewn on clothes, etc.
****

Flight Dynamics
16600 SW 72nd Ave.
City: Portland
State: OR
Postal Code: 97224
Country: United States of America
Voice Phone: (1) 503 684 5384
Fax Phone: (1) 503 684 0169
Web: www.fltdyn.com
Description: Manufacturer of HOEs and HeadUp Displays.
****

Floating Images, Inc.
95 Post Avenue
City: Westbury
State: NY
Postal Code: 11590
Country: United States of America
Voice Phone: (I) 516 338 5000
Fax Phone: (1) 516 338 5008
Email: info@floatingimages.com
Web: http://www.floatingimages.com
Contact # 1: Gene Dolgoff
Description: Floating Images, Inc. has developed the software and hardware for a new, patent pending, ""floating 3D, off-the-screenexperience:'" (non-holographic) display technology. Floating Images actually produces images at different depths on any display, such as CRT and LCD, for television, computer, projection, and other formats .
****

Focal Image Ltd.
2 St John's Place
City: London
State: England
Postal Code: ECIM 4DE
Country: United Kingdom
Voice Phone: (44) 171 250 1101
Fax Phone: (44) 17l 2S0 3750
Email: kaveh@focal.demon .co.uk
Web: http: //www.focalimage.com
Contact #1: Kaveh Bazargan
Description: Research and development in holographic recording and display systems. Consultant on holographic projects.
****

Foil Stamping and Embossing Association
P.O. Box 12090
City: Portland
State: OR
Postal Code: 97212
Country: United States of America
Voice Phone: (1) 503 331 6221
Fax Phone: (I) 503 331 6928
Email: fseamail@aol.com
Web: http ://www.fsea.com
Fi - Fo
Contact # 1: Heather Wade
Description: A non-profit international trade association of the foil stamping, embossing, die cutting and other graphic finishing industries. Its purpose is to develop a cohesive alliance within the trade for the advancement of the entire finishing industry.
****

Foilmark Holographic Images
(a division of Foilmark, Inc.)
5 Malcolm Hoyt Drive
City: Newburyport
State: MA
Postal Code: 01950
Country: United States of America
Voice Phone: (I) 978 462 7300
Voice Phone: (1) 800468 7826
Fax Phone: (1) 978 462 0831
Email: flmkmgr@aol.com
Web: http://www.foilmark.com
Contact # 1: John Halotek
Description: FoilMark Holographic Images, a division of Foilmark, Inc. , is a manufacturer of diffraction embossed films. These films are printable and are available in many different mediums; for example, unsupported film, film laminated to paper or board, pressure sensitive material, and hot stamping foils.
****

Fong Teng Technology
No 41 , Lane 63 , Hwa Chen Road
City: Hsin Chuang, Taipei
Country: Taiwan
Voice Phone: (886) 2 2 998 4760
Fax Phone: (886) 2 2 992 1240
Contact #1: Mark Chiang
Description: 60 inch hologram and dot-matrix pattern foil manufaCturer, service from origination to finished product.
****

Foreign Dimension (The)
The Peak Galleria
Level 2, Shops 29 & 42, The Peak
City, Province: Hong Kong, Hong Kong
Country: China
Voice Phone: (852) 2 849 6361
Fax Phone: (852) 2 541 60 II
Email: schvarzy@netvigator.com
Web: http ://www.dimension.com.hk
Contact # 1: Frederic Schvartzman
Description: Holography shop/showroom offering all varieties of holograms for sale to the public. We also offer holograms for sale wholesale to other businesses.
SEE OUR ADVERTISEMENT
****

Foreign Dimension (The)
190 I Manley Commercial Bldg.
367-375 Queen's Road, Central
City, Province: Hong Kong, Hong Kong
Country: China
Voice Phone: (852) 2 542 0282
Fax Phone: (852) 2 541 60 II
Email: schvarzy@netvigator.com
Web: http ://www.dimension.com.hk/
Contact # 1: Frederic Schvartzman
Description: Specialists in manufacturing all kinds of holographic and illusion products (watches, keyrings, etc.). If you are a hologram manufacturer, we can also make top quality products at unbeatable prices using your holograms'
SEE OUR ADVERTISEMENT
****

Fornari, Arthur David
195 Garfield Place
City: Brooklyn
State: NY
Postal Code: 11215
Country: United States of America

Voice Phone: (1) 71 89653956
Email: afornari@mindspring .com
Contact # 1: Arthur David Fornari
Description: Artistic holographer; silver halide transmission & reflection holograms.
****

Forth Dimension Holographies
2759 Helmsburg St.
City: Nashville
State: IN
Postal Code: 47448
Country: United States of America
Voice Phone: (1) 812 988 8212
Fax Phone: (I) 812 988 9211
Email: hologram4d@aol. com
Web: http://members.aol.comlhologram4d
Contact # 1: Rob Taylor
Description: U.S. Distributor for Slavich holographic emulsions. Full-line distributor of display holograms and other related holographic products. Hologram shop, gallery, & museum. Pulsed holographic studio & lab.
SEE OUR ADVERTISEMENT
****

Foshan Holosun Packaging Co. Ltd
Zhang-Cha-Zhen
Zhang-Cha-Yi-Lu, Yu-Dai-Kai-Fa-Qu
City, Province: Foshan, Guangdong
Postal Code: 528000
Country: China
Voice Phone: (86) 757 22 12368
Fax Phone: (86) 757 22 11 228
Contact # 1: Qin Yijun
****

Frank J. Deutsch Inc.
17 Spielman Road
City: Fairfield
State: NJ
Postal Code: 07004
Country: United States of America
Voice Phone: (1) 800 394 77 13
Fax Phone: (I) 973 808 1168
Description: High speed precision web presses for hot stamping bf holographic images, die cutting, lamination sheeting and rewinding of holograms.
****

Frank M. Schenker's Aquarius-Vertrieb
Crailsheimer Strasse I
City: Kirchberg/Jagst
Postal Code: D-74592
Country: Germany
Voice Phone: (49) (0)7954 222
Contact # 1: Frank M. Schenker
Description: Retail and wholesaler.
****

Free University Of Brussels
Faculty Of Applied Sciences
Alna-Tw Pleinlaan 2
City: Brussels
Postal Code: B-I 050
Country: Belgium
Voice Phone: (32) 2 629 3452
Fax Phone: (32) 2 629 3450
Contact # 1: Erik Styns
Description: Academic and Scientific research on diffractive elements and HOE's.
****

Fresnel Technologies Inc.
101 West Morningside Drive
City: Fort Worth
State: TX
Postal Code: 76110
Country: United States of America
Voice Phone: (1) 817 926 7474
Fax Phone: (I) 817 926 7 146
Email: info@fresneltech.com
Web: http ://www.fresneltech.com
Contact # 1: Linda H. Claytor

Description: Manufactures plastic Fresnel lenses & lens arrays from its POLY IR plastics for use into the infrared; also other optical products for use into the ultraviolet from acrylic & other plastics.
****

Fringe Research Holographies
Interference Hologram Gallery
1179A King Street West, Suite 010
City, Province: Toronto, Ontario
Postal Code: M6K 3C5
Country: Canada
Voice Phone: (I) 41 6 535 2323
Contact # 1: Michael Sowdon
Description: Gallery of Arti stic holography; silver halide holograms; pulse portraits; gallery; workshops; traveling exhibit.
****

Fuji Electric Co. Ltd
Mecatronics DivisIon
1-12-1 Yuraku-Cho Chiyoda-Ku
City: Tokyo
Postal Code: 100
Country: Japan
Voice Phone: ((81» 3 211 7 111 ..
Description: Manufactures C02 lasers and related equipment.
****

Fujitsu Laboratories Ltd.
Peripheral Systems Laboratories
10-1 Morinosato-Wakamiya
City: Atugi, Kanagawa
Postal Code: 2430124
Country: Japan
Voice Phone: ((8 1» 462 50-8821
Fax Phone: ((81» 462 48-3233
Email: nakashim@fl ab.fuj itsu.co.jp
Web: http://www.fujitsu.com
Contact #1: Masato Nakashima
Description: Research and development applications to computer i/o systems.
****

G. Franck OptroniK GmbH - GFO
Kiihnerstrasse 75 Haus 2
City: Hamburg
Country: Germany
Voice Phone: (49) (0)40 669622-0
Fax Phone: (49) (0)40 669622-30
Contact # 1: Gerhard Franck
Description: Distributor of optical equipment.
****

G.M. Vacuum Coating Lab, Inc.
882 Production Place
City: Newport Beach
State: CA
Postal Code: 92663
Country: United States of America
Voice Phone: (I) 949 642 5446
Fax Phone: (I) 949 642 7530
Contact # 1: Dan Coursen
Description: Custom manufacturing only, usually on your substrate . Will do coatings for front surface mirrors, beamsplitters, etc. for holographic use.
****

Galerie 3D
Goldbacher Strasse 31
City: Aschaffenburg
Postal Code: D-63 739
Country: Germany
Voice Phone: (49) (0)6021 26447
Contact #1: Stefan Merget
Description: Retailer of j ewelry, embossed holograms, and holograms up to 9 x 12 cm.
****

Galerie Illusoria
Schwarztorstrasse 70
City: Bern
Postal Code: CH-3 007
Country: Switzerland

Voice Phone: (41) 31 381 773 1
Fax Phone: (4 1) 31 381773 1
Contact # 1: Sandro del -Prete
Description: Gallery featuring holograms.
****

Galerie Illusoria
Schwarztorstrasse 70
City: Bern
Postal Code: CH-3007
Country: Switzerland
Voice Phone: (41) 31 381 7731
Fax Phone: (41) 31 381 773 1
Contact # 1: Sandro de l-Prete
Description: Gallery featuring holograms.
****

Galerie WesterlandiSylt
S trandstrasse
City: Westerland / Sylt
Postal Code: D-25980
Country: Germany
Voice Phone: (49) (0)465 1 21313
Contact #1: Martin Hofmann
Description: Gallery
****

Galvoptics Ltd.
Harvey Road
Burnt Mills Industrial Estate
City, Province: Basildon, Essex
State: England
Postal Code: SSl3 IES
Country: United Kingdom
Voice Phone: (44) 1268 728 077
Fax Phone: (44) 1268 590 445
Contact # 1: R. D. Wale
Description: Optics; mirrors, lenses.
****

GEHOL sarl
28 quai des Bateliers
City: Strasbourg
Postal Code: F-67000
Country: France
Voice Phone: (33) 88 52.17. 16
Fax Phone: (33) 88 52.17.44
Email : gehol@calvanet.calvacom.fr
Contact # 1: Jean-Luc Perreau
Description: I have an holography shop and a laboratory with a Denisyuk table.
****

General Design
2005 - 18th Street
City: San Francisco
State: CA
Postal Code: 94107
Country: United States of America
Voice Phone: (1) 415 550 9193
Email : bk@cronos.net
Web: http ://cronos.netl-bklgd/
Contact #1: Brian Kane
Description: Creative services - Computer graphics for print, video and holography. 3D computer modeling and 2D computer composition. General image design and construction.
****

General Holographics, Inc.
P.O. Box 82247
City, Province: Burnaby, British Columbia
Postal Code: V5C 5P7
Country: Canada
Voice Phone: (I) 604 685 6666
Voice Phone: (I) 800 667 9669
Fax Phone: (I) 604 685 6678
Email: bsimson@capcollege.bc.ca
Web: http ://www2.capcollege.bc.cal-bsimsonl
Contact # 1: Paula Simson
Description: Distributor of dichromate & embossed gift and jewelry items, silver halide wall and desk decor, and photopolymer for the Canadian market. Custom and stock.
****

General Optics Pty. Ltd.
PO Box 8
City, Province: Hove, SA
Postal Code: 5048
Country: Australia
Voice Phone: (61) 8 829 69708
Fax Phone: (61) 8 829 69708
Email: austholo@camtech.net.au
Description: Distributor of Siavich holographic emulsions.
****

Geola
P.O. Box 343
City: Vilnius
Postal Code: 2006
Country: Lithuania
Voice Phone: (370) 2 232737
Fax Phone: (370) 2 232838
Email: sales@geola.com
Alternate Emails:techinal@geola.com;
geola@post.omnitel.net
Web: http://www.geola.coml
Contact #1: Dr. Stasys Zacharovas
Description: Manufacturer of Pulsed Neodymium lasers for holography and compact automated Holoportraiture systems. International sales of Soviet holographic materials. Optics sales. Hologram studio rental. Reflection hologram stock images. Pulsed holography jobs to Ix2m. Equipment and hologram rental.
SEE OUR ADVERTISEMENT
****

Gerhard Winopal Forschungsbedarf
Echtemfeld 25
City: Hannover
Postal Code: D-30657
Country: Germany
Voice Phone: (49) (0)511 65444
Contact #1: Gerhard Winopal
Description: Isolation tables for holographyHolopal system.
****

GESA - Arbeitskreis fur optische Verfahren
FG 1.1 Holografie
Reichenhainer Strasse 88
City: Chemnitz
Postal Code: D-09126
Country: Germany
Voice Phone: (49) (0)211 6214-224
Fax Phone: (49) (0)211 6214-161
Email: gma@vdi.de
Contact #1: Dr. R. Hiifling
Description: Association
****

Gesellschaft fur Holografie mbH - GfH
A10is Wohlmuth Strasse 25
City: Munich
Postal Code: D-81545
Country: Germany
Voice Phone: (49) (0)89 6253117
Fax Phone: (49) (0)89 6253728
Contact #1: Dipl Phys. 1. Akhmedjanov
Description: Distributor of all types of Russian holograms.
****

Gigahertz-Optik
Fischerstrasse 4
City: Puchheim
Postal Code: D-82178
Country: Germany
Voice Phone: (49) (0)89 890159 0
Fax Phone: (49) (0)89 890159 50
Contact #1: Wolfgang G. O. Dahn
Description: Optics retailer
****

Gile Foil Securities, Inc.
II Caldwell Dr.
City: Amherst

State: NH
Postal Code: 03031
Country: United States of America
Voice Phone: (1) 603 880 6217
Fax Phone: (1) 603 882 6590
Description: Holographic securities for Continuous Form Documents such as checks, vouchers, stock certificates, event tickets & other protected items. Foil stamping, MICRI OCR encoding, embossed images, & imprinting from ribbon.
****

Glaser - Technical Consulting
24 Hashnayim Street
City: Givatayim
Postal Code: 53239
Country: Israel
Voice Phone: (972) 3 673 2734
Fax Phone: (972) 3 6732734
Email: feglaser@weizmann.ac.il
Web: http://www.weizmann.ac.ill
Contact #1: Shelly Glaser
Description: Technical consulting on holography (HOE, display, etc.), diffractive optics (DOE and systems containing DOEs), nonconventional optical systems (lens let array based etc.), and image processing (specifically imaging optics for image processing). Services include feasibility studies, system design and evaluation, courses, etc.
****

Glass Mountain Optics
9517 Old McNeil Road
City: Austin
State: TX
Postal Code: 78758-5225
Country: United States of America
Voice Phone: (1) 512 339 7442
Fax Phone: (I) 512 339 0589
Email: hardyhar@ix.net.com
Web: http://www.glassmountain.com
Contact #1: Don Conklin
Description: Specialize in custom manufacturing front surfaced collimating mirrors. Emphasis on massive optics. See our web site for surplus mirrors .
****

Global Images
I Northumberland Ave
City: Lond6n
State: England
Postal Code: WC2N 5BW
Country: United Kingdom
Voice Phone: (44) 171 872 5452
Fax Phone: (44) 171 872 5611
Contact #1: Walter Clarke
Description: Specialists in high volume, low cost, quality embossing equipment. ISO 9002 Qualification.
****

GOM Gesellschaft fiir optische Messtechnik
mbH
Rebenring 33
City: Braunschweig
Postal Code: D-38106
Country: Germany
Voice Phone: (49) (0)531 3804-330
Fax Phone: (49) (0)531 3804-152
Description: Industrial holographic interferometry.
****
*****

Gorglione, Nancy
2047 Blucher Valley Road
City: Sabastopol
State: CA
Postal Code: 95472
Country: United States of America
Voice Phone: (1) 707 823 7171
Fax Phone: (I) 707 823 8073
Email: gorglione@aol.com
Contact #1: Nancy Gorglione

Description: Holographic artist specializing in architectural installations and public art environments. Extensive portfolio of one-of-a kind fine artworks.
****

Grafix Plastics
Graphix inc.
19499 Miles Rd.
City: Cleveland
State: OH
Postal Code: 44128
Country: United States of America
Voice Phone: (1) 800 447 2349
Voice Phone: (1) 216 581 9050
Fax Phone: (1) 216 581 9041
Email: info@grafixplastics.com
Web: www.grafixplastics.com
Contact #1: Jordan Katz
Description: Full service converter of plastic film and sheet. Including Holographic Film, Polyester, Acetate, Rivid PVC, Pressure Sensitive Film, Drafting Film, Masking Film, and other substrates. Converting capabilities include: slitting/rewinding, sheeting/trimming, custom coating, packaging, custom sourcing.
****

Gresser E. KG
An der Warth 10
City: Ochsenfurt
Postal Code: D-97199
Country: Germany
Voice Phone: (49) (0)933 22 77
Fax Phone: (49) (0)933 78 41
Contact #1: Joachim Mueller
Description: Laser measurement techniques.
****

Gsiinger Optoelektronik GmbH
Robert Koch Strasse I a
City: Planegg
Postal Code: D-82152
Country: Germany
Voice Phone: (49) (0)89 859 -5621
Fax Phone : (49) (0)89 859-7875
Email: sales@gsaenger.de
Web: www.gsaenger.de
Description: Distributor of optical holographic equipment.
****

GTO Lasertechnik GmbH
1m Lindenbusch 37
City: Baden Baden
Postal Code: D-76534
Country: Germany
Voice Phone: (49) (0)7223 589-15
Fax Phone: (49) (0)7223 589-16
Contact # 1: Ulf Wilzner
Description: Distributor of holographic material.
****

Guangdong Dongguan South Holoprint Co.
Dongguan Cheng-Qu Qi-Feng-Lu #77
City, Province: Dongguan, Guangdong
Postal Code: 511700
Country: China
Contact #1: Fan Cheng
Description: Hologram distribution.
****

Guangzhou Chuntian Industrial
Techniques Inc.
Dong-Shan-Qu Jiang-Ling-Dong #14
1st Floor
City, Province: Guangzhou, Guangdong
Postal Code: 510080
Country: China
Contact #1: Xu Chuntian
****

Guangzhou Inst.of Electronics
Lab for Holography & Optoelectronic Tech
Xian-Lie-Zhong-Lu # 100
City, Province: Guangzhou, Guangdong

INTERNATIONAL BUSINESS DIRECTORY H - Holo

Postal Code : 510070
Country: China
Voice Phone: (86) 20 7668176
Fax Phone: (86) 20 7668176
Contact #1: Wang Tianji
Description: Holography research and classes.
****

H. Kallenbach - H.M.V
Holographie Marketing Vertrieb
Friedrich Breuer Strasse 79
City: Bonn
Postal Code: D-53225
Country: Germany
Voice Phone: (49) (0)228 478675
Fax Phone: (49) (0)228 467096
Contact #1: Heinz Kallenbach
Description: Wholesale of holograms on film, glass, and embossed holograms, holographic gifts.
****

HAKRO
Hologramm Pragegerate
Gladbacherstrasse 58
City: Langenfeld
Postal Code: D-40764
Country: Germany
Voice Phone: (49) (0)2173 21700
Fax Phone: (49) (0)2173 24368
Contact #1: Hans Krosel
Description: Distributor for holography mass production equipment.
****

HAM Kristall Technologie
City: Schwendi-Harenhausen
Postal Code: D-88477
Country: Germany
Voice Phone: (49) (0)7353 760
Contact #1: Andreas Maier
Description: Distributor of optical mechanical and chemical holographic equipment.
****

HB Laserkomponenten
Bergstrasse 15
City: Schwabisch GmUnd
Postal Code: D-73525
Country: Germany
Voice Phone: (49) (0)7171 61107
Fax Phone: (49) (0)7171 64679
Email: info@hb-Iaser.com
Web: www.hb-Iaser.com
Contact # 1: Harald Bohlinger
Description: Distributor of optical/mechanical holographic equipment.
****

Heiss, Peter Dr. Priv. Doz.
An der Bleiche 2
City: Korschenbroich
Postal Code: D-41352
Country: Germany
Description: Hobby holographer, teacher. Books and seminars on holography.
****

Hellenic Institute Of Holography
28 Dionyssou Street
City: Chalandri
Postal Code: GR-15234
Country: Greece
Voice Phone: (30) 1 684 6776
Fax Phone: (30) 1 685 0807
Contact # 1: Alkis Lembessis
Description: Established in 1987, the Institute aims at the overall introduction and promonon of holography in Greece. Exhibitions, OUIses. vocational training and mastering laboratory.
****

Heptagon Oy
Otaniemi Science and Technology Park
Telmllkantie 12
City: Espoo

Postal Code: FIN-02150
Country: Finland
Voice Phone: (358) 9 4354 2041
Fax Phone: (358) 9 4354 2041
Email: info@heptagon.fi
Web: http://www.heptagon.fi
Contact # 1: Jyrki Saarinen
Description: Heptagon provides complete design services for designing diffi'active optical elements (DOEs) to customer requirements. Also consulting services and engineering assistance to the fabrication and exploitation of DOEs.
****

Hi Tech Tips
Bachstrasse 24
City: Ettlingen
Postal Code: D-76275
Country: Germany
Voice Phone: (49) (0)7243 32640
Description: Holography lab.
****

Hiat Image Technology Group, Inc.
2F, No. 16 Lane 6, Sec. I, Hang Chou .S. Rd.
City: Taipei
Country: Taiwan
Voice Phone: (886) 2 393 0306
Fax Phone: (886) 2 395 8122
Contact #1: Billy Chou
****

Highlite
Alexanderstrasse 63-65
City: Aachen
Postal Code: D-52062
Country: Germany
Voice Phone: (49) (0)241 35402
Voice Phone: (49) (0)241 31407
Contact # 1: Hans J. Bose
Description: Manufacturer of dichromate holograms up to 30x40cm.
****

HMS-Elektronik
Hans M. Strassner GmbH
Tannenweg 7
City: Leverkusen
Postal Code: D-51381
Country: Germany
Voice Phone: (49) (0)2171 3814
Voice Phone: (49) (0)2171 3815
Contact # 1: Hans M. Strassner
Description: Manufacturing and sale of hightech electronics. Lock-In amplifier-pre-amplifier- optical chopper-optical componentsultrasonic devices-broadband amplifiers.
****

Hofmann-Lange Brigitte
Kaiserstrasse 12
City: Weinheim
Country: Germany
Voice Phone: (49) (0)6201 58571
Contact #1: Dr rer. nat. Brigitte HofmannLange
Description: Seminars on holography.
****

Hahere Grafische Bundes Lehr- und
Versuchsanstalt
Leysalstrasse 6
City: Wien
Postal Code: A- 1140
Country: Austria
Voice Phone: (43) (0)1 9823914
Contact #1: Prof. Dr. Haslauer
****

HOL 3-Galerie fur Holographie GmbH
Europa Center
City: Berlin
Postal Code: D-l0789
Country: Germany
Voice Phone: (49) (0)30 261 4490
Fax Phone: (49) (0)30 344 6379

Contact #1: Valeska Cordner-Guled
Description: Exhibition of mainly stock holograms of various producers. Sometimes feature one man shows. Sales of holograms and all sorts of related holographic items.
****

Holar Seele KG
Wasserwerksweg 16a
City: Aurich
Postal Code: D-26603
Country: Germany
Voice Phone: (49) (0)4941 10005
Fax Phone: (49) (0)4941 63644
Contact #1: Gerd Dipl.-Ing. Seele
Description: Large format holograms up to 1 x3m for use in architecture-display holograms also possible. Portrait holograms with pulsed laser.
****

Holarium Museum fur Holografie
Kirchplatz
City: Esens
Postal Code: D-29042
Country: Germany
Voice Phone: (49) (0)4971 3088
Email: digi_art@t-online.de
Contact # 1: Erik Speer
Description: Museum for holography.
****

Holman Technology, Inc.
5B Marlen Drive
South Gold Industrial Park
City: Hamilton
State: NJ
Postal Code: 08691
Country: United States of America
Voice Phone: (1) 609 8904320
Fax Phone: (1) 609 890 4322
Email: holmtec@erols.com
Contact # 1: John Dixon
Description: Holographic technology and holographic machine sales including: Turnkey holographic lab, dot matrix Light Machines, electro forming equipment, embossing machines, some finishing equipment and related hardware. Also training and in-house manufacturing services.
****

Holo 3D S.p.A.
Trieste Area Science Park
Padriciano, 99
City: Trieste
Postal Code: 1-34012
Country: Italy
Voice Phone: (39) 40 226 327
Voice Phone: (39) 40 364 725
Fax Phone: (39) 40 226 431
Email: hol03d@com.area.trieste.it
Web: www.area.trieste.it/area/centri/hol03d
Contact # 1: Dr Glauco Miniussi
Description: In-house manufacturer of holographic products. Design, mastering, electroforming, embossing and converting. Security and product authentication projects.
****

Holo Art
Island Holographics
City: Northport
State: NY
Postal Code: 11768
Country: United States of America
Voice Phone: (1) 516 757 3866
Email: holoart@aol.com
Contact #1: Dave Battin
Description: Supplier of educational diffraction/holographic art kits.
****

Holo Images Tech Co., Ltd.
17, Alley 20, Lane 7, Jong Hwa Road
City: Yung Kang City, Tainan County

Rainbow reflectivity at its best in these stock transfer designs, much more in catalog. Custom designs and 3D holograms available, too. Display alone or as a focal point with screen printing or embroidery. Sealed edges ... washing / drying 150 times ... no delamination.

Industry's leader for decorative rainbow transfers for any substrate.

Patented product and process.
Distributors sought. Licensing available.

Holography Presses On ®

Jan Bussard
Box 193 Spring Lake, MI 49456 USA
Phone 616/842-5626 Fax 616/842-5653

Can you or your customers *tell the difference between* *your authentic property and a*

Counterfeit?

Absolutely!

Yours is the one with the
hologram security label

secure . . . permanent . . . washable

Use the power of hologram security labels, applied
with heat or stickers with pressure, to all products.
Security holograms and custom shapes for all uses . . .
guaranteed to remain bright and effective after many launderings.

- ✓ Durable construction
- ✓ Manufacturer verifiable
- ✓ Difficult to replicate

- ✓ Quick recognition
- ✓ Easy application
- ✓ Cost effective

Patented product and process.

**Holography
Presses
On**

Jan Bussard
Box 193 Spring Lake, MI 49456 USA
Phone 616/842-5626 Fax 616/842-5653

Country: Taiwan
Voice Phone: (886) 66 237 3896
Fax Phone: (886) 66 238 4641
Contact #1: Craig Chiou
Description: Embossed holograms and products.
****

Holo Impressions Inc
47-1 Wu Chuan Rd
Wu-Ku Industrial Park
City: Taipei Shein
Country: Taiwan
Voice Phone: (886) 2 299 7576
Fax Phone: (886) 2 299 7050
Contact #1: Jonathan Hsu
Description: Embossed holography.
****

Holo Sciences, LLC
480 East Rudasill Road
City: Tucson
State: AZ
Postal Code: 85704
Country: United States of America
Voice Phone: (1) 520 293 9393
Fax Phone: (1) 520 696 0773
Email: deck_O@azstarnet.com
Contact #1: Chuck Hassen
Description: HI Mastering and stereogram creation from video or computer graphic source imagery. Special capabilities to produce full-parallax, animated stereograms. Custom 3-D computer modeling and animation services on request.
SEE OUR ADVERTISEMENT
****

Holo Service/Service-Druck
MorsestraBe 5
City: Duesseldorf
Postal Code: D-40215
Country: Germany
Voice Phone: (49) (0)221 370917
Contact #1: Klaus Kleinherne
Description: Retailer of holograms. Orig. printing shop.
****

Holo Time Gericke
SchertlinstraBe 32
City: FreiberglNeckar
Postal Code: D-71691
Country: Germany
Voice Phone: (49) (0)7141 76850
Fax Phone: (49) (0)7141 73050
Contact # 1: W K. Gericke
Description: Hologram full service: all kinds of holograms, consulting, engineering, sale, mediation.
****

Holo-Idee Reiner Kleinherne
Chopinstrasse 2a
City: Meerbusch-Struemp
Postal Code: D-40670
Country: Germany
Voice Phone: (49) (0)2159 8733
Contact # 1: Reiner Kleinherne
Description: Consultant for industry and advertising, organizer of exhibitions, manufacturing of sample holograms on film before embossing process.
****

Holo-Laser
12 rue de Vouille, Ecole
City: Paris
Postal Code: F-75015
Country: France
Voice Phone: (33) 1 45 31 52 75
Fax Phone: (33) I 48 33 17 07
Contact #1: Dr.Jean Louis Tribillon
Description: Embossed holography and equipment; artistic holography; buying and selling; education.
****

Holo-OrLtd

PO Box 1051
Kiryat Weizmann
City: Rehovot
Country: Israel
Voice Phone: (972) 89 469 687
Fax Phone: (972) 8 9466 378
Contact #1: Uri Levy
Description: Manufactures computer-generated diffractive optical elements by VLSI techniques. Catalogue elements and custom designs. Substrates include AnSe, GaAs, various glasses. DOE work station - dedicated workstation for element design, mask generation.
****

Holo-Source Corporation
11930 Farmington Rd
City: Livonia
State: MI
Postal Code: 48150
Country: United States of America
Voice Phone: (1) 734 427 1530
Fax Phone: (1) 734 525 8520
Email: sales@holo-source.com
Web: http://www.holo-source.com/
Contact #1: Lee Lacey
Description: Paperboard sheets of holographic film and paper. Holographic image mastering of all types and finished flexo printed holographic labels.
****

Holo-Spectra
7742 Gloria Ave.
City: Van Nuys
State: CA
Postal Code: 91406
Country: United States of America
Voice Phone: (1) 818 994 9577
Fax Phone: (1) 818 9944709
Email: bill@lasershs.com
Web: http ://www.lasershs.com
Contact #1: Bill Arkin
Description: Embossed hologram production. Embossed mastering equipment, laser repairs. Optical table and holographic equipment resold.
****

Holoart Studio
NO.13 II Lane
Shin-Fu S.treet ,Shihlin
City: Taipei
Country: Taiwan
Voice Phone: (886) 2 8323843
Fax Phone: (886) 2 8339445
Email: linyow@tcts.seed.net.tw
Contact #1: Yow-Snin Lin
Description: Artistic reflection hologram up to 30cm*40cm,stock and custom design.
****

HoloCom
401 Carver Road
City: Las Cruces
State: NM
Postal Code: 88005
Country: United States of America
Voice Phone: (1) 505 527 9184
Fax Phone: (1) 505 527 9927
Email: arts@holo.com
Web: http://www.holo.com
Contact #1: Ken Harris
Description: Holographic web site and web provider. Holographic technology transfer and holographic photoresist training. Custom originations in photoresist.
****

Holocrafts
Canadian Holographic Developments Ltd.
Box 1035
City, Province: Delta, British Columbia
Holo - Holografie

Postal Code: V4M 3T2
Country: Canada
Voice Phone: (1) 604 946 1926
Fax Phone: (1) 604 946 1648
Email: canholo@interramp.com
Web: www.ghiweb.com
Contact #1: Karoline Cullen
Description: Holocrafts manufacturers dichromate holograms. Offering stock and custom production in a variety of formats such as plain discs, watches, keychains, pendants and 3 "" x 3 '"" plates. Providing a tradition of excellence since 1979.
SEE OUR ADVERTISEMENT
****

Holocrafts Europe Limited
Barton Mill House
Barton Mill Road
City, Province: Canterbury, Kent
State: England
Postal Code: cn !BY
Country: United Kingdom
Voice Phone: (44) 1227 463 223
Fax Phone: (44) 1227 450 399
Contact #1: Chris Luton
Description: Specialists in manufacture of dichromate reflection holograms. Also produce holographic gift products as well as selling photopolymeL
SEE OUR ADVERTISEMENT
****

Holodesign
Fiichtenbusch & Hufmann GbR
Am Forst 38
City: Wesel
Postal Code: D-46485
Country: Germany
Voice Phone: (49) (0)i81 52837
Contact #1: Annette Fuechtenbusch
Description: Manufacturing of holograms.
****

Holografia Polska
ul.sw.Mikolaja 16117
City: Wroclaw
Postal Code: PL-50-128
Country: Poland
Voice Phone: (48) 71 343 46
Fax Phone: (48) 71 339 48
Contact #1: Boguslaw Stich
Description: Practical application of holography and importers of holographic products.
****

Holografie Galerie
Leipziger Chaussee 147
City: Halle
Postal Code: D-06112
Country: Germany
Contact #1: Stephan Rauschenbach
Description: Holography giftshop and gallery.
****

Holografic Hofmann
Schidiickerstrasse 13
City: Salach
Postal Code: D-73084
Country: Germany
Voice Phone: (49) (0)7162 44064
Fax Phone: (49) (0)7162 43900
Contact # 1: Martin Hofmann
Description: Gallery-manufacturer of holograms up to 30x40cm-custom made up to lxlm.
****

Holografie Illusion mit Licht
Turmstrasse 17
City: Pattensen
Postal Code: D-30982
Country: Germany
Voice Phone: (49) (0)5101 1871
Email: bei@imLuni-hannover.de
Web: www.imLuni-hannover.de/-bei

Holografie - Holograms
Contact #1: Günter Beichert
Description: Organizer of holography courses.
****

Holografie Manufaktur
Lychener Strasse 19
City: Berlin
Postal Code: D-10437
Country: Germany
Voice Phone: (49) (0)304417710
Email: geotz.greiner@gestaltung.uniweimar.
de
Web: www.uni-weimar.de/~greiner
Contact #1: Gbtz Greiner
Description: artist
****

Holografie Studio Niirnberg
mind electronics
MAXIMUM - Farberstrasse 11
City: Niirnberg
Postal Code: 0-90402
Country: Germany
Voice Phone: (49) (0)911 241074
Fax Phone: (49) (0)9105 1604
Web: www.eurosmile.comlbauer
Description: Holography lab
****

Holografie Vertrieb
Ritterfelddamm 213
City: Berlin
Postal Code: 0-14089
Country: Germany
Voice Phone: (49) (0)30 3658312
Description: Distributor of all types of holograms.
****

Hologram Industries
Edewechter Land-28
City: Oldenburg
Postal Code: D-26131
Country: Germany
Voice Phone: (49) (0)441 52484
Description: Holography lab.
****

Hologram Center Holmby
Holmby gamla skola
City: Flynge
Postal Code: S-24032
Country: Sweden
Voice Phone: (46) 46 524 30
****

Hologram Company RAKO GmbH
Moellner Landstrasse 15
City: Witzhave
Postal Code: D-22969
Country: Germany
Voice Phone: (49) (0)410 693 250
Fax Phone: (49) (0)410 693 249
Contact #1: Wilfried Schipper
Description: Specializing in production of embossed
holograms and the sale of embossing equipment.
****

Hologram Development Corp.
37 Standish Ave.
City, Province: Toronto, Ontario
Postal Code: M4W 3B2
Country: Canada
Voice Phone: (1) 416 925 5569
Contact #1: Ed Burke
Description: General hologram services.
****

Hologram Industries
22 Ave de I'Europe
Parc Gustave Eiffel
City, Province: Bussy Saint-Georges, Marne
La Vallee Cedex 3
Postal Code: F-77606
Country: France
Voice Phone: (33) I 6476 3100
Fax Phone: (33) I 6476 3570

162 Holography MarketPlace - 8'h Edition
Contact #1: Hughes Souparis
Description: Embossed holography. Security holo-
grams.
****

Hologram Land
284 E. Broadway
Mall Of America
City: Bloomington
State: MN
Postal Code: 55425
Country: United States of America
Voice Phone: (1) 612 854 9344
Fax Phone: (1) 612 854 7857
Contact #1: Sue Rickert
Description: Retail store specializing in everything ho-
lographic. Product range includes: artwork, watches,
jewelry, T-shirts, small gift items and optical novel-
ties. Framing and lighting accessories provided. Also
carry various scientific novelties and glow-in-the-dark
merchandise.
****

Hologram Research, Inc.
P. O. Box 377
City: Locust Valley
State: NY
Postal Code: 11560
Country: United States of America
Voice Phone: (1) 516 922 2560
Fax Phone: (1) 516 624 8687
Email: jburns@nassau.cv.net
Web: http:\\www.hologramres...com
Contact #1: Joseph Burns
Description: Exclusive source for Ilford silver halide
plates and film. Custom holograms to 42"" x 72"", ste-
reograms and limited editions in silver halide, photore-
sist, embossed and injection -molded from our argon
and he-ne laser labs. Exhibition services available from
our extensive hologram collection; brokering services
for the New York Metro area. Fully equipped studio
for rent.
SEE OUR ADVERTISEMENT
****

Hologram Varga Miklos
Kiraly u.102.
City: Budapest
Postal Code: H-1068
Country: Hungary
Voice Phone: (36) 1 351 4725
Voice Phone: (36) 20 342 076
Fax Phone: (36) 1 227 354
Contact #1: Miklos Varga
Description: Stock and custom made holograms on
best quality photopolymer.
****

Hologram World, Inc.
1860 Berkshire Lane North
City: Plymouth
State: MN
Postal Code: 55441
Country: United States of America
Voice Phone: (1) 612 559 5539
Voice Phone: (1) 800 882 4656
Fax Phone: (1) 612 559 2286
Contact #1: Jim Paletz
Description: One of the largest wholesale distributors
of holographic novelties. We represent over 50 holo-
graphic manufacturers. Specialize in helping the new
retail store owner in all stages of development from
start to finish.
****

Hologramas, S.A. de C.V.
(Holograms of Mexico)
Pino 343 Local 3
Col. Atlampa
City: 06450 Mexico City
State: D.F.
Country: Mexico

Voice Phone: (52) 5 5411791
Voice Phone: (52) 5 5479046
Fax Phone: (52) 5 5474084
Email: holomex@holomex.com.mx
Web: http://www.holomex.com.mx/
Contact #1: Laura Sosa
Description: Manufacturer of embossed holograms
since 1984. Our services include security labels, over-
laminates, hot- stamping foil, wide web materials for
flexible, and rigid packaging up to 43" wide, and tech-
nology transfers.
SEE OUR ADVERTISEMENT
****

Holograms 3D
286 Earl's Court Road
City: London
State: England
Postal Code: SW5 9AS
Country: United Kingdom
Voice Phone: (44) 171 3702239
Fax Phone: (44) 171 3732511
Email: jross@holograms.deamon.co.uk
Web: www.holonet.khm.de/jross
Contact #1: Jonathan Ross
Description: Jonathan Ross has a personal holography
collection available for touring shows. He also deals
privately in holographic art and consults on commer-
cial applications.
****

Holograms and Lasers International
Hologram Production Facilities
1200 McKinney, Suite 433
City: Houston
State: TX
Postal Code: 77010
Country: United States of America
Voice Phone: (1) 713 650 9204
Fax Phone: (1) 713 650 9204
Email: felix@electrotex.com
Web: http://www.holoshop.coml
Contact #1: Perry Felix
Description: A full service Ruby Pulse and CW laser
hologram production lab providing complete origina-
tion services for reflection, transmission and mass pro-
duction holograms, specializing in fine quality large
format hologram portraits and trade show exhibits.
****

Holograms and Lasers International
Retail Merchandise Division
P.O. Box 41259
City: Houston
State: TX
Postal Code: 77242-2159
Country: United States of America
Voice Phone: (1) 281 498 0235
Fax Phone: (1) 281 498 6337
Email: felix@electrotex.com
Web: http://www.holoshop.coml
Contact #1: Perry Felix
Description: Holograms and Lasers International oper-
ates the largest retail hologram shop in Houston, Texas
with a complete library of over 750 hologram images
and products to view on the Internet's World Wide Web
located at http://www.holoshop.com.
****

Holograms Fantastic and Optical Illusions
P.O. Box 130
City: Buddina
State: Queensland
Postal Code: 4575
Country: Australia
Voice Phone: (61) 18 776 226
Voice Phone: (61) 414 776 226
Fax Phone: (61) 3 9729 6020
Contact #1: Trevor McGaw
Description: Glass, film and foil (opp, PET & PVC)
2D, 2D/3D, 3D & multi images & patterns. Services to
printers, hot-stampers, pack

aging, label, security marketing, sales promotion and advertising. Specialists in foil holography.
****

Holograms International
211 18th Street
City: Huntington Beach
State: CA
Postal Code: 92648
Country: United States of America
Voice Phone: (1) 714 536 0608
Fax Phone: (1) 714 536 0608
Email: holograms@worldnet.att.net
Contact #1: Dave Krueger
Description: Distributor of all kinds of holograms to retail stores and wholesale accounts. We are known for our fast delivery, friendly consulting and factory-direct prices. Call or write for quote or catalogue.
****

Holograms Unlimited
(U.K. Gold Purchasers, Inc.)
110 Central Park Mall
City: San Antonio
State: TX
Postal Code: 78216
Country: United States of America
Voice Phone: (1) 210 530 0045
Voice Phone: (1) 800 722 7590
Fax Phone: (1) 210 530 0048
Email: marvinuram@hotmail.com
Web: http://www.eden.com/-mainlinkltx/sat/artlrailindex.htm
Contact #1: Marvin Uram
Description: Full line distributor of hologram and related products for specialty retailers - representing more than 80 firms. One stop shopping at competitive prices.
SEE OUR ADVERTISEMENT
****

Holographia
Kaiserstrasse 45
City: Lahr
Postal Code: D-77933
Country: Germany
Voice Phone: (49) (0)7821 38921
Fax Phone: (49) (0)7821 3 89 24
Contact #1: Thomas Krautter
Description: Distributor of all types of holography.
****

Holographic Applications, Inc
21 Woodland Way
City: Greenbelt
State: MD
Postal Code: 20770-1728
Country: United States of America
Voice Phone: (1) 301 345 4652
Fax Phone: (1) 301 345 4653
Contact #1: Suzanne St. Cyr
Description: Design and product engineering services for consumer products and licensed promotions using 3-D imaging technologies. Product specifications and quality assurance. Vendor selection and product management. General contractor delivering finished, packaged product.
****

Holographic Consulting Agency
Jampot Cottage, WestHill,
Elstead, Godalming,
City: Surrey
State: England
Postal Code: GU8 6DQ
Country: United Kingdom
Voice Phone: (44) 1252 702781
Fax Phone: (44) 973 763121
Email: Holoconsul@aol.com
Contact #1: Mark Dicker
Description: Equipment & Technology for the holographic industry.
****

Holographic Design Systems
1134 West Washington Blvd.
City: Chicago
State: IL
Postal Code: 60607
Country: United States of America
Voice Phone: (1) 312 829 2292
Fax Phone: (1) 312 829 9636
Email: hologram@flash.net
Web: www.museumofholography.com
Contact #1: John Hoffman
Description: Unrivaled creativity, combining artistic imagination with complete technical mastery of all forms of holography resulting in a worldwide reputation for excellence. The most complete labs in the industry with the most powerful and advanced lasers and computers. Our clients include the most innovative and sophisticated companies in the US and abroad.
SEE OUR ADVERTISEMENT
****

Holographic Dimensions, Inc. (USA)
7503 N.W. 36th street
City: Miami
State: FL
Postal Code: 33166
Country: United States of America
Voice Phone: (1) 305 994 7577
Fax Phone: (1) 305 994 7702
Email: sales@hgrm.com
Web: http://www.hgrm.com
Contact #1: Kevin Brown
Description: Holographic Dimensions, Inc. is a vertically integrated manufacturer of holographic imagery, with extensive experience in high volume security and authentication holograms. A recently, traded Public Company, it has complete in-house facilities from artwork origination to embossing.
SEE OUR ADVERTISEMENT
****

Holographic Dimensions - Panama
5201 Zone 5
City: Panama
Country: Panama
Voice PhDne: (507) 212 0177
Voice Phone: (507) 212 0173
Fax Phone: (507) 212 0177
Email: jimmywoo@sinfo.net
Web: http://www.sinfo.netlholographic
Contact #1: Jimmy Woolford
Description: HDI Security Holograms
****

Holographic Dimensions, Poland S.A.
UI. Traugutta 25
City: Lodz
Postal Code: PL-90-950
Country: Poland
Contact #1: Grace Golen, VP
Description: Subsidiary of Holographic Dimensions, Inc.
****

Holographic Display Artists & Eng. Club (HOmC)
c/o Seiki Tsushinsha Co., Ltd.
3F 1-1 -8, Shinsenri-Nishimachi,
City: Tokyo
Postal Code: 169-0073
Country: Japan
Voice Phone: (81) 3 3367 0571
Fax Phone: (81) 3 3368 1519
Contact #1: Ms. Nonoko Kita
Description: Regular meetings 4 times a year with oral presentations on holography given. Publishes HODIC circular. Membership open to all.
****

Holograms - Holographic
Holographic Finishing, Inc.

501 Hendricks Causeway
(P.O .Box 597)
City: Ridgefield
State: NJ
Postal Code: 07657
Country: United States of America
Voice Phone: (1) 20 I 941 4651
Fax Phone: (1) 201 941 4453
Contact #1: Michael Vuleano
Description: Finishing operation for printing trade. Hologram application; stamping, embossing; die cutting; gluing.
****

Holographic Identity
Kleine Rosenstrasse 8
City: Hamburg
Postal Code: D-20095
Country: Germany
Contact #1: Behnam Keyaniyam
Description: Distributor of all types of holography.
****

Holographic Images Inc .
521 Michigan Ave.
City: Miami Beach
State: FL
Postal Code: 33139
Country: United States of America
Voice Phone: (1) 305 531 5465
Fax Phone: (1) 305 532 4090
Contact #1: Matthew Schrieber
Description: Mastering and replication facility dedicated to the production of multi-color I full color limited-edition holographic artworks produced on silver halide film.
****

Holographic Impressions
96 North Almaden Blvd.
City: San Jose
State: CA
Postal Code: 95110-2490
Country: United States of America
Voice Phone: (1) 408 292 8901
Fax Phone: (1) 408 292 0417
Email: smithmckay@aol.com
Contact #1: Dave McKay
Description: Manufacturer and distributor of unique line of greeting cards, stationary and business cards. Stock products available.
****

Holographic Industries, Inc .
P.O. Box 1109
City: Libertyville
State: IL
Postal Code: 60048
Country: United States of America
Voice Phone: (1) 847 680 1884
Fax Phone: (1) 847 680 0505
Contact #1: Robert Pricone
Description: Consulting and distribution for giftware industry.
****

Holographic Label Converting (HLC)
9675 Hamilton Rd. Suite 100
City: Eden Prtairie
State: MN
Postal Code: 55344
Country: United States of America
Voice Phone: (1) 612 944 7408
Fax Phone: (1) 612 944 7210
Email: inquire@hle.com
Web: hle.com
Contact #1: Scott Labelle
Description: Full service capabilities, 2D/3D holography, designing, embossing, hot-stamping, precision die-cutting, wide variety of foils. Custom holographic labeling, magnetic holograms, packaging and more ... You think of it, and we can put it together.

INTERNATIONAL BUSINESS DIRECTORY Holographic - Holography

Holographic Laserdesign
Kunst- und Geschenkartikel
Trendelbuscher Weg 1
City: Ganderkesee
Postal Code: D-27777
Country: Germany
Voice Phone: (49) 4221 89298
Contact #1 : Tilmann Eimers
Description: Custom made holograms .

Holographic Materials Distributors
38 Arlington Road Southgate
City: London
Postal Code: N14 5AS
Country: United Kingdom
Voice Phone: (44) 181 3686425
Fax Phone: (44) 181 361 8761
Email: hmd@eivd.globalnet.co.uk
Contact #1: AJ Ranalli
Description: We supply film and plates for holographers from Siavich and HRT.

Holographic Optics Inc.
358 Saw Mill River Rd
City: Millwood
State: NY
Postal Code: 10546
Country: United States of America
Voice Phone: (1) 914 762 1774
Fax Phone: (1) 914 762 2557
Email: hoptics@hol-optics.cnchost.com
Contact #1: Jose R. Magarinos
Description: Manufacturer of holographic optical elements, particularly holographic filters, holographic mirror and beamsplitters. Design and manufacture of prototypes.

Holographic Products
1711 St. Clair Ave.
City: St. Paul
State: MN
Postal Code: 55105,
Country: United States of America
Voice Phone: (1) 65 I 698 6893
Fax Phone: (1) 65 I 698 1619
Email: sugarOOI@maroon.tc.umn.edu
Contact #1: Stephen Sugarman
Description: Holographic Products is actively pursuing new product development in educational toys, intermedia print design, ad specialties, promotions, and premiums. Specialists in security laminates (for ID card application) and tamper evident labeling. Also conducts hands-on elementary school workshops, and instructional presentations.

Holographic Security Marking Systems PVT.
LTD.
2, Ashoka Tower, Kulupwadi Road
Borivli -(East),
City: Mumbai
Postal Code: 400 0066
Country: India
Voice Phone: (91) 887 095 I
Fax Phone: (91) 885 2482
Email: hsms@bom3.vsnl.net.in
Contact #1: Rohit Mistry
Description: Holographic Security Marking Systems Pvt. Ltd. , the choice of banks for authentication marking for use on important documents, cheques and bonds. The company with experience and capability of developing products to customer's specifications. Tamper Evident lables, Induction cap sealing wads, Hot stamping foils etc. We have the confidence of our clients because we have the capability and the ability to service all their holographic security marking requirements.

Holographic Studios
Diisseldorfer Strasse 18
City: Willich
Postal Code: D-47877
Country: Germany
Voice Phone: (49) (0)2154 427-633
Fax Phone: (49) (0)2154 427-583
Contact #1: Ralf Rosowski
Description: Holography lab

Holographic Studios
240 East 26th Street
City: New York
State: NY
Postal Code: 10010
Country: United States of America
Voice Phone: (1) 212 686 9397
Fax Phone: (1) 212481 8645
Email: drlaser@interport.net
Web: http://www.holostudios.com
Contact #1: Jason Sapan
Description: New York's only gallery and commercial holographic lab. Custom and stock holograms. Single or mass-produced. Integral portrait cinematography, mastering, and scan copies from small to large format. Computer generated holograms. Classes.

Holographics (Uk) Ltd.
12 Whidborne St
City: London
State: England
Postal Code: WCIH 8EU
Country: United Kingdom
Voice Phone: (44) 777 552 2550
Voice Phone: (44) 0171 4627548
Fax Phone: (44) 171 4627696
Email: j.john@virgin.net
Contact #1: Jon Vogel
Description: Holographic & 3-D multimedia, design origination and production specialists (est. 1982) providing comprehensive service for the corporate, retail, & leisure sectors.

Holographics Inc.
44-0 I Eleventh Street
City: Long Island City
State: NY
Postal Code: 11101
Country: United States of America
Voice Phone: (1) 718 784 3435
Fax Phone: (1) 718 706 0813
Contact #1: Fred Nicholas
Description: Company involved in 3 main areas of research and development primarily involving pulsed holography; NDT for government and corporate applications (Dr. John Webster, Tim Schmidt); and laser development (Peter Nicholson) .

Holographics North Inc.
444 South Union Street
City: Burlington
State: VT
Postal Code: 05401
Country: United States of America
Voice Phone: (1) 802 658 2275
Fax Phone: (1) 802 658 5471
Email: jp @holonorth.com
Web: http://www.holonorth.com
Contact #1: John Perry
Description: Designers/manufacturers of large format holograms for commercial, museum and fine art applications. Multicolor, animated holograms up to 44x72 inches (I.I m x 1.8m). Design, model building, production, installation and consulting services.

Holographie & Design
Kalk-Miilheimer-Strasse 124
City: Cologne
Postal Code: D-5 1J 03
Country : Germany
Voice Phone: (49) (0)221 857386
Contact #1: Peter Ludwig
Description: Custom made holograms, exhibition for rent, modeling. Large collection.

Holographie Anubis
Oberer Kaulberg 37
City: Bamberg
Postal Code: D-96049
Country: Germany
Voice Phone: (49) (0)951 57951
Fax Phone: (49) (0)95 I 59529
Email: holographie.anubis@t-online.de
Contact #\: M.T. Frieb
Description: We are producer and distributor of all formats of holograms. Import and export. Consulting and education. Mass production. Full pulse laser facilities. Fully pictured wholesale catalog (122 pages).

Holographie Fachstudio Bad Rothenfelde
Postfach 1304
Niederesch 28
City: Bad Rothenfelde
Postal Code: D-49214
Country: Germany
Voice Phone: (49) (0)542 5365
Fax Phone: (49) (0)542 5359
Contact #1: Giinther Deutschmann
Description: Expert consultancy for integration of holographic products into finished advertising media-including application-overprinting etc. Founder and office of the AHT (Arbeitskreis Hologramm-Techniken & 3D Medien).

Holographie Konzept GmbH
Rudolf Diesel Strasse 20b
City: Eschborn
Postal Code: D-65760
Country: Germany
Voice Phone: (49) (0)6173 320889
Fax Phone: (49) (0)6173 320887
Contact #1: Gabriele Chmielewski
Description: Deliver all kinds of holograms.

Holographie Labor
Bertelsmann AG
Auf dem Eickholt 47
City: Giitersloh
Postal Code: D-33334
Country: Germany
Voice Phone: (49) (0)5241 580192
Fax Phone: (49) (0)5241 580549
Contact #1: Saurda Uwe
Description: Holograms, related projects.

Holographie Roth
Schmale Strasse 5
City: Reutlingen
Postal Code: D-72764
Country: Germany
Voice Phone: (49) (0)172 7300581
Fax Phone: (49) (0)7121 44687
Contact #1: Ulrich G. Roth
Description: Hologram production, sale, rent, exhibition.

Holography and Media Institute of Quebec
1139 Ave des Laurentides
City, Province: Quebec City, Quebec
Postal Code: GIS 3C2
Country: Canada
Voice Phone: (1) 418 687 2985
Voice Phone: (1) 418 656 3095
Fax Phone: (1) 418 687 2985
Email: Marie-Andree.Cossette@arv.ulaval.ca

Web: http://www.ulaval.caJ
Contact #1: Marie-Andree Cossette
Description: Established in 1990 as an international centre for the study and production of holographic art. Workshops and private tutorials. Residency programs offered on an invitation only basis. Director Marie-Andree Cossette is Associate Professor of Visual Art at Laval University, Quebec.
****

Holography Center of Austria (Holo.Trend)
Kahlenbergstrasse 6
City: Wiirmla
Postal Code: A-3042
Country: Austria
Voice Phone: (43) (0)2275 8210
Fax Phone: (43) (0)2275 82105
Contact #1: Irrnfried Wober
Description: Our holography laboratoryfounded in 1985-is the first in Austria and the most comprehensive in the region. We offer high quality pulsed and CW laser originationmastering and production (transportable pulsed laser systems for portrait-stereograms up I x I meter and computer generated holograms available). In addition, we organize exhibitions in Austria and Germany and sell embossed holograms.
SEE OUR ADVERTISEMENT
****

Holography Development Co.Ltd.
News building No.2 Room622
Shen Nan Zhong Road #2
City, Province: Shenzhen, Guangdong
Postal Code: 518027
Country: China
Voice Phone: (86) 755 2271973
Fax Phone: (86) 755 2271973
Email: Holohot@nenpub.szptt.net.cn
Web: http://www.holoworld.com/www/hdc/
Contact #1: Hunter Wang
Description: The Holography Development Company is located in the P.R. China. Our aim is to serve holographic companies throughout the world. We can make contact and find outlets for your product all over China. Some of the potential customers would be the department stores, hotels and restaurants, museums, advertising and exhibition companies, etc. We are not a holographic manufacturer. Our job is to market your product as your agent.
****

Holography Group TEMOO
Nordala 3031P
City: Angelholm
Postal Code: S-262 73
Country: Sweden
Voice Phone: (46) 431 82577
Fax Phone: (46) 431 82577
Email: nordala @algonet.se
Description: 10 years old organization for enthusiasts. Lab for holographers.
****

Holography Institute of San Francisco
Email: hologram@well.com
Contact #1: Jeffrey Murray
Description: Specializing in one-on-one instruction in display holography (design, optics, recording, processing, etc.). Courses can be customized to the required balance of theory and practice for artistic, technical or production applications. Classes for artists, scientists, young and old. No prerequisites for beginners. (Classes may be suspended for the 1999 year. Please Email for details.)
****

Holography Israel
21 Hakomemiut Str.
City: Herzlia

Postal Code: 46683
Country: Israel
Voice Phone: (972) 09 572 387
Voice Phone: (972) 09 559 766
Fax Phone: (972) 09 570 569
Contact #1: Hameiri Shimon
Description: Holography Israel specializes in exhibitions-lectures and demonstrations to pupils and students-advertising, commission, sales and production of art holograms.
****

Holography Marketplace
(c/o Ross Books)
P.O.Box 4340
City: Berkeley
State: CA
Postal Code: 94704
Country: United States of America
Voice Phone: (1) 510 841 2474
Voice Phone: (1) 800 367 0930
Fax Phone: (1) 510 841 2695
Email: staff@rossbooks.com
Web: http:\\www.holoinfo.com
Contact #1: Alan Rhody
Description:
The world's most comprehensive and informative book for, and about, the holography industry. Published annually, each edition of THE HOLOGRAPHY MARKETPLACE includes: a worldwide corporate directory; chapters of useful reference material; interviews with industry experts; and a sample kit of actual holograms produced by major manufacturers. Eight editions available.
ORDER YOUR OWN COPY TODAYl
****

Holography Presses On (HPO)
(see Textile Graphics)
201 North Fruitport Road, Box 193
City: Spring Lake
State: MI
Postal Code: 49456-0193
Country: United States of America
Voice Phone: (1) 616 842 5626
Fax Phone: (1) 616 842 5653
Contact #1: Jan Bussard
Description: Holographic stock or custom shapes aqd sizes as decorative or security labels applied with heat or pressure for adhesion to all substrates. Specialize in stickers and textile applications. Sealed edges prevent delamination in all environments; washable. Worldwide distributors sought.
SEE OUR ADVERTISEMENT
****

Holographyx Inc.
99 Ronald Court
City: Ramsey
State: NJ
Postal Code: 07446
Country: United States of America
Voice Phone: (1) 201 327 4414
Voice Phone: (1) 954 345 5001
Fax Phone: (1) 201 327 2606
Email: sdhgx@sprynet.com
Contact #1: Scott Devens
Description: Full service creators and producers of holographic promotional programs and materials for the consumer product marketplace.
****

Holographyx Inc.
10661 NW 43rd Court
City: Pompano Beach
State: FL
Postal Code: 33065-2320
Country: United States of America
Voice Phone: (1) 954 345 5001
Fax Phone: (1) 954 345 2991
Email: rdarnott@aol.com

Contact #1: Bob Arnott
Description: Full service creators and producers of holographic promotional programs and materials for the consumer product marketplace.
****

Holo1and S.c.
Batumi 6 m 43
City: Warszawa
Postal Code: PL-02-760
Country: Poland
Voice Phone: (48) 22 427 463
Fax Phone: (48) 2 625 5567
Email: stepien@if.pw.edu.pl
Contact #1: Pawel Stepien
Description: Low volume holographic labels holographic consultancy, security CGH research & development.
****

Hololaser Gallery
PO Box 23386
City: Dubai
Country: United Arab Emerates
Voice Phone: (971) 4 518 989
Fax Phone: (971) 4 528 015
Contact #1: Abdul Wahab Baghdadi
Description: Holography Gallery and holographic items. We are the first and only gal lery in the Gulf Countries and we produce laser shows.
****

HoloMedia Ab/Hologram Museum
PO Box 45012
City: Stockholm
Postal Code: S-10460
Country: Sweden
Voice Phone: (46) 8, 411 1108
Fax Phone: (46) 8 107 638
Contact #1: Mona Forsberg
Description: Broker for embossed and custom made artistic holography; buying & selling holograms; holography education; gallery. Display unit available.
****

Holomedia France
16 rue Maurice Fontvielle
City: Toulouse
Postal Code: F-31000
Country: France
Voice Phone: (33) 62 27 17 04
Fax Phone: (33) 62 27 17 04
Description: Wholesale and distribution of silver halide, jewelry and fine art holograms .
****

Ho10media France
4 r St Jean
City: Lyon
Postal Code: F-69005
Country: France
Voice Phone: (33) 4 7240 2840
Description: Wholesale and distribution of silver halide, jewelry and fine art holograms.
****

Holomex
4 Borrowdale Avenue
City, Province: Harrow, Middlesex
State: England
Postal Code: HA3 7PZ
Country: United Kingdom
Voice Phone: (44) 181 427 9685
Contact #1: Mike Anderson
Description: Supplier of film processing kits and safe1ights. Designs and manufacturers a "holographic camera and viewer" .
****

Holonix
Box 45577
City: Seattle
State: WA
Postal Code: 98145

Country: United States of America
Email: jkollin@holonix.com
Web: http://www.holonix.coml
Contact #1: Joel Kollin
Description: Optical systems design and consultation and for 3-D imaging, holography, scanning systems, lighting and image projection. Please contact by Email.
****

Holophile, Inc.
56 Abner Lane
City: Killingworth
State: CT
Postal Code: 06419
Country: United States of America
Voice Phone: (1) 860 663 3030
Voice Phone: (1) 212 840 1540
Fax Phone: (1) 860 663 3067
Email: info@holophile.com
Web: http://www.holophile .com
Contact #1: Paul D. Barefoot
Description: Founded in 1975, Holophile provides consulting services in holography and "spectral imagery"(3-D projection of moving images) to corporations, museums and display builders. Our company is a premier producer of holography exhibitions for museums, science centers and children's museurns.
****

Holopress Holographic Techn. GmbH
Deichstrasse 9
City: Hamburg
Postal Code: D-20459
Country: Germany
Voice Phone: (49) (0)40 3743551
Description: Mass production of holograms .
****

Holoptics
Heidjerweg 13
City: Oldenburg
Postal Code: D-261~3
Country: Germany
Voice Phone: (49) (0)441 45166
Fax Phone: (49) (0)441 4860928
Email: schweer@t-online.de
Contact #1 : Jiirg Schweer
Description: Production of silver halide holograms- origination of SHH (F.E.: Eye in Pyramid)- mass production of small size SHH-distribution of holograms and holographic articles- sophisticated display-systems. Mobile exhibitions-listed supplier of holograms for fastidious department stores.
****

Holopublic Unbehaun
Hirschstrasse 84
City: Wuppertal
Postal Code: D-42285
Country: Germany
Voice Phone: (49) (0)202 84118
Contact #1: Klaus Unbehaun
Description: Consulting-education-newsletters Holography 3D Software and AHT Reflexionen-fine arts (Holofotografik) book Holo Show International-founding member AHT -Association for Holography and New Media.
****

Holos An Galerie
4 Place Grenus
City: Geneva
Postal Code: CH-1201
Country : Switzerland
Voice Phone: (41) (0)22 1 7328551
Fax Phone: (4 1) (0)221 7325191
Contact #1: Pascal Barre
Description: Gallery-retail sales.
****

Holosco, Ernest Barnes
Bajada de Viladecols, 2
City: Barcelona
Postal Code: 08002
Country: Spain
Voice Phone: (34) 3 3107113
Fax Phone: (34) 3 319 1676
Description: Holography Lab. Reflection and transmission, transfer to photoresist - embossing facilities. Consulting services.
****

Holosta Holographie-Galerie
Hofkamp 51
City: Wuppertal (Elberfeld)
Postal Code: D-42103
Country: Germany
Contact #1: Brigitta Staiger
Description: Retaile; of holograms on glass or film-multiplex holograms-embossed holograms-dichromate-holographic jewelry.
****

Holostar
College de Maisonneuve
3800 Sherbrooke Est
City, Province: Montreal, Quebec
Postal Code: HIX 2A2
Country: Canada
Voice Phone: (1) 514 254-7131 #4509
Fax Phone: (1) 514 253-8909
Email: holostar@cmaisonneuve.qc.ca
Web: http://www.cmaisonne"uve.qc.ca/holostar.html
Contact #1: Eric Bosco
Description: Center dedicated to the promotion of sciences through holography. Education, services, workshops, research and production. Since 1988.
****

Holostik India Pvl. Ltd.
50, Adhchini
Sri Aurobindo Marg.
City: New Delhi
Postal Code: 110017
Country: India
Voice Phone: (9 1) 11 665 690
Voice Phone: (91) 11 669 725
Fax Phone: (91) 11 686 8828
Contact #1: Govind Sharma
Description: We are among the first to have set up a fully automated plant for manufacture of security and promotional holograms and films in India. Soon setting up master lab and 40 inch wide web machine for holographic packaging.
****

Holostudio Beate Krengel
Mariawalder StraBe 2
City: HeimbachlEifel
Postal Code: D-52396
Country: Germany
Voice Phone: (49) (0)2446 3592
Contact #1: Beate Krengel
Description: Distribution of optics and components for holography.
****

Holotec
Heilwigstrasse 19
City: Munich
Postal Code: D-81825
Country: Germany
Voice Phone: (49) (0)89 429741
Contact #1: M. Wagensonner
Description: Activities: - Display holography (max. 30x40 cm) - Portrait holography (PulseMaster) - Holography seminars.
****

HOLOTECH -Texel
Boodtlaan 41
City, Province: De Koog - Texel, Holland
Postal Code: 1796

Country: Netherlands
Voice Phone: (31) 22 2027352
Fax Phone: (31) 22 202 7429
Contact #1: Dave Platts
Description: Gallery
****

Holotek
(a division of ECRM Inc.)
205 Summit Point Drive
City: Henrietta
State: NY
Postal Code: 14467
Country: United States of America
Voice Phone: (1) 716 321 6000
Voice Phone: (1) 888 465 6832
Fax Phone: (1) 716 321 6001
Contact #1: John Hart
Description: Engineering and design of sub systems for laser optic scanning devices. Commercial and industrial applications.
****

Holotopia II
Kleiner Schlossplatz 5
City: Stuttgart
Postal Code: D-70173
Country: Germany
Voice Phone: (49) (0)711 2263727
Contact #1: Martin Hofmann
Description: Gallery
****

Holovision AB
Box 70002
City: Stockholm
Postal Code: S-10044
Country: Sweden
Voice Phone: (46) 8 331 186
Fax Phone: (46) 8 331 186
Contact #1: Jonny Gustafsson.
Description: Specializing in silver halide holography with pulsed lasers. Denisyuk and transferred-type reflection holograms up to 30 x 40 cm. Rainbow holograms with pulsed laser up to 2 x I m.
****

Holovision Systems Inc.
119 South Main SI.
City: Findlay
State: OH
Postal Code: 45840
Country: United States of America
Voice Phone: (1) 419 422 3604
Fax Phone: (1) 419 422 4270
Email: atl@brighl.net
Contact #1: Ronald L. Kirk
Description: Holovision Systems, Inc. specializes in technological development for holographic and 3-dimensional image display. Holovision currently manufactures and markets Real Image (TM) displays which produce live full color 3-dimensional video projections into 3-D space for point of sale, trade show and exhibit applications. It is also developing a higher level of technology called Holoview(TM) or real time holographic displays of medical CADCAM and cinemagraphic applications.
****

HoloWebs, LLC
9475 Chesapeake Drive, Suite A
City: San Diego
State: CA
Postal Code: 92123
Country: United States of America
Voice Phone: (1) 619 576 1778
Fax Phone: (1) 619 576 2181
Email: holowebs@holowebs.com
Web: www.holowebs.com
Contact #1: Dan Lieberman
Description : Description: Manufacturer of embossed holograms since 1984. Our services

include security labels, overlaminates, hotstamping foil, wide web materials for flexible, and rigid packaging up to 72" wide, and technology transfers.
SEE OUR ADVERTISEMENT
****

Holoworld.com
Internet Webseum of Holography
815 Allen Street
City: Allentown
State: PA
Postal Code: 18102
Country: United States of America
Voice Phone: (1) 800 458 3525
Voice Phone: (1) 610 770 0341
Email: director@holoworld.com
Web: http://www.holoworld.coml
Contact #1: Frank DeFreitas
Description: Designer of the Internet Webseum of Holography - a multi-award winning web site dedicated to amateur and hobbyist holography.
****

Honeywell Technology Center
MN 65 - 2500
3660 Technology Drive
City: Minneapolis
State: MN
Postal Code: 55418
Country: United States of America
Voice Phone: (1) 612 951 7738
Fax Phone: (1) 612 951 7438
Email: cox@src.honeywell.com
Web: http://www.honeywell.com
Contact #1: Dr. J. Allen Cox
Description: Diffractive optics. Micromachining.
****

HOPSec/dii
P.O. Box 765
City: Bayswater
State: Victoria
Postal Code: 3153
Country: Australia
Voice Phone: (61) 39 729 6337
Voice Phone: (61) 18 776226
Fax Phone: (61) 39 729 6020
Contact #1: Trevor McGaw
Description: Security, Optical Holography and Lenticulars.
****

Hotek Holografie Full service
Am Trieb 5
City: Neu Isenburg
Postal Code: D-63263
Country: Germany
Voice Phone: (49) (0)6102 326520
Fax Phone: (49) (0)6102 329312
Description: Coordinator
****

HRT GmbH
Holographic Recording Technologies
Am Steinaubach 19
City: Steinau
Postal Code: D-36396
Country: Germany
Voice Phone: (49) (0)6663 7668
Fax Phone: (49) (0)6663 7463
Email: hrt.birenheide@t-online.de
Web: http://www.holographic-materials.de/
Contact #1: Dr. Richard Birenheide
Description: Own production and distribution of silver halide emulsions with low noise and high diffraction efficiency.
SEE OUR ADVERTISEMENT
****

HSM
Holographic Systems Munich
Melchior Huber Strasse 25
City: Ottersberg/Pliening
Postal Code: D-85652

Country: Germany
Voice Phone: (49) (0)8121 99250
Fax Phone: (49) (0)89 992599
Contact #1: GUnther Dausmann
Description: Display holography (custom made and series, editions, portraits), Embossed holography (self adhesive und hot stamped holo grams), Holographic cameras (Holomatic, Holomatic, Dental, Secumatic, Micromatic) Security (entrance control systems, copysafe holograms)
****

H Space'
Web: www.hspace.com
Description: Holographic Services, Products and Creative Environments, for the lifestyle of the twentyfirst century.
****

Humboldt University Berlin
Mathematische Naturwissenschaftliche
Fakultat I
Universitatsstrasse
City: Berlin
Postal Code: D-10117
Country: Germany
Voice Phone: (49) (0)302095-7857
Fax Phone: (49) (0)30 2095-7666
Email: wernicke@albert.physik.hu-berlin.de
Contact #1: Prof. Dr. GUnther Wernicke
Description: R&D in sign recognition,
****

Hyogo Prefectual Museum of Modern Art
Art Curator
Kobe-3-8-3 Harada-Dori
City: Nada-Ku Kobe, Hyogo Ken
Postal Code: 657
Country: Japan
Voice Phone: ((81)) 778 801 1591
Fax Phone: ((81)) 78 8614731
Contact #1: Hitoshi Yamazaki
Description: 20th century Art, History of Art and Holography, Art and Optics, curating a exhibition of holography into Art.
****

I.S. Gill
214 Kailash Hills
East of Kailash
City: New Delhi
Postal Code: 110065
Country: India
Voice Phone: (91) II 1684 0377
Voice Phone: (91) 11 68470377
Description: Bindry - application of holographic foil and stickers.
****

IBM Almaden Research Center
K03/G2 - 650 Harry Road
City: San Jose
State: CA
Postal Code: 95120
Country: United State~ of America
Voice Phone: (1) 408 927 1283
Fax Phone: (1) 408 927 30 II
Email: mikeross@almaden.ibm.com
Contact #1: Michael Ross
Description: Research; holographic storage.
****

Ibsen Micro Structures AlS
CAT, Frederiksborgvej 399
P.O. Box 30
City: Roskilde
Postal Code: DK-4000
Country: Denmark
Voice Phone: (45) 46 75 40 14
Fax Phone: (45) 46 75 40 12
Email: ibsen@risoe.dk
Web: http://www.ibsen .dk!
Description: Producer of photoresist plates for holography and diffractive optics.

iC Holographics
8 Flitcroft St.
City: London
State: England
Postal Code: WC2H 8DJ
Country: United Kingdom
Voice Phone: (44) 171 2406767
Fax Phone: (44) 171 2406768
Email: 100413.3406@compuserve.com
Contact #1: Chris Levine
Description: Holographic design and digital mastering.
****

Illinois Institute Of Technology
MechanicallMaterials & Aerospace Engring.
Engineering Building #1 Rm 252-B
City: Chicago
State: IL
Postal Code: 60616
Country: United States of America
Voice Phone: (I) 312 567 3220
Fax Phone: (1) 312 567 7230
Email: mesciammarella@mimna.iit.ezu
Contact #1: Cesar Sciammarella
Description: Holographic interferometry; industrial holographic research; NDT
****

Illuminations
1252 7th Avenue
City: San Francisco
State: CA
Postal Code: 94122
Country: United States of America
Voice Phone: (1) 415 664 0694
Email: Ibrill@slip.net
Contact #1: Louis Brill
Description: Involved in developing & expanding market & sales efforts for holographic retail/wholesale product lines. Assist in preparation of promotions and collateral sales materials, identify potential sales markets & implementation of sales.
****

Image Engine
1535 East 1260 North
City: Logan
State: UT
Postal Code: 84341
Country: United States of America
Voice Phone: (1) 435 258 0709
Fax Phone: (1) 435 258 0109
Email: hutch9@cache.net
Contact #1: George Sivy
Description: Specialists in the design and creation of models and sculptures for holographic imaging, including digital origination services for stereograms. Consultant services offered. Extensive portfolio. Samples available upon request. Computer Stereo grams for holography a specialty.
****

Image Technical Development Co.
Huazhong University of Science & Tech.
City, Province: Wuhan, Hubei
Postal Code: 430074
Country: China
Voice Phone: (86) 27 7547655
Fax Phone: (86) 27 7547655
Email: Megpei@sever20.hust.edu.cn
Contact #1 : Zhang Zhaoqun
Description: Holography classes.
****

Imagenes Holograficas De Columbia
Raul Delgado Z - Manager
AVDA 5 Norte #17 - 23
City: Cali
Country: Columbia
Voice Phone: (57) II 572 6684032
Fax Phone: (57) II 572 6685450
****

INTERNATIONAL BUSINESS DIRECTORY Im - In

Images Company
39 Seneca Loop
City: Staten Island
State: NY
Postal Code: 10314
Country: United States of America
Voice Phone: (1) 718 698 8305
Fax Phone: (1) 718 982 6145
Email: images@imagesco.com
Web: www.imagesco.com
Description: Sells holographic equipment and materials to educational institutions, students and private holographers. Equipment and materials available: lasers, film, development kits, mounting kits for lenses, beamsplitters, mirrors, spatial filters, safelights, filters and display lights.
****

Imagination Plantation
2650 18th street, 2nd floor
City: San Francisco
State: CA
Postal Code: 94110
Country: United States of America
Voice Phone: (1) 415 487 0841
Description: 3D imaging and content creation for all media. Experienced in modeling for holographic applications, including direct output to master.
****

ImEdge Technology
2123 Fountain Court
City: Yorktown Heights
State: NY
Postal Code: 10598
Country: United States of America
Voice Phone: (1) 914 962-1774
Fax Phone: (1) 914 962 4922
Email: mmetz@imedge.com
Contact #1 : Michael Metz
Description: Research, development and manufacturing of edge-lit holograms; creative and innovative h010graphy and optics problem solving; custom display volume holograms; edge-lit holographic optical elements; consulting; edge-lit fingerprint imaging device and other industrial holographic products.
****

Imperial College Of Science
Optics Section, Blackett Laboratory
City: London
State: England
Postal Code: SW7 2BZ
Country: United Kingdom
Voice Phone: (44) 171 5895111
Web: http://www.ic.ac.uk/
Description: Courses in holography; holography research; particle measurement.
****

Industrial Technology Research Inst.
Holography Department
Bldg 44, 195 Chung Hsing Road, Section 4
City, Province: Chutung, Hsinchu
Postal Code: 31015
Country: Taiwan
Voice Phone: (886) 35 917 482
Fax Phone: (886) 35 917 479
Contact #1 : Dr. J.J. Su
Description: Research in HUD (Head Up Display) and Dot Matrix Hologram systems.
****

INETI - Institute of Information Tech.
LAER - Aerospace Laboratory
Estrada do Paco do Lumiar
City: Lisboa Codex
Postal Code: P-1699
Country: Portugal
Voice Phone: (351) 1 7165181
Fax Phone: (351) 1 7166067

Email: xana@laer.ineti.pt
Web: http://www.laer.ineti.ptl
Contact #1: Ana Alexandra Andrade
Description: R&D in Holography and OVDs. Special interests in security features. Consultants on technology and customized production projects of embossed holograms . Inhouse design and origination.
****

Infinity Laser Laboratories
6811 Flanders Station
City: Polk City
State: FL
Postal Code: 33868
Country: United States of America
Voice Phone: (1) 941 984 3108
Fax Phone: (1) 941 984 4244
Email: infinityll@earthlink.net
Contact #1: Thad Cason
Description: Full service company from concept and design to finished product. Laboratories include photopolymer capability with pulsed, continuous wave and solid sttate lasers including Argon, Krypton, Nd:YAG:KTP and HeCd.
****

Infox Corporation
3rd floor No 283
Sec2 Fu-hsing South Road
City: Taipei
Country: Taiwan
Voice Phone: (886) 2 70566~9
Fax Phone: (886) 2 7551800
Contact #1: Alex C.T. Chen
Description: Injected-molded holograms .
****

Infrared Optical Products, Inc.
PO Box 292
City: Farmingdale
State: NY
Postal Code: 11735-0664
Country: United States of America
Voice Phone: (1) 516 694 6035
Fax Phone: (1) 516 694 6049
Contact #1: Barry Bassin
Description: Front surface optical coatings for mirrors. A variety of IR optics.
****

Innolas (UK) Ltd,
67 Somers Road
City, Province: Rugby, Warwickshire
State: England
Postal Code: CV22 7DG
Country: United Kingdom
Voice Phone: (44) 1788 550 777
Fax Phone: (44) 1788 550 888
Email: InnoLasUK@aol.com
Contact #1: William Brown
Description: Innolas Ltd. is a newly formed company and originates from Laser Techniques, United Kingdom and InnoLas GmbH., Planegg-Steinkirchen, (Munich) Germany. It is largely created on the sale of the Lumonics Ltd. Solid State Scientific Products business to both InnoLas (UK) Ltd. and InnoLas GmbH. Innolass Ltd. Continues to manufacture scientific Q switched ND:YAG lasers and ruby holographic lasers. Laser Technical Services is the agent for InnoLas Ltd. in North & South America.
****

Inrad, Inc.
181 Legrand Avenue
City: Northvale
State: NJ
Postal Code: 07647
Country: United States of America
Voice Phone: (1) 201 767 1910
Fax Phone: (1) 201 767 9644
Contact #1: Maria Murray

Description: Manufacturer of nonlinear materials, harmonic generation systems, electrooptic and acoustic-optic devices and drivers . Also provides optical components, assemblies and optical coatings for the UV, visible and IR.
****

Inside Finishing Magazine
P.O. Box 12090
City: Portland
State: OR
Postal Code: 97212
Country: United States of America
Voice Phone: (1) 503 331 6221
Fax Phone: (1) 503 331 6928
Email: fsea@aol.com
Web: www.fsea.com
Contact #1: Jeff Peterson
Description: Only trade publication specifically targeting the graphic finishing industry. Editorial focus is on foil stamping, embossing, holograms, die cutting, folding/gluing, and coatings. Published quarterly by the Foil Stamping & Embossing Assn.
****

Inspeck, Inc.
360 rue Franquet, Suite 20
City, Province: St. Foy, Quebec
Postal Code: G IP 4N3
Country: Canada
Voice Phone: (1) 418 650 2112
Fax Phone: (1) 418 650 2141
Email: inspeck@riq.qc.ca
Web: http://www.inspeck.coml
Contact #1: Li Song
Description: Manufacturers a 3-D color digitizer (scans into computer the coordinates of object and color). Data is imported as DXF or other system independent data format. Works on PC and can be ported to SGI, SUN, etc. for use with any rendering program.
****

Inst. flir Licht- und Bautechnik - ILB
an der Fachhochschule Kaln e.v.
Gremberger Strasse 151a
City: Cologne
Postal Code: D-51105
Country: Germany
Voice Phone: (49) (0)221 831094-97
Fax Phone: (49) (0)221 835513
Email: gregorh@gauss.fo.fh-koeln.de
Contact #1: Prof. Jarg Gutjahr
Description: Holography lab working on the architectural field.
****

Institute for Applied Physics
TH Darmstadt
Schlossgartenstrasse 7
City: Darmstadt
Postal Code: D-64289
Country: Germany
Voice Phone: (49) (0)6151 162786
Voice Phone: (49) (0)615 1 162022
Fax Phone: (49) (0)6151 164534
Email: Andreas.Billo@Physik.THDarmstadt.de
Web: www.physik.th-darmstadt.de/andreas
Contact #1: Andreas Billo
Description: Investigation of cavitation bubbles-sprays and droplets with high speed holographic particle image velocimetry (HPIV). HOEs-computer generated holograms-holographic interferometry.
****

Institute for Holographic Tech.
Central Univ. of Nationalities
Dept. of Physics
City: Beijing
Postal Code: 100081
Country: China

Voice Phone: (86) 769 2467775
Fax Phone: (86) 769 2478473
Contact #1: Zhu Weili
Description: Holography class.
****

Institute of Applied Optics
National Academy of Sciences of Ukraine
10-G Kudryavaskaya St.
City: Kiev
Postal Code: 254053
Country: Ukraine
Voice Phone: (380) 44 212 21 58
Fax Phone: (380) 44 21248 12
Email: vmarkov@iao.freenet.kiev.ua
Contact #1: Vladimir B. Markov
Description: R&D in the areas of: display holography (reflection, transmission, mastering) and color holography for museum items reproduction; holographic interferometry and its application in industry and cultural heritage protection; diffraction on the volume grating; laser physics, including properties of the cavity, tunable lasers, etc.; non-linear optical effects and multibeam interaction.
****

Institute Of Optical Science
Central University
City: Chung-Li
Postal Code: 32054
Country: Taiwan
Voice Phone: (886) 3 425 7681
Fax Phone: (886) 3 425 8816
Contact #1: Tang Yaw Tzong
Description: HOEs, academic research.
****

Integraf
P.O. Box 586
745 N. Waukegan Rd.
City: Lake Forest
State: IL
Postal Code: 60045
Country: United States of America
Voice Phone: (1) 847 234 3756
Fax Phone: (1) 847 615 0835
Email: tjeong@aol.com
Contact #1: T.H. Jeong
Description: Distributor of holographic films and plates, including Russian emulsions. Agfa materials still in stock. Expert consultation on HOEs, NDT, system designs and other holographic projects. Also, educational materials and stock holograms.
****

Interactive Industries, Inc.
40 Todd Rd
City: Shelton
State: CT
Postal Code: 06484
Country: United States of America
Voice Phone: (1) 203 929 9000
Fax Phone: (1) 203 929 9001
Contact #1: Ron Phillips
Description: Custom, Stock, 3-D, 2-D, Key Chains, Bookmarks, Mugs, Posters, Calendars, Signs, Rulers, Gambling Specialties, Labels, Tags, Souvenirs, Buttons, Portfolios.
****

Interferens Holografi D.A.
Museum , Gallery, Studio
Halvor Hoels Gt 6
City: Hamar
Postal Code: N-2300
Country: Norway
Voice Phone: (47) 62 25050
Voice Phone: (47) 62 30659
Fax Phone: (47) 62 30659
Contact #1: Olav Skipnes
Description: Ongoing exhibition of NorwayJEs largest collection of holograms. Makes glass (mainly reflection) holograms of museum exhibits. Continuous wave laser. Norway's largest collection of holograms. Our specialty: museum exhibits.
****

International Data Ltd.
127 Kingfischer Indst estae
west moors
City, Province: windborne, dorset
State: England
Postal Code: VH21 6US
Country: United Kingdom
Voice Phone: (44) 1202 875 555
Fax Phone: (44) 1202 871 166
Email: 101763.1273@compuserve.com
Contact #1: Dawn Dreelaw
Description: Manufacturer of plastic cards. Capable of hot stamping holographic materials.
****

International Hologram Manufacturers
Association (IHMA)
Runnymede Malthouse
Runnymede Road
City: Egham
State: England
Postal Code: TW20 9BD
Country: United Kingdom
Voice Phone: (44) 1784497008
Fax Phone: (44) 1784 497 001
Email: 100142.1164@compuserve.com
Description: IMHA was founded in 1993 to promote the interests of hologram manufacturers and the holography industry worldwide. It is a non profit membership organization open to all producers of holograms, suppliers of equipment and material for the manufacture of holograms and hologram converters and finishers. Annual meeting in November.
SEE OUR ADVERTISEMENT
****

International Holographic Paper
300 High Point Drive
City: Chalfont
State: PA
Postal Code: 18914
Country: United States of America
Voice Phone: (1) 215 572 8600
Fax Phone: (1) 215 572 8154
Email: hoklgramer@aol.com
Web: http://www.holoprism.com
Contact" #1: Brian Monaghan
Description: Manufacturer of ""Prismatic Illusions"" holographic paper line. Holographic paper and board suitable for high volume printing and packaging applications. 5 stock patterns as well as custom images available (utilizing computer-generated dot matrix origination) Wide web up to 30 in. x 40 in. image size.
****

Intrepid World Communications
2777 Product Drive, Suite 100
City: Rochester Hills.
State: MI
Postal Code: 48309
Country: United States of America
Voice Phone: (1) 248 299 4040
Fax Phone: (1) 248 299 6206
Email: dea@intrepidworld.com
Web: www.intrepidworld.com
Contact #1 : James Mies
Description: Distributor of display holograms, in particular true-color holograms. Currently working on archiving Vatican treasures.
****

Introduct GmbH
Siemensstrasse 21-23
City: Neu- Anspach
Postal Code: D-61267
Country: Germany
Voice Phone: (49) (0)6081 402-0
Fax Phone: (49) (0)6081 -90
Description: Distributor of mass produced holograms.
****

Ishii, Ms. Setsuko
#404,
1-23 26 Kohinata,Bunkyo-Ku
City: Tokyo
Postal Code: 102
Country: Japan
Voice Phone: ((81)) 3 3945 9017
Fax Phone: ((81)) 3 3945 9068
Contact #1: Setsuko Ishii
Description: Holographic Fine Artist.
****

Island Graphix
22 Bayview Ave.
Ward's Island
City, Province: Toronto, Ontraio
Country: Canada
Voice Phone: (1) 416 203 7243
Fax Phone: (416) 203 7243
Email: page@astral.magic.ca
Contact #1: Michael Page
Description: Production of all formats. Computer Generated Holograms. Mass Production. Consulting. Stock images.
****

J.B. Consultants
11 Apple Tree Grove, Burwell
City: Cambridge
Postal Code: CB5 OBF
Country: Great Britain
Voice Phone: (44) 1223 420405
Fax Phone: (44) 1223 420797
****

James River Products
211 East German School Road
City: Richmond
State: VA
Postal Code: 23224-1460
Country: United States of America
Voice Phone: (1) 804 231 0600
Fax Phone: (1) 804 231 0900
Email: jrp@richmond.infi.net
Web: www.hmt.comlholography/jrp
Contact #1: Mike Florence
Description: World leader in embossed hologram machinery. Products include: origination lab equipment, photoresist plate spinning, electroform facilities, embossing machines, die cutting equipment, supporting technology and training .
SEE OUR ADVERTISEMENT
****

Japan Communication Arts Co.
Yonezawa Bldg2F
2-37 Suehirocho, Kita-Ku
City: Osaka .
Postal Code: 530
Country: Japan
Voice Phone: ((81)) 6 314 1919
Fax Phone: ((81)) 6 315 1900
Contact #1: Mineko Fukuma
Description: Sales of cards with hologram.
****

Jayco Holographics
29-43 Sydney Road
City, Province: Watford, Herts
State: England
Postal Code: WDI 7PY
Country: United Kingdom
Voice Phone: (44) 1923 246 760
Fax Phone: (44) 1923 247 769
Contact #1: Rohit Mistry
Description: Complete production service for embossed holograms. Embossing masters through to fully finished product. Our years of experience enables Jayco to offer outstand

INTERNATIONAL BUSINESS DIRECTORY Je - Ki

ing quality of product and service at competitive prices.
****
Jeffery Murray Custom Holography
Email: hologram@well.com
Contact #1: Jeffery Murray
Description: Museum displays, artists collaborations, commercial advertising, image research, HOE custom optics for visual display. Research specialties: high quality display holography, holographic optics, holographic recording systems. One-offs, limited editions and unusual projects . Able to produce HI masters, reflection and transmission image plane transfers, and Denisyuk holograms up to 50x60 cm.
****
Jiangsu Sida Images, Inc.
Nanjing Normal University
City, Province: Nanjing, Jiangsu
Postal Code: 210024
Country: China
Voice Phone: (86) 25 3318848
Fax Phone: (86) 25 7718174
Contact #1: Sun Hangjia
Description: Holography class.
****
Jodon Inc.
62 Enterprise Drive
City: Ann Arbor
State: MI
Postal Code: 48103
Country: United States of America
Voice Phone: (1) 734 761 4044
Voice Phone: (1) 800 989 jodon
Fax Phone: (1) 734 761 3322
Email: johng@wwn.com
Contact #1: John Wernenski
Description: Manufacturer of HeNe lasers, laser systems, specialty laser tubes, optical and electro-optical instruments and systems. Holographic films, plates and chemicals.
****
John, Pearl
Voice Phone: (1) 573 449 6739
Contact #1: Pearl John
Description: Artist with portfolio, including pulsed and multiplex holograms. M.A. degree awarded from Royal College of Art, London. Also provides educational, design, and consulting services. Currently teaching in USA.
****
Junker, Timo
Otto Hahn Strasse 7
City: Gebrunnen
Country: Germany
Voice Phone: (49) (0)3931 7059147
Fax Phone: (49) (0)3931 7059 149
Email: junker@holografie.com
Web: www.holografie.com
Description: Holography course organizer.
****
K. Thielker & T. Rost-Holographievertrieb
Vermeerweg 15
City: Wesseling
Postal Code: D-50389
Country: Germany
Voice Phone: (49) (0)2236 43138
Contact #1: Klaus Thielker
Description: Wholesale of holograms, traveling exhibitions, holograms for rent.
****
K.c. Brown Holographics
17 Salisbury Road
New Malden
City: Surrey
State: England
Postal Code: KT3 3HZ
Country: United Kingdom

Voice Phone: (44) 181 9428294
Fax Phone: (44) 181 877 3400
Email: 100306.2015@compuserve.com
Contact #1: Kevin Brown
Description: Pulse portraits; artistic works.
****
Ka-Lor Cubicle & Supply Co.
P.O. Box 804 H
City: Fair Lawn
State: NJ
Postal Code: 07410
Country: United States of America
Voice Phone: (1) 201 891 8077
Fax Phone: (1) 201 891 6331
Description: We design and manufacature high quality black-out curtains and related hardware for your studio or laboratory. An easily installable track system custom-designed to meet your business· needs.
SEE OUR ADVERTISEMENT
****
Kaiser Optical Systems, Inc.
PO Box 983
371 Parkland Plaza
City: Ann Arbor
State: MI
Postal Code: 48106
Country: United States of America
Voice Phone: (1) 734 665 8083
Fax Phone: (1) 734 665 8199
Web: http://www.kosi.comlProdlicts/Raman/5s1cover.htm
Contact #1: Karen Mitchie
Description: Compact fiber coupled Raman spectrometers for routine quality control and remote, real time, in line process monitoring applications with microscope accessory for line applications. Fast fl1.8 volume transmission grating based Holographic Imaging Spectrographs for visible, fluorescence and Raman applications. Holographic Notch and Laser Bandpass fliters for Raman laser induced fluorescence spectroscopy.
****
Kaliningrad State University
Physics Dept.
14, A. Newsky
City: Kaliningrad
Postal Code: 236041
Country: Russia - CIS
Voice Phone: (7) 112 469805
Fax Phone: (7) 112 465813
Email: gusev@phys.ksu.kern.ru
Web: www.ksu.kern.ru\MPHS\index.html
Description: Holographic mounting for nondestructive testing is a mobile autonomous complex, carrying out contact-less measuring of locations constructions deformations under the influence of different operational loadings.
****
Karas Studios Holografia
Ave Maria, 46
City: Madrid
Postal Code: 28012
Country: Spain
Voice Phone: (34) 1 530 89 88
Fax Phone: (34) I 530 89 88
Email: karasrb@ddnet.es
Web: http://www.webvent.comlkaras
Contact #1: Ramon Benito
Description: Established in 1988, Karas owns a collection of holographic pieces selected under a sole criteria: art. Among others, the most relevant works of the Spanish art holographers are in the Karas Collection. 3 Galleries. Publications. Art curators.
****
Kauffman, John
Box 477
City: Point Reyes Station

State: CA
Postal Code: 94956
Country: United States of America
Voice Phone: (1) 415 663 1216
Fax Phone: (1) 415 663 1216
Contact #1: John Kauffman
Description: Holographic Fine Artist. Specializes in multi color reflection holograms. Extensive portfolio.
****
Keio University
Dept Of Electrical Engineering
3-14-1 Hiyoshi Kohoku-Ku
City: Yokohama
Postal Code: 223
Country: Japan
Voice Phone: (81) 45 563 1141
Fax Phone: (8 1) 45 563 2773
Contact #1: Dr. Masato Nakajima
Description: Research using HNDT.
****
Kempter
Kreuzackerweg 2
City: Konstanz
Postal Code: D-78465
Country: Germany
Voice Phone: (49) (0)7531 45997
Fax Phone: (49) (0)7531 43707
Description: Manufacturer of lasers.
****
Kendall Hyde Ltd.
Stroudley Road
City, Province: Basingstoke, Hants
State: England
Postal Code: RG24 8UG
Country: United Kingdom
Voice Phone: (44) 125 684 0830
Fax Phone: (44) 125 684 0443
Contact #1: Clive Birch
Description: Optical coating specialists. Front surface mirrors, and any type of film coating.
****
Keystone Scientific Co.
1460 Sawmill Road
City: Downington
State: PA
Postal Code: 19335
Country: United States of America
Voice Phone: (1) 610 269 9065
Fax Phone: (1) 610 269 4855
Email: EKELLYI460@aol.com
Description: Manufacturer of complete holography kits (recording materials, optics, lasers, etc.) and automatic processors for schools, hobbyists and others.
****
Kimmon Electric Co., Ltd.
TM21 Building
1-53-2 Itabashi, Itabashi-ku
City: Tokyo
Postal Code: 173
Country: Japan
Voice Phone: (81) 3 5248 4811
Fax Phone: (81) 3 5248-0021
Email: lasers@kimmon.com
Web: http://www.kimmon.com
Contact #1: Shinichi Fukuda
Description : Manufacturer of Helium Cadmium lasers which are used in holography. Kimmon manufacturers the highest powered HeCd laser in the world. The model IK4171IG is the laser of choice for holographers because of its 180mW @442 TE-Moo specified output power and long lifetime.
****
Kinetic Systems, Inc.
20 Arboretum Road
City: Boston
State: MA

Postal Code: 02131
Country: United States of America
Voice Phone: (1) 617 522 8700
Voice Phone: (1) 800 992 2884
Fax Phone: (1) 617 522 6323
Email: sales@kineticsystems.com
Web: http://www.kineticsystems.com
Contact #1: Moss Blosvem
Description: Manufacturers of Vibraplane standard and special Honeycomb optieal tables in four grades up to 5' x 16' x 24'. Larger sizes available by butt splicing. Also vibration isolation support systems.
****

Koch Herbert
Schulstrasse 28
City: Bad Lauterberg
Country: Germany
Voice Phone: (49) (0)5524 80366
Fax Phone: (49) (0)5524 89197
Description: Distributor of all types of holograms.
****

Kolbe-Druck
1m Industriegeliinde 50
City: Versmold
Postal Code: D-33775
Country: Germany
Voice Phone: (49) (0)5423 9670
Fax Phone: (49) (0)5423 41230
Contact #1 : Roland Pahnke
Description : Bindry: Hot stamping holographic foil-Lenticular-Printing-3-D-motionchange.
****

KomSa Kommunikationstechnik GmbH
August Bebel Strasse 23
City: Gelenau
Postal Code: D-09423
Country: Germany
Voice Phone: (49) (0)37297 4000
Fax Phone: (49) (0)37297 4001
Email: 0372974000-0001@t-online.de
Contact #1: Ronald Dittrich
Description : Holography lab working Dicromate plates and a stereogram printer.
****

Korn, Michael
B urggasse 103/8
City: Wien
Country: Austria
Voice Phone: (43) (0)1 5242191
Email: e9526820@student.tuwien.ac.at
Description: artist
****

Kremo
Industriepark E
Alte Neckarelzer 24
City: Mosbach
Postal Code: D-74821
Country: Germany
Voice Phone: (49) (0)6261 14805
Fax Phone: (49) (0)6261 18471
Contact #1: Karl Kretschmer
Description: Dist. mass production equipment.
****

Krystal Holographies France SARL
I, rue de la Republique
City: Lyon
Postal Code: F-69001
Country: France
Voice Phone: (33) 4 72 10 9099
Fax Phone: (33) 4 72 109094
Email: e.vertongen@Krysaltech.de
Web: http://www.khiinc.com
Contact #1: Emmanuel Vertongen
Description: All products supplied by Krystal Holographics International and especially physical and digital mastering and mass replication on photopolymer film.
****

Krystal Holographies International Inc.
U.S. Holographics Division
365 North 600 West
City: Logan
State: UT
Postal Code: 84321
Country: United States of America
Voice Phone: (1) 435 753 5775
Voice Phone: (1) 800 998 5775
Fax Phone: (1) 435 753 5876
Email: krystal@sunrem.com
Web: http://www.khiinc.com
Contact #1: Dave Rayfield
Descript ion: Marketing/manufacturing for mass-produced and custom photopolymer and dichromate holograms for retail and ad specialty needs. Stock and custom products available. KHI is also a manufacturer of mastering and mass replication equipment.
SEE OUR ADVERTISEMENT
****

Krystal Holographies International Inc.
555 West 57th Street, Suite 1750
City: New York
State: NY
Postal Code: 10019
Country: United States of America
Voice Phone: (1) 212 261 0400
Voice Phone: (1) 801 753 5775
Fax Phone: (1) 212 262 0414
Email: hologram@krystaltech.com
Web: http://www.khiinc.com
Contact #1: Marion Baker
Description: World headquarters of Krystal Holographics International Inc ., a manufacturer and marketer of photopolymer holograms produced by proprietary technology. Experienced with large volume orders of custom holograms for commercial applications including packaging, giftware, security/authentication, advertising premiums, etc.
SEE OUR ADVERTISEMENT
****

Krystal Holographics Vertriebs-GmbH
Birnenweg 15
City: Reutlingen
Postal Code: D-72766
Country: Germany
Voice Phone: (49) 7121 9461 58
Fax Pholle: (49) 7121 9461 55
Email: R.Stooss@krystaltech.de
Web: http://www.khiinc.com
Contact #1: Richard Stooss
Description: Eye-Catching and brand enhancement at its best! Distribution of innovative KrystalGram products in Western Europe: High quality 3D-Holograms on DuPont Omnidex photopolymer film-aimed for high volume promotional & OEM applications. Custom and stock designs available. Krystalmark security systems for high level security. You need to see it!
****

Kurz Foils
3200 Woodpark Blvd.
City: Charlotte
State: NC
Postal Code: 28206
Country: United States of America
Voice Phone: (1) 800 950 3645
Voice Phone: (1) 704 596 9091
Fax Phone: (1) 704 596 3321
Contact #1: Donald Tomking
Description: Provides holographic hot stamp foil to the industry. Custom designs done to customer specifications. Also provide bindery hot stamping service. Sales offices in Germany, Canada and throughout USA.
****

L.O.T. Oriel GmbH & Co. KG

1m Tiefen See 58
City: Darmstadt
Postal Code: D-64293
Country: Germany
Voice Phone: (49) (0)6151 8806-0
Fax Phone: (49) (0)6151 896667
Email: info@lot-oriel.com
Web: www.lot-oriel.comlde
Description: Distributor of opticaUmechanical holographic equipment.
****

Label Systems
56 Cherry Street
City: Bridgeport
State: CT
Postal Code: 06605
Country: United States of America
Voice Phone: (1) 203 333 5503
Fax Phone: (1) 203 336 8570
Contact #1: Richard Roule
Description: Label Systems is a manufacturer of complex holographic products. We offer a wide variety of products for security, promotional, collectable, and packaging applications. The company specializes in the development of security devices for anti counterfeiting and identification uses.
****

Laboratories of Image Information
Science and Technology
5-10, Hongo 4-Chome, Bunkyo-ku
City: Toyonaka, Osaka
Postal Code: 565-0083
Country: Japan
Voice Phone: (81) 6 836 0256
Fax Phone: (81) 6 871 5733
Description: R&D on 'Holographic Display, Dynamic Holography, HOEs.
****

Laboratory for Optical Data Processing
Carnegie Mellon University
Dept. of Electrical And Computer Enginrg.
City: Pittsburg
State: PA
Postal Code: 15213
Country: United States of America
Voice Phone: (1) 412 268 2464
Fax Phone: (1) 412 268 6345
Email: casasent@ece.cmu.edu
Web: http://www.ece.cmu.edu/
Contact #1 : David Casasent
Description: Research in optical data processing, pattern recognition, product inspection, neural nets. Processors, filters and feature extractors using computer generated holograms.
****

Laboratory Vinckiner
Holography Workshop Univ Gent
41 St Pietersnieuwstraat
City: Gent
Postal Code: B-9000
Country: Belgium
Voice Phone: (32) 9 264 3242
Voice Phone: (32) 91 626 384
Fax Phone: (32) 9 223 7326
Contact #1: Pierre Boone
Description: Consultancy, education, problem- solving for display holography. Museum applications and (mainly!) non-destructive testing.
****

Lacher Peter
Freileiten 7
City: Vocklabruck
Country: Austria
Voice Phone: (43) (0)7672 29876
Fax Phone: (43) (0)662 626415
Email: lacherpelo@netway.at
Description: artist
****

Laczynski & Angus Creative Consultancy
4 Wheeler Ave.
City, Province: Toronto, Ontario
Postal Code: M4L 3V2
Country: Canada
Voice Phone: (1) 416 691 2016
Fax Phone: (1) 416 691 2383
Email: andrewl@passport.com
****

Lake Forest College
Center for Photonics Studies
555 N. Sheridan Road
City: Lake Forest
State: IL
Postal Code: 60045
Country: United States of America
Voice Phone: (1) 847 735 5160
Fax Phone: (I) 847 615 0835
Email: jeong@lfc.edu
Web: http://www.lfc.edu/physics/holography/
Contact #1: T.H. Jeong
Description: Lake Forest College offers a hands-on workshop for participants who have no prior experience in holography and also advanced workshops. Due to building renovation, during the 1999 year workshops will be offered by arrangement "off-site". Regular college level courses also available.
****

Langen R.
Stemstrasse 4
City: Bonn
Country: Germany
Voice Phone: (49) (0)228 694525
Description: Giftshop with holograms.
****

Larry Lieberman Holography
5101 Collins Ave., #5-D
City: Miami Beach
State: FL
Postal Code: 33140
Country: United States of America
Voice Phone: (1) 305 994 7577
Voice Phone: (1) '305867 1321
Fax Phone: (1) 305 994 7702
Email: lieber741@aol.com
Contact #1: Larry Lieberman
Description: Mastering facility for full color holograms and stereogram portraiture. A collection of full color limited edition artworks available for sale and exhibition. Call for details.
****

Lasart Ltd.
2911 San Isidro Ct.
City: Santa Fe
State: NM
Postal Code: 87501
Country: United States of America
Voice Phone: (1) 505 438 8224
Fax Phone: (1) 505 438 8224
Email: aurorean@nets.com
Contact #1: August Muth
Description: Lasart Ltd. is a manufacturer of production holographic jewelry and gifts as well as one-of-a-kind holographic glass sculpture. We also specialize in unique corporate gifts, taking the project from modeling through completion of the product in our in house facilities.
****

Laser 2000 GmbH
Argelsrieder Feld 14
City: Wessling
Postal Code: D-82234
Country: Germany
Voice Phone: (49) (0)8153 405-0
Fax Phone: (49) (0)8153 405-33
Web: www.laser2000 .de
Description: Manufacturer of lasers.

****

Laser Affiliates
2047 Blucher Valley Road
City: Sebastopol
State: CA
Postal Code: 95472
Country: United States of America
Voice Phone: (1) 707 823 7171
Fax Phone: (1) 707 823 8073
Contact #1: Nancy Gorglione
Description: Laser Affiliates is an award-winning non-profit organization that designs innovative holographic and laser theatrical productions, in stallations and exhibitions. Services include curatorial guidance, videotapes and media lectures.
****

Laser and Motion .Development Company
3101 Whipple Road, Bld.22
City: Union City
State: CA
Postal Code: 94587
Country: United States of America
Voice Phone: (1) 5104291060
Fax Phone: (1) 510 429 1065
Email: em@lasermotion.com
Web: http ://www.lasermotion.com
Contact #1: Ed Monberg
Description: LMDC is a buyer and seller of lasers, motion and optical equipment. We also integrate laser processing systems at significant savings to our customers. We offer experience and engineering in galvanometer, X-Y table, and laser based optical systems.
****

Laser Art Studio Ltd
Room 12, 3/F., Blk. A2, Yau Tong Ind. City
17 Ko Fai Road, Yau Tong
City, Province: Kowloon, Hong Kong
Country: China
Voice Phone: (852) 2349 1193
Fax Phone: (852) 2349 0923
Email: laserart@laser-art.com
Contact #1: Joyce Kwok
Description: Full line of hologram products manufacturing in Hong Kong. Such as stickers, Hot stamping foil, security stickers. We can also numbering your labels in series. Products from Hong Kong giving you security, time save as well as money save. Clients are our lives, we guarantee all our clients 100% satisfaction.
****

Laser Arts Society For Education and Research
Country: United States of America
Email: hologram@well.com
Web: http://www.hmt.comlholography/laser/index.html
Description: Volunteer staffed non-profit organization dedicated to holography and laser education and research.
****

Laser Focus World
(a division of Penwell Publishing)
10 Tara Boulevard - 5th floor
City: Nashua
State: NH
Postal Code: 03062
Country: United States of America
Voice Phone: (1) 603 891 0123
Voice Phone: (l) 800 331 4463
Fax Phone: (1) 603 891 0574
Email: barbaraw@pennwell.com
Web: http://www.lfw.com/WWW/home.htm
Contact #1: Debbie Bourgault
Description: Trade publication covering the field of optics, lasers , electro-optics, and related imaging research, as well as commercial applications.
****

Laser Innovations

668 Flinn Ave. #22
City: Moorpark
State: CA
Postal Code: 93021
Country: United States of America
Voice Phone: (1) 805 529 5864
Fax Phone: (1) 805 529 6621
Contact #1: R. Eric King
Description: Laser Innovations offers sales, repair, and support of ion laser systems. Specializing in the repair and service of COHERENT lasers ; Laser Innovations stocks remanufactured INNOVA plasma tubes for fast and reliable support of your ion laser system.
****

Laser Institute Of America
12424 Research Parkway #125
City: Orlando
State: FL
Postal Code: 32826
Country: United States of America
Voice Phone: (1) 407 380 1553
Fax Phone: (l) 407 380 5588
Email: lia@mail.creol.ucf.edu
Web: http://www.laserinstitute.org/
Contact #1: Jackie Thomas
Description: Laser safety training courses. Publishes ""Journal of Laser Applications ..... Hosts annual International Congress on Applications of Lasers and Electro-Optics (ICALEO), including holographic applications and International Laser Safety Conference. Call for membership details and publication catalog.
****

Laser International
19 Normanton Rise
Holbeck Hill
City, Province: Scarborough, N Yorks
State: England
Postal Code: YO 11 2XE
Country: United Kingdom
Voice Phone: (44) 172 336 4452
Contact #1: Keith Dutton
Description: Specializing in laser display systems for exhibitions. Auto & manual control available. Full range price, features.
****

Laser Las Vegas
4300 N. Pecos, Unit 18
City: Las Vegas
State: NV
Postal Code: 89115
Country: United States of America
Voice Phone: (1) 702 644-4224
Fax Phone: (1) 702 644-4001
Email: laserlv@ aol.com
Contact #1: Bill Aymar
Description: Laser sales, repairs and rentals. Specialists in high power laser systems.
****

Laser Light Designs
241 2 Kennedy Way
City: Antioch
State: CA
Postal Code: 94509
Country: United States of America
Voice Phone: (1) 925 754 3144
Contact #1: Michael Malott
Description: New product designs using embossed foil, tinsel and holographic films . I specialize in jeweled novelty designs. Designer of the original Rainbow Flasher, inventor of original Rainbow Sparkler.
****

Laser Optics, Inc.
111 Wooster St.
City: Bethel
State: CT

Postal Code: 06801
Country: United States of America
Voice Phone: (1) 203 744 4160
Fax Phone: (1) 203 798 7941
Contact #1 : Jim Larim
Description: A complete line of laser and optical components for ultraviolet, visible and infrared applications from 250 nm to 16 microns, including focusing lenses, windows, cavity components, prisms , beamsplitters, mirrors and coatings.
****

Laser Reflections
589 Howard St.
City: San Francisco
State: CA
Postal Code: 94105
Country: United States of America
Voice Phone: (1) 415 896 5958
Fax Phone: (1) 4158965171
Email: hologram@laser-reflections.com
Web: http: //www.laser-reflections.com
Contact #1: Ron Olson
Description: Highest quality holographic portraiture. Advanced Nd:YAG laser recording technology producing visibly superior holograms of living subjects up to 14"" x 24"". Self-standing displays and custom installations. Unique commercial applications for signage and advertisements. Extensive portfolio of limited fine art editions. Stock images available on silver halide and photopolymer.
SEE OUR ADVERTISEMENT
****

Laser Resale Inc.
54 Balcom Road
City: Sudbury
State: MA
Postal Code: 01776
Country: United States of America
Voice Phone: (1) 978 443 8484
Fax Phone: (1) 978 443 7620
Email: LaseResale@aol.com
Web: http://www.laserresale.com
Contact #1: Tom Van Duyne
Description: Laser Resale provides a marketplace for buying and selling pre-owned lasers, laser systems, optical tables and associated equipment for holographers.
****

Laser Solution Management
Ingenieur-Biiro flir Laser-Systeme
Barkhorstruecken 5a
City: Essen
Postal Code: D-45239
Country: Germany
Contact #1: Vera Hartmann
Description: Development of lasers for air traffic control-light-shows and holography
****

Laser Technical Services
1396 River Road, Box 248
City: Upper Black Eddy
State: PA
Postal Code: 18972
Country: United States of America
Voice Phone: (1) 610 982 0226
Fax Phone: (1) 610 982 0226
Email: lasertek@ptd.net
Contact #1: Dan Morrison
Description: Technical consultant and field repair of lasers. Full customer service of laser equipment - Specifically Pulsed Ruby holographic lenses. Specialize in Lumonics lasers.
****

Laser Technology, Inc.
1055 West Germantown Pike
City: Norristown
State: PA
Postal Code: 19403

Country: United States of America
Voice Phone: (1) 610 631 5043
Fax Phone: (1) 610 631 0934
Contact #1: John Newman
Description: Manufacture equipment for laser-based NDT; Holography and Shearography equipment and inspection services. Portable and production units available.
****

Laser Trend
Uferstrasse 72
City: Kiel
Postal Code: D-24106
Country: Germany
Voice Phone: (49) (0)431 330085
Fax Phone: (49) (0)431 330086
Description: Distributor of all types of holograms.
****

Laserfilm
Milchstrasse 12
City: Munich
Postal Code: D-81667
Country: Germany
Voice Phone: (49) (0)89 4807714
Fax Phone: (49) (0)89 485666
Contact #1: Eckard Knuth
Description: 120 I 360 degree Multiplex holograms (stereograms). Moving images. Diameter 45 cm or 65 cm. Most representati ve work in 1996 - a 360 degree hologram with an imperial crown in original size for the exhibition Austria 996 -1996.
****

Laserfilm
Milchstrasse 12
City: Miinchen
Postal Code: D-81667
Country: Germany
Voice Phone: (49) (0)89 4807714
Fax Phone: (49) (0)89 485666
Contact #1: Eckard Knuth
****

LaserMax, Inc.
3495 Winton Place,Building B
City: Rochester
State: NY
Postal Code: 14623
Country: Urtited States of America
Voice Phone: (1) 716 272 5420
Fax Phone: (1) 716 272 5427
Email : blarabee@ lasermax-inc.com
Web: http://www.lasermax-inc.coml
Description: LaserMax is a manufacturer of diode laser systems and is the industry leader for reliability and technical support. Our engineers have drawn upon years of experience in laser design and manufacturing in order to develop the most robust and complete miniature diode drive circuits currently on the market. We offer the most advanced optical systems and circuitry, capable of operation in harsh environments from space flight to mining.
****

Lasermetrics, Inc.
(a division of Fastpulse Technology, Inc.)
220 Midland Ave.
City: Saddlebrook
State: NJ
Postal Code: 07663
Country: United States of America
Voice Phone: (1) 973 478 5757
Voice Phone: (I) 800449 FAST
Fax Phone: (1) 973 478 6115
Email: rglase@aol.com
Web: http ://lasermetrics.com
Contact #1: Robert Goldstein
Description: Laser components and electronic drivers

for lasers.
****

Laserworks
PO Box 2408
City: Orange
State: CA
Postal Code: 92859
Country: United States of America
Voice Phone: (1) 714 832 2686
Fax Phone: (1) 714 832 1451
Email: 110344.2454@compuserve.com
Web: http :\\www.laser-works.com
Contact #1: Selwyn Lissack
Description : Company manufactures programmable laser scanning equipment for sign age and display applications. Selwyn Lissack has been a holographic artist and researcher since 1969. He has produced numerous holographic exhibitions in addition to the "Salvador Dali " holograms.
****

Lasing S.A.,
Marques De Pico Velasco 64
E-28027
City: Madrid
Country: Spain
Voice Phone: (34) 01 268 3643
Description: Branch office of Newport Corporation
****

Lasiris Inc.
Main Office
3549 Ashby
City, Province: Ville St Laurent, Quebec
Postal Code: H4R 2K3
Country: Canada
Voice Phone: (1) 514 335 1005
Voice Phone: (1) 800 814 9552
Fax Phone: (1) 514 33S 4576
Email: sales@lasiris.com
Web: http://www.lasiris.coml
Contact #1: Alain Beauregard
Description: HOE optics in stock gratings. HOE special projects. Beamsplitters . Laser pattern projectors for industrial inspection and machine vision.
****

LASOS
Schwedenweg 25 T03/II
City: Ebersberg
Postal Code: D-85560
Country : Germany
Voice Phone: (49) (0)8092 8259-18
Fax Phone: (49) (0)8092 8259-19
Description: Distributor of optical holographic equipment.
****

Lauk & Partner GmbH
(a division of Lauk Kommunikation)
Augustinusstrasse 9B
City: Frechen
Postal Code: D-50226
Country: Germany
Voice Phone: (49) (0)2234 963220
Fax Phone: (49) (0)2234 65013
Email: 101624.331@compuserve.com
Contact #1: Matthias Lauk
Description: Holograms-holographic projectshologram museum.
****

Lawrence Berkeley Laboratory
1 Cyclotron Road
City: Berkeley
State: CA
Postal Code: 94720
Country: United States of America
Voice Phone: (1) 510 486 4000
Description: Industrial & academic holography research . Will do commercial research projects.
****

INTERNATIONAL BUSINESS DIRECTORY Laz - Li

Laza Holograms Ltd.
Unit 7 Wilton Rd.
St. Georges Industrial Estate
City, Province: Camberley, Surrey
State: England
Postal Code: GUI5 2QW
Country: United Kingdom
Voice Phone: (44) 1276 683000
Fax Phone: (44) 1276 683000
Email: sjkyle@cableol.co.uk
Contact #1: Stephen Kyle
Description: Laza Holograms is one of the worlds largest suppliers of stock imagery of silver halide and photo polymer film holograms. We have an extensive stock image range constantly being updated. Volume distributor of holographic glasses.
****

Lazart Holographics
22 Erina Valley Road
City: Erina
State: NSW
Postal Code: 2250
Country: Australia
Voice Phone: (6 1) 2 4367 6245
Fax Phone: (61) 2 4367 2306
Email: lazart@ozEmail.com.au
Web: http://www.ozEmail.com.au/-lazart/
Contact #1 : Brett Wilson
Description: Artistic holography; buying & selling holograms. Wholesale distribution and retail sales of artist editions and stock images. Production of jewelry items and novelty products from embossed images. Gallery exhibition open 7 days .
****

Lenox Laser
12530 Manor Road
City: Glen Arm
State: MD
Postal Code: 21057
Country: United States of America
Voice Phone: (1) ,410 592 3106
Fax Phone: (1) 410 592 3362
Email: sales@lenoxlaser.com
Web: http://www.lenoxlaser.com
Contact #1 : Joseph P. D'Entremont
Description: Laser-systems laboratory specializing in laser drilling and etching. Manufactures pinhole kits and spatial filters suitable for holography and precision industrial/optical applications. Also sells holographic components, used laser systems, and low cost optical kits for holography.
****

Leonhard Kurz GmbH
Schwabacher Strasse 482
City: FUrth
Postal Code: D-90763
Country: Germany
Voice Phone: (49) (0)911 71410
Fax Phone: (49) (0)911 7141507
Email: chri stian.hermann@kurz.de
Web: www.kurz.de
Contact #1: Christian Hermann
Description: Manufacturer of embossing equip.; broker for hologram embossing.
****

Leseberg Dr. Detlef
Kamener Strasse 172
City: Liinen-Beckinghausen
Postal Code: D-44532
Country: Germany
Voice Phone: (49) (0)2306 1794
Fax Phone: (49) (0)2306 1793
Contact #1: Detlef Leseberg
Description: Scientific holography researchHOE. computer-generated holography.
****

Letterhead Press, Inc.
W226 N880 Eastmound Drive
City: Waukesha
State: WI
Postal Code: 53186-1689
Country: United States of America
Voice Phone: (1) 414 5741717
Fax Phone: (1) 414 5741718
Email: let@execpc.com
Contact #1: Mike Graf
Description: Full service trade finisher with 24-hour, 7-days/week manufacturing. Featuring 19 x 25 inch and 40 inch formats for holographic stamping. Complete projects from print to final bindery assuring single-source responsibility.
****

Lexel Laser, Inc. .
48503 Milmont Drive
City: Fremont
State: CA
Postal Code: 94538
Country: United States of America
Voice Phone: (1) 510 770 0800
Voice Phone: (1) 800 527 3795
Fax Phone: (1) 510 651 6598
Email: lexel@aol.com
Contact #1: Ben Graham
Description: Lexel produces the highest quality Argon, Krypton and mixed gas laser systems. In particular, Lexel specializes in production of single frequency systems which are very stable over a variety' of environmental situations.
****

LG Chern
overseas business team
20,yoido-dong, youngdungpo-gu
City, Province: Seoul, Kyungi
Postal Code: 150-721
Country: Korea
Voice Phone: (82) 2 3773 3416
Fax Phone: (82) 2 3773 3428
Email: jwleeq@mail.lgchem.co.kr
Description: Raw material (plastic) supplier for hologram sheets.
****

LichtBlicke Classen & Voss
Strahlenberger Weg 8
City: Frankfurt
Postal Code: D-60599
Country: Germany
Voice Phone: (49) (0)69 628492
Fax Phone: (49) (0)69 610176
Contact #1: Walter ClaBen
Description: Custom made holograms-small exhibition.
****

Lichtwelt Gratings
Ladgraf Friedrich Strasse 25
City: Offenbach
Postal Code: D-63075
Country: Germany
Voice Phone: (49) (0)69 867810-27
Fax Phone: (49) (0)69 867810-29
Description:Hologram distributor.
****

LiCONiX, A Melles Griot Company
3281 Scott Boulevard
City: Santa Clara
State: CA
Postal Code: 95054
Country: United States of America
Voice Phone: (1) 408 496 0300
Voice Phone: (1) 800 825 2554
Fax Phone: (1) 408 492 1303
Email: sales@liconix.com
Web: http :\\www.liconix.com
Contact #1: Ken Ibbs

Description: LiCONiX has long been recognized as the leader in Helium Cadmium laser technology and offers the worlds most extensive range of HeCd lasers for commercial, OEM and scientific users. Small, light and robust, LiCONiX HeCd lasers require little maintenance, no water cooling and have lifetimes up to 10,000 hours.
****

Light Dimension, Inc.
Sunfamily Hongo #403
2-16-13 Hyakunin-cho, Shinjuku-ku
City: Tokyo
Postal Code: 113-0033
Country: Japan
Voice Phone: (81) 3 38129201
Fax Phone: (81) 3 3812 9422
Contact #1: Mariko Oishi
Description: Handling the whole range of holograms/ holographic products from embossed holograms to fine art images; also focusing on exhibitions on holography.
****

Light Dimension, Inc.
Sunfamily Hongo #403
1-33, Yayoi-cho
City: Tokyo
Postal Code: 113-0033
Country: Japan
Voice Phone: ((81)) 3 38129201
Fax Phone: ((81)) 3 3812 9422
Contact #1: Mariko Oishi
Description: Handling the whole range of holograms/ holographic products from embossed holograms to fine art images; also focusing on exhibitions on holography.
****

Light Impressions International, Ltd.
5 Mole Business Park 3
City, Province: Leatherhead, Surrey
State: England
Postal Code: KT22 7BA
Country: United Kingdom
Voice Phone: (44) 1372 386 677
Fax Phone: (44) 1372 386 548
Email: postbox@lightimpressions.co.uk
Web: http://www.lightimpressions.co.ukl
Contact #1: John Brown
Description: A full service manufacturer offering the highest quality custom mastering. Embossed product available in hot stamp foil or pressure sensitive labels for security or promotional application. Stock images and holographic equipment available. Office representatives.
****

Lightgate, Ltd
Novy Svet 21
P.O.Box 2, Post 012 HRAD
City: Prague
Postal Code: 11900
Country: Czech Republic
Voice Phone: (42) 2 352325
Fax Phone: (42) 2 352479
Email: light@boohem-net.cz
Contact #1: Jana Vancurova
Description: Description of services: I) Mastering of silver halide display holograms a copies on classic materials up to 50 x 50 cm. 2) Mastering for embossed holograms for copies up to 15 x15 cm of label. 3) Stock holograms of Lightgate and European partners. 4) Manufacturing of embossed holograms, label and hot foils .
****

Lightrix, Inc.
2132 Adams Avenue
City: San Leandro
State: CA
Postal Code: 94577
Country: United States of America

Voice Phone: (1) 510 577 7800
Fax Phone: (1) 510 577 7816
Email: dir@lightrix.com
Web: http://www.lightrix.com
Contact #1: Deborah Robinson
Description: Lightrix manufactures and designs high quality holographic toys, gifts and wall decor. Lightrix offers state of the art holographic product development and graphic design. Lightrix holograms are available to the wholesale trade. We offer a full custom program for embossed and photopolymer holograms.
SEE OUR ADVERTISEMENT
****

LightVision Confections
8 Faneuil Hall
City: Boston
State: MA
Postal Code: 02109
Country: United States of America
Voice Phone: (1) 513 469 0330
Fax Phone: (1) 513 489 8222
Email: info@lightvision.com
Web: http://www.lightvis ion.com
Description: LightVisions markets a process that embosses holographic images in candy. The process is owned by Dimensional Foods Corp.
****

Lightwave Electronics
2400 Charleston Rd.
City: Mountain View
State: CA
Postal Code: 94043-1630
Country: United States of America
Voice Phone: (1) 650 962 0755
Voice Phone: (1) 888 544-4892
Fax Phone: (1) 650 962 1661
Email: info@lwecorp.com
Web: www.lwecorp.com
Description: Lightwave Electronics is the premier manufacturer of diode-pumped solidstate lasers, systems and accessories. We produce a broad range of lasers including CW, Q-switched and mode-locked, with infrared and visible output.
****

Likom
Poslovni Center Ledina
Kotnikova 5
City: Ljubljana
Country: Slovenia
Description: Holograms on display and sale.
****

Linda Law Holographics
P.O. Box 434
City: Centerport
State: NY
Postal Code: 11721
Country: United States of America
Voice Phone: (1) 516 754 6121
Fax Phone: (1) 516 754 9227
Email: llholo@i-2000.com
Contact #1: Linda Law
Description: State of the art computer graphics for holography. Using Mac and SGI computers, artwork can be created for mass production holograms or large scale display holograms.
****

Linea - Bregaglia
Hauptstrasse 30
City: Castasegna
Postal Code: CH- 7608
Country: Switzerland
Voice Phone: (41) (0)81 8221718
Fax Phone: (4 1) (0)81 8221268
Contact #1: H. & R. Gutekunst
Description: Museum
****

Liti Holographies
11834 Canon Blvd., Suite H
City: Newport News
State: VA
Postal Code: 23606
Country: United States of America
Voice Phone: (1) 757 873 6460
Fax Phone: (1) 757 873 6461
Email: rnrbob@media.mit.edu
Web: www.litiholographics.com
Contact #1: Paul Christie
Description: High Quality holograms of people and objects using pulsed laser and stereogram technology. Company has invested heavily in the technology and techniques behind these holograms to make them more friendly and affordable. We want YOU to be able to make a hologram of the things YOU like. (Company founder is a graduate of the MIT Media Lab's Spatial Imaging Group).
SEE OUR ADVERTISEMENT
****

London Holographic Image Studio
9 Warple Mews
Warple Way
City: London
State: England
Postal Code: W3 ORF
Country: United Kingdom
Voice Phone: (44) 181 7405322
Fax Phone: (44) 181 740 1733
Email: easyhologram@easy net
Contact #1: Martin Richardson
Description: Commissioned holograms up to I x 2 meters, pulse-portraiture, movie stereograms and mass production of silver halide holograms. Catalogues on request.
****

Lone Star Illusions
2901 Capital Of Texas Highway, #191
City: Austin
State: TX
Postal Code: 78746
Country: United States of America
Voice Phone: (1) 512 328 3599
Fax Phone: (1) 512 328 3599
Email: webmaster@lonestarillusions.com
Web: http://www.lonestarillusions.com
Contact #1: Alan Lifshen
Description: Austin's only hologram gallery and retail shop which features a full range of holographic giftware and related optical novelties.
****

Loughborough Univ. Of Tech.
Dept Of Physics - Dept of Mechanical Engineering
City, Province: Loughborough, Leicestershire
State: England
Postal Code: LEI 1 3TU
Country: United Kingdom
Voice Phone: (44) 1509263 171
Fax Phone: (44) 1509 '219702
Web: http://www.lboro.ac.uk
Description: Scientific and industrial R&D.
****

Louis Paul Jonas Studios, Inc.
304 Miller Road
City: Hudson
State: NY
Postal Code: 12534
Country: United States of America
Voice Phone: (1) 518 851 2211
Fax Phone: (1) 518 851 2284
Email: dmerrit@jonasstudios.com
Web: http://www.jonasstudios.com
Contact #1: Dave Merritt
Description: Jonas Studios specializes in making models and miniatures for the museum

and film industries. Services include: sculpting, model making, dioramas, EDM machining, CAD for rapid prototyping, laser cutting, mold making and casting in a wide variety of materials.
****

LST - Laser & Strahl Technik GmbH
Testarellogasse 1113
City: Wien
Postal Code: A- 1130
Country: Austria
Voice Phone: (43) (0)1 8764694
Fax Phone: (43) (0)1 876469422
Description: Retailer of Laser optics - new and second hand
****

Lumenx Technologies, Inc.
PO Box 219
City: New Durham
State: NH
Postal Code: 03855
Country: United States of America
Voice Phone: (1) 603 859 3800
Fax Phone: (1) 603 859 2501
Email: eneister@zet.com
Contact #1: Ed Neister
Description: We manufacture and repair a variety of laser equipment, mostly for scientific and medical applications. Call for more details.
****

Luminer Printing and Converting
1925 Swarthmore Ave.
City: Lakewood
State: NJ
Postal Code: 08701
Country: United States of America
Voice Phone: (1) 732 886 6557
Fax Phone: (1) 732 886' 6692
Description: Innovative printer and converter; labels and promotional materials. Expertise and technology for adhesive coating imaged holographic materials, including zone and patterned areas. Overprint, laminate, fold multiple webs . Complete design and origination services.
****

Lumonics Ltd.
Cosford Lane
Swift Valley
City, Province: Rugby, Warwickshire
State: England
Postal Code: CV21 lQN
Country: United Kingdom
Voice Phone: (44) 1788 570 321
Fax Phone: (44) 1788 579 824
Email: info@lumonics.com
Web: http://www.lumonics.com/
Contact #1: George Synowiec
Description: Due to recent business changes, see Laser Technical Services in USA for sales and service of Lumonics pulsed ruby lasers. Lumonics manufactures a variety of lasers and laser based systems for industrial applications. These include a range of pulsed ruby lasers specifically designed for holography with output energies spanning from 30 mJ per pulse to greater than 10 Joules per pulse.
****

Lund Institute Of Tech
Department Of Physics
Box 118
City: Lund
Postal Code: S-221
Country: Sweden
Voice Phone: (46) 046 222 7656
Fax Phone: (46) 046 222 4017
Email: seven-goran.pattersson@fysik.l th.se
Description: Color H-1; holography education; academic research.

INTERNATIONAL BUSINESS DIRECTORY M - Me

M.I.T. Spatial Imaging Group
Massachusetts Institute of Technology
20 Ames Street # E15-416
City: Cambridge
State: MA
Postal Code: 02139 - 4307
Country: United States of America
Voice Phone: (1) 617 253 0632
Fax Phone: (1) 617 253 8823
Email: sab@media.mit.edu
Web: http://www.media.mit.edu/groups/spi/
Contact #1: Stephen A. Benton
Description: Education, research and development center. Current areas of research include: computer generated holography, holographic hard copy printers, and edge lit holograms.

M.I.T. Museum
265 Massachusetts Ave.
City: Cambridge
State: MA
Postal Code: 02139-4307
Country: United States of America
Voice Phone: (1) 617 253 4462
Fax Phone: (1) 617 253 8994
Web: http://web.mit.edu/museum/home/index.html
Contact #1: Diego Garcia
Description: Museum has approximately 50 holograms on display from a inventory of approximately 1,000 holograms. Some of the most historically-significant holograms ever made are on display here. Museum shop has holograms for sale.

M.O.M. Inc.
2436 Forest Green Rd.
City: Baltimore
State: MD
Postal Code: 21209
Country: United States of America
Contact #1: Alan Evan
Description: Maryland Optical Manufacturing. Highest quality. 40 years experience.

Macquarie Holographics
Macquarie University
City, Province: Macquarie, NSW
Postal Code: 2109
Country: Australia
Voice Phone: (61) 2 9850 8715
Fax Phone: (61) 2 9850 8128
Contact #1: John Tobin
Description: "Hands-On" classes in holography. Please call us for our class schedule. We also do commissioned research projects in holography and mastering.

MacShane Holography
C/O Laser Arts Productions
512 West Braeside Drive
City: Arlington Heights
State: IL
Postal Code: 60004-2060
Country: United States of America
Voice Phone: (1) 847 398 4983
Contact #1: Jim MacShane
Description: MacShane Holography/Laser Arts Productions goes to schools and sets up holography class programs.

Magic Laser
105 r Moines
City: Paris
Postal Code: F-75017
Country: France
Voice Phone: (33) 1 40 33 17 49

Description: Importer and wholesaler of all holographic products-travelling exhibition.

Magick Signs Holografie
Isenburg-Zentrum
Shopteil West
City: Neu-Isenburg
Postal Code: D-63263
Country: Germany
Voice Phone: (49) (0)610 328404
Fax Phone: (49) (0)610 45548
Contact #1: Andreas Wollenweber
Description: Sells holograms of all kinds to the public.

Magick Signs Holografie
August-Bebel-Strasse 40
City: Egelsbach .
Postal Code: D-63329
Country: Germany
Voice Phone: (49) (0)610 45544
Fax Phone: (49) (0)61045548
Contact #1 : Andreas Wollenweber
Description: PRODUCTION of embossed holograms: Artwork-models-origination-embossing- many stock images (Trade mark). RETAIL in our four holographic magic gift stores (Frankfurt).

Man/Environment, Inc.
2251 Federal Avenue
City: Los Angeles
State: CA
Postal Code: 90064
Country: United States of America
Voice Phone: (1) 310477 7922
Voice Phone: (1) 310 477 8960
Fax Phone: (1) 310477 4910
Email: holograms@armchair.com
Web: http:\\www.armchair.com
Contact #1: Gary Fisher
Description: Silver halide and photopolymer R&D projects. Design and manufacture optical printers and holographic systems. Complete website development. Check our website for additional info.

Mannheim
Lehrstuhl flir Informatik V
Quadrat B6126
City: Mannheim
Postal Code: D-68131
Country: Germany
Voice Phone: (49) (0)621 292-0
Email: noehte@ti .uni-mannheim.de
Web: www-mp.informatik.uni-mannheim.de
Contact #1: Arbeitsgruppe Optisches Rechnen
Description: R&D in optical holographic computer and storage.

mario liedtke pro design
PixelerstraBe 34
City: Rheda-Wiedenbrueck
Postal Code: D-33378
Country: Germany
Contact #1: Mario Liedtke
Description: modeling for holographic setups.

Marita Schlemmer GmbH
Friedrich Bergius Strasse 15c
City: Hohenbrunn
Postal Code: D-85662
Country: Germany
Voice Phone: (49) (0)8102 5108
Fax Phone: (49) (0)8102 6199
Description: Hologram distributor.

Marks, Gerald
29 West 26th Street
City: New York

State: NY
Postal Code: 10010
Country: United States of America
Voice Phone: (1) 212 889 5994
Fax Phone: (1) 212 889 5926
Email: pulltime3d@aol.com
Web: http://www.pulltime3d.com
Contact #1: Gerald Marks
Description: Artist specializing in stereoscopic 3D of every type for over twenty years. He is best known for the 3D music videos he created for the Rolling Stones and his 3D museum exhibits, anaglyph prints and books, lenticulars, random dot stereograms, computer multimedia and computer generated holography.

Markus Studio d.o.o.
B. Magovca 16a
City: Zagreb
Country: Croatia
Voice Phone: (38 5) 1 694 008
Fax Phone: (385) 1 515 824

Mazda Motor Corp.
Technical Research Center
POBox 18
City: Hiroshima
Postal Code: 730 91
Country: Japan
Voice Phone: «81)) 82 282 1111
Fax Phone: «81)) 82 252 5343
Web: http://www.mazda.com/
Description: Holographic Interferometry

McMahan Electro-Optic
2160 Park Avenue
(Orlando Division)
City: Winter Park
State: FL
Postal Code: 32789
Country: United States of America
Voice Phone: (1) 407 645 1000
Fax Phone: (1) 407 644 9000
Email: bobmcmahn@aol.com
Contact #1: Robert McMahan
Description: McMahan Electro-Optics manufactures a laser-based NDT system for testing composite aerospace components and assemblies. Mobile unit.

Media Interface, Ltd.
215 Berkeley Place
City: Brooklyn
State: NY
Postal Code: 11217
Country: United States of America
Voice Phone: (1) 718 398 1136
Fax Phone: (1) 718 622 1753
Email: ronholog@bway.net
Web: http://www.bway.net/-ronholog
Contact #1: Ronald R. Erickson
Description: Consulting in holographic applications and mass produced and commercial holographic image design and production. Medical holography. Custom holographic optical configurations. Computer assisted holographic image design and production - research or commercial.

Mefoma Fototechnik GmbH
Ulmer Strasse 1
City: Elchingen
Postal Code: D-73450
Country: Germany
Voice Phone: (49) (0)731 266355
Description: Equipment for holographic development processes

MeiLing - Lasertechnik
Dorfstrasse 24
City: Meiling I Seefeld
Postal Code: D-82229
Country: Germany
Voice Phone: (49) (0)8153 4825
Fax Phone: (49) (0)8153 4825
Contact #1 : Harald Richter
Description: Distributor of optical/mechanical holographic equipment.
****

Meisinger Verlag
Lucile - Grahn - Strasse 39
City: MUnchen
Postal Code: D-81675
Country: Germany
Voice Phone: (49) (0)89 4194020
Fax Phone: (49) (0)89 4701081
Contact #1: Mrs. M. Zeller
Description: Publisher of books for children with holographic elements
****

Melissa Crenshaw Holography Studio International
Jl RRI
City, Province: Bowen Island, British Colombia
Postal Code: VON lGO
Country: Canada
Voice Phone: (1) 604 645 2019
Voice Phone: (1) 604 645 2019
Email: mcrensha@capcollege.bc.ca
Web: http://www.capcollege.bc.ca/dept/physicsl
Contact #1: Melissa Crenshaw
Description: Holographic fine artist with experience integrating holographic elements into commercial projects, including architectural and lighting design. Color reflect ion hologram mastering and mass production (l2x 16, 4x5 film) services offered. Extensive portfolio.
****

Melles Griot
16542 Millikan Ave.
City: Irvine
State: CA
Postal Code: 92606
Country: United States of America
Voice Phone: (1) 949 261 5600
Voice Phone: (1) 800 835 2626
Fax Phone: (1) 949 261 7589
Email: mgtech@irvine.mellesgriot.com
Web: http:\\www.mellesgriot.com
Description: Melles Griot, worldwide laser and photonics components manufacturer, offers a broad spectrum of helium neon, helium cadmium, argon ion and krypton argon ion lasers covering the blue to infrared range. Ideal for material analysis, testing and inspection, as well as interferometric measurement, surface inspection, scatter measurement, holography, high resolution metrology, optical disk mastering and more.
SEE OUR ADVERTISEMENT
****

Melles Griot GmbH
Lilienthalstrasse 30-32
City: Bensheim
Postal Code: D-64625
Country: Germany
Voice Phone: (49) (0)6251 84060
Fax Phone: (49) (0)625 840618
Email: 106223.1243@compuserve.com
Web: www.mellesgriot.com
Contact #1: Dieter KUrsten
Description: Condensers (optics),fiber optics construction components, laser diodes,laser optics, filters optical lenses.
****

Melles Griot Laser Group
2051 Palomar Airport Rd., 200
City: Carlsbad
State: CA
Postal Code: 92009
Country: United States of America
Voice Phone: (1) 7604382131
Voice Phone: (1) 800 645 2737
Fax Phone: (1) 760 438 5208
Email: sales@carlsbad.mellesgriot.com
Web: http:\\www.mellesgriot.com
Description: Melles Griot, worldwide laser and photonics components manufacturer, offers a broad spectrum of helium neon, helium cadmium, argon ion and krypton argon ion lasers covering the blue to infrared range. Ideal for material analysis, testing and inspection, as well as interferometric measurement, surface inspection, scatter measurement, holography, high resolution metrology, optical disk mastering and more.
SEE OUR ADVERTISEMENT
****

Menning, Melinda
171 Hopetoun Ave., Vaucluse
City: Sydney
State: NSW
Postal Code: 2030
Country: Australia
Voice Phone: (61) 2 9850 896 ..
Fax Phone: (61) 2 9850 8983
Email: mmenning@mpce.mq.edu.au
Web: hnp://www.comp .mq.edu .au/gen/perso
n/mmenning
Cont act #1: Melinda Menning
Description : Practicing Artist. Holographic Consultant. Producing medium format limited edition Art pieces. Producing Laser Transmission masters suitable for mass replication via photopolymer and embossed material. Producer of White light reflection and transmission holograms for displays.
****

Meredith Instruments
5420 West Camelback Rd., Suite 4
City: Glendale
State: AZ
Postal Code: 85301
Country: Uni,ted States of America
Voice Phone: (1) 602 934 9387
Fax Phone: (1) 602 934 9482
Email: info@mi-lasers.com
Web: http://www.mi-lasers.com
Contact #1: Lee Toland
Description: Specializing in surplus inventories of HeNe lasers as well as argon and diode lasers, Meredith Instruments is the USA's largest laser discount dealer. Free catalogue. Laser repair.
****

Merrick, Michael G.
1605 Bensington Court
City: Normal
State: IL
Postal Code: 61761-4811
Country: United States of America
Voice Phone: (1) 309 452 5228
Email: merrick@dave-world.net
Contact #1: Michael Merrick
Description: Production of master, image plane and reflection holograms using pseudocolor and shadow gram techniques . Experience giving educational programs to elementary and adult groups. Also coordinating holography shows for fundraising . Currently working with children's museum.
****

Metrologic Instruments GmbH
Dornierstrasse 2
City: Puchheim
Postal Code: D-82178
Country: Germany
Me-Mi

Voice Phone: (49) (0)89 89019 0
Fax Phone: (49) (0)89 89019 200
Contact #1: Benny Noems
Description: see Metrologic Instruments Inc.USA. Please fax us .
****

Metrologic Instruments, Inc.
Coles Road at Route 42
City: Blackwood
State: NJ
Postal Code: 08012
Country: United States of America
Voice Phone: (1) 609 228 8100
Voice Phone: (1) 800 IDMETRO
Fax Phone: (1) 609 228 6673
Email: marketing@metrologic.com
Web: http://www. metrologic.com/
Contact #1: Christen Kendall
Descripti on: Metrologic In struments manufacturers holographic laser bar code scanners or HoloTrak (TM). The Holotrak is omnidirectional with a 40 inch depth of field and 26 inch scan width. Applications include pallet scanning, unattended scanning, truck unloading and conveyor belts.
****

Meulien Odile
Madam Weg 77
City: Braunschweig
Postal Code: D-38120
Country: Germany
Voice Phone: (49) (0)531 352816
Fax Phone: (49) (0)531 352 816
Email: holodietma@aol.com
Web: www.holonet.khm.de/holodietma/od/
index.htm
Contact #1: Odile Meulien
Description: Collector and analyst of the holographic trend since 10 years in the US and Europe. Manage a private collection. Conduct studies on future uses of holography. Coordinate holographic and light happenings.
****

MGM Converters Inc.
16604 Edwards Road
City: Cerritos
State: CA
Postal Code: 90703
Country: United States of America
Voice Phone: (1) 562 404 3779
Fax Phone: (1) 5624047408
Contact #1: Steve Meyer
Description: Full service converting services for the holography market. Foil hot stamping, including continuous application.
****

Micos GmbH & Co KG
Am Laidholzle 1
City: Umkirch
Postal Code: D-79224
Country: Germany
Voice Phone: (49) (0)7665 51052
Fax Phone: (49) (0)7665 51737
Contact #1: Hr. Ammelung
Description: Distributor of optical/mechanical holographic equipment.
****

Microplan GmbH
Am Holzbrunnen 4
City: SaarbrUcken
Postal Code: D-66121
Country: Germany
Voice Phone: (49) (0)681 813137
Description: Distributor of optical/mechanical and chemical holographic equipment.
****

CONTACT US TODAY TO INCLUDE
YOUR BUSINESS IN THESE LISTINGS.

Midwest Laser Products
PO Box 262
City: Frankfort
Staie: TL
Postal Code: 60423
Country: United States of America
Voice Phone: (1) 815 464 0085
Fax Phone: (1) 815 4640767
Email: mlp@midwest-Iaser.com
Web: http://www.midwest-laser.comi
Contact #1: Steve Garrett
Description: We carry HeNe, Argon, HeCd, Nd:YAG and visible diode lasers. We sell complete holography kits, including low-cost HeNe lasers and related materials. Lasers for holography starting under $100.
SEE OUR ADVERTISEMENT
****

Miller, Neal
The Career Center
4203 South Providence Road
City: Columbia
State: MO
Postal Code: 65203
Country: United States of America
Voice Phone: (1) 573 886 2610
Fax Phone: (1) 573 886 2904
Email: nemiller@columbia.k12.mo.us
Contact #1: Neal Miller
Description: Physics technology instructor. Optics related courses for vocational instruction.
****

Miller, Peter
136 Clinton Road
City: Newfoundland
State: NJ
Postal Code: 07435
Country: United States of America
Voice Phone: (1) 973 697 1773
Contact #1: Peter Miller
Description: Professional holographer - 20 years experience.
****

Ministry Of International Trade
Electrotechnical Laboratory
Optical Information Section
City: Tsukuba Science City
Postal Code: 305
Country: Japan
Voice Phone: «8 1» 298 58 5625
Fax Phone: ((81) 298 58 5627
Contact #1: Dr. Satoshi Ishihara
Description: Research using HOEs.
****

Minjian Laser Holography
Anticounterfeit Label Factory
Tong-Bei-Lu #540
City: Shanghai
Postal Code: 200082
Country: China
Contact #1: Wang Rumo
****

Mitsubishi Heavy Industries Ltd.
Nagasaki Technical Institute
I-I Akunoura-Machi
City: Nagasaki
Postal Code: 850-91
Country: Japan
Voice Phone: ((8 1» 3 3218 2111
Email: info@mitsubishi.com
Web: http://www.mitsubishi.comi
Contact #1: M. Murata
Description: Holographic non-destructive testing; industrial research.
****

Mitutoyo Measuring Instruments (MTI Corp.)
965 Corporate Blvd.
City: Aurora
State: IL

Postal Code: 60504
Country: United States of America
Voice Phone: (1) 630 820 9666
Fax Phone: (1) 630 820 1393
Contact #1: Bill Naaman
Description: Manufacturers of precision measuring instruments including highly accurate holographic linear tracking systems suitable for precision industrial and research applications.
****

Modern Marketing
Olgastrasse 2
City: Boennigheim
Postal Code: D-74357
Country: Germany
Voice Phone: (49) 7143 22909
Contact #1: Wilfried Moedinger
Description: Marketillg of holography in the area of youth education.
****

Molins PLC
13-13A Westeood Way
Westwood Business Park
City: Coventry
State: England
Postal Code: CV4 8HS
Country: United Kingdom
Voice Phone: (44) 1203 421 100
Fax Phone: (44) 1203 421 255
Email: Mike.Cahill@Molins.com
Web: http://www.molins.comi
Description: Makes foil applicator machines to work on currency and other paper substrates.
****

Moonbeamers
8 CuI goa Ave
City: Eastwood
State: NSW
Postal Code: 2122
Country: Australia
Voice Phone: (61) 2 9878 6427
Fax Phone: (61) 2 9878 6427
Contact #1: John Tobin
Description: Since 1984, we have been producing commercial holograms and diffractions for security and display applications. We offer a complete service from artwork creation through application & printing within Australia and Asia.
****

Morning Light Holograms
106 Xi Huan Middle Street
Cang Zhou
City: He Bei
Postal Code: 061001
Country: China
Voice Phone: (86) 317 226 164
Fax Phone: (86) 317 226 167
Contact #1: Chen Guo Tong
Description: Hologram distributor.
****

Mu's Laser Works
1328 Dunsterville Avenue
City, Province: Victoria, British Colombia
Postal Code: V8Z 2XI
Country: Canada
Voice Phone: (1) 250 479 4357
Contact #1: Ron Meuse
Description: Holographic and 3-D photographic services, Can provide lab rental and technical assistance. Laser light show production and rental.
****

Mulhern, Dominique
I, Residense les Camelias,
7, rue du 18 Juin 1940
City: Asnieres
Postal Code: F-92600
Country: France
Voice Phone: (33) 1 47 94 82 42

Fax Phone: (33) I 47 94 82 42
Email: dwm@mail.dotcom.fr
Web: http://www.alphapix.com
Description: Holographic artist.
****

Multiplex Moving Holograms
746 Treat Street
City: San Francisco
State: CA
Postal Code: 94110
Country: United States of America
Voice Phone: (1) 415 285 9035
Fax Phone: (1) 415 206 1622
Contact #1: Peter Claudius
Description: We are the originators of the Multiplex Hologram. We produce white light viewable moving holograms for trade shows and exhibits. 120, 360 degree and flat format white light viewable holograms made to your specifications. Stock images also available. Ask for our catalogue! In business since 1973!
****

Miinchner Volkshochschule e. V.
Am Gasteig-KeUerstrasse 6
City: Munich
Postal Code: D-81667
Country: Germany
Voice Phone: (49) (0)89 418060
Description: Seminars on holography
****

Museu D' Holografia
Jaume I, 1
City: Barcelona
Postal Code: 08002
Country: Spain
Voice Phone: (343) 3 102 172
Fax Phone: (343) 3 319 1676
Email: museuholos@rnx3 .redestb.es
Web: http://www.museuholos.com
Description: Holographic Gallery, Itinerant exhibitions, sale of holograms. Teaching. Holographic courses. Holographic laboratory.
****

Museum 3. Dimension
Stadtmiihle / Nbrdlinger Tor
City: Dinkelsbuehl
Postal Code: D-91S50
Country: Germany
Voice Phone: (49) (0)9851 6336
Fax Phone: (49) (0)69 787777
Contact #1: Gerhard Stief
Description: Museum for holography and stereography.
****

Museum 3. Dimension
N brdlinger Tor
City: Dinkelsbiihl
Postal Code: D-91550
Country: Germany
Voice Phone: (49) (0)9851 6336
Fax Phone: (49) (0)9851 2882
Contact #1: Gerhard Stief
Description: German private museum. Holography collection on display.
****

Museum Of Holography / Chicago
1134 West Washington Blvd.
City: Chicago
State: IL
Postal Code: 60607
Country: United States of America
Voice Phone: (1) 312 226 1007
Fax Phone: (1) 312 829 9636
Email: museum@concentric.net
Web: http://www.cris.comi-museumh
Contact #1: Loren Billings
Description: Founded in 1978, the MOHC is now the world's oldest institution devoted to

the display, acquisition and maintenance of holography as well as education and research in the field. Permanent collection is now the largest in the world. At least two major exhibitions a year featuring artists from around the world.
****

MWK Industries
1269 West Pomona Road #112
City: Corona
State: CA
Postal Code: 91720
Country: United States of America
Voice Phone: (1) 909 278 0563
Voice Phone: (1) 800 356 7714
Fax Phone: (1) 909 278 4887
Email: mkennyl989@aol.com
Web: http://www.mwkindustries.comi
Contact #1: Mike Kenny
Description: Large selection of surplus and used lasers from major manufacturers. Save 30% to 60% on brand name laser purchases. We offer the beginning, intermediate and advanced holographer a large selection of lasers suitable for holography (including HeNe lasers ranging up to 25 mw), as well as other related materials.
SEE OUR ADVERTISEMENT
****

N.Y.K. Holographie
New Vision Konzept
Otto Wehrle Strasse 9
City : Emmedingen
Postal Code: 0-78312
Country: Germany
Voice Phone: (49) (0)764 1 52854
Fax Phone: (49) (0)7641 570913
Description: Distributor for all types of holograms.
****

Nakamura, Ikuo
864 President SI. #3
City: Brooklyn
State: NY
Postal Code: 11215
Country: United States of America
Voice Phone: (1) 718 6369112
Email: ikuo@spacelab.net
Web: http://www.spacelab.net/-ikuo/
Contact #1: Ikuo Nakamura
Description: Artist with portfolio.
****

Nanjing Sanle Laser Technology R&D
Nanjing Pukou New Technology Development Zone
PO Box 62
City, Province: Nanjing, Jiangsu
Postal Code: 210032
Country: China
Voice Phone: (86) 25 8840357
Fax Phone: (86) 25 3304991
Contact #1: Lu Zhongming
****

NEC Electronics (Europe) GmbH
Oberrather Str. 4
City: Duesseldorf
Postal Code: 0-40472
Country: Germany
Voice Phone: (49) (0)211 65 03-01
Fax Phone: (49) (0)211 65 03-488
Description: Manufacturer of lasers-di stribution argon ion laser-he-ne laser.
****

Neo Vision Productions
PO Box 74277
City: Los Angeles
State: CA
Postal Code: 90004
Country: United States of America
Voice Phone: (1) 323 255 3597

Contact #1: Bill Hillard
Description: Traveling show. Fine art originals. Produce holograms for home and industry. Consulting.
****

New Dimension Holographics
23 Victoria Ave.
City: Concord West
State: NSW
Postal Code: 2138
Country: Australia
Voice Phone: (61) 2 9743 3767
Voice Phone: (6 1) 15 435 076
Fax Phone: (6 1) 2 9743 3241
Contact #1: Tony Butteriss
Description: Retail shop. Wholesale distribution. Origination consultant. Educational consultant. Gallery.
****

New Focus, Inc
2630 Walsh Ave.
City: Santa Clara
State: CA
Postal Code: 95051-0905
Country: United States of America
Voice Phone: (1) 408 980 8088
Fax Phone: (1) 408 980 8883
Email: Contact@NewFocus.com
Web: http://www.newfocus.comi
Contact #1: Milton Chang
Description: New Focus is a supplier of photonics tools for laser applications. Products include narrow-linewidth tunable diode lasers, ultrafast photo detectors (DC-60 GHz), electro-optic modulators, wavelength meters, mechanical positioners, motorized positioners, and high-performance optics.
****

New Light Industries
West 9715 Sunset Hwy.
City: Spokane
State: WA
Postal Code: 99224
Country: United States of America
Voice Phone: (1) 509 456 8321
Fax Phone: (1) 509 456 8351
Email: stevem@iea.com
Web: http://www.iea.comi-nli
Contact #1: 'Steve McGrew
Description: Extensive experience with technology transfer, consulting and R&D for embossed holography. Complete origination and production system installations, worldwide.
****

New York Hall Of Science
47-01 111 Th Street
City: Corona
State: NY
Postal Code: 11368
Country: United States of America
Voice Phone: (1) 718 699 0005
Email: mweiss@nyhallsci.org
Web: http://www.nyhallsci.org/
Contact #1: Beth Weinstein
Description: The New York Hall of Science is New York's only hands-on science and technology museum. Lasers and optics demonstrated daily. Color hologram depicting quantum atom is on display.
****

New York Holographic Laboratories
P.O. Box 20391
Thomkins Square Station
City: New York
State: NY
Postal Code: 10009
Country: United States of America
Voice Phone: (1) 212 674 1007

Fax Phone: (1) 212 677 6304
Email: dschwitzer@mindspring.com
Web:www.dschwitzer.home.mindspring.com
Contact #1: Dan Schweitzer
Description: Fine art editions. Tutorial courses, lectures and consultations.
****

Newport Corporation
1791 Deere Ave.
City: Irvine
State: CA
Postal Code: 92606
Country: United States of America
Voice Phone: (1) 800 222 9980
Voice Phone: (1) 949 863-3144
Fax Phone: (1) 949 253 1800
Email: sales@newport.com
Web: http://www.newport.com
Contact #1: Gary Spiegel
Description: Designer and manufacturer of E/O components, optics, spatial filters, optical & beamsteering instruments, magnetic bases, fiber optic components, vibration isolation systems , and holographic recording materials.
****

Newport GmbH
European Headquarters
Holzhofallee 19
City: Darmstadt
Postal Code: 0-64295
Country: Germany
Voice Phone: (49) (0)615 362 10
Fax Phone: (49) (0)615 362 152
Description: Designer and manufacturer of laser/holographic systems-E/O componentsoptics- spatial filters-optical & beamsteering instruments-magnetic bases-fiber optic components- vibration isolation systems-and holographic recording material s.
****

Nihon University
Dept Electronic Engineering
7-24-1 Narashinodai
City: Funabashi-Shi, Chiba
Postal Code: 274-0063
Country: Japan
Voice Phone: (81) 474 69 5391
Fax Phone: (81) 474 67 9683
Contact #1: Dr. Hiroshi Yoshikawa
Description: Research NOT
****

Nika Prazisionsoptik GmbH
Vertriebs Service & Beratungsgesellschaft
Schulstrasse 20
City: Winningen Mosel
Postal Code: 0-56333
Country: Germany
Voice Phone: (49) (0)2606 363
Description: Distributor of optical/mechanical holographic equipment.
****

Nimbus Manufacturing, Inc.
(a division of Nimbus CD International)
P.O. Box 7427
City: Charlottesville
State: VA
Postal Code: 22906
Country: United States of America
Voice Phone: (1) 800 231 0778 x457
Voice Phone: (1) 804985-1100
Fax Phone: (1) 804985 4625
Email: Ihaney@nimbuscd.com
Web: http://www.nimbuscd.comi
Contact #1: Lorri Haney
Description: Nimbus manufactures holographic CDs, CD ROMs, CDIs, enhanced CDs, and DVD' s, as well as providing packaging, pre-press, print procurement, and spine labels

Nippon Polaroid K.K.
Business Development Div. - Mori Bldg No.30
3-2-2 Toranomon, Minato-ku
City: Tokyo
Postal Code: 105-0001
Country: Japan
Voice Phone: (81) 3 3438 8883
Fax Phone: ((1) 3 5473 8637
Contact #1: Makoto ide
Description: Subsidiary of Polaroid Corp., Cambridge, MA USA
****

Nippondenso Co., Ltd.
System Development Engineering
1-1 Showa-Cho Kariya-Shi
City: Aichi-Ken
Postal Code: 448-0029
Country: Japan
Voice Phone: ((81)) 566 25 6 948
Contact #1: Hiroshi Ando
Description: Manufacture HeadsUp Display. Also Mr. Toru Mizuno, Mr. Tatsuya Fujita, or Mr. Shinji Nanba.
****

Nissan Motor
Central Research Lab
Natsushima Machi
City: Yokosuka
Postal Code: 237
Country: Japan
Voice Phone: ((81)) 468 625 182
Fax Phone: ((81)) 46 654 183
Description: Hologram manufacturer headup display.
****

Norland Products, Inc.
695 Joyce Kilmer Avenue
City: North Brunswick
State: NJ
Postal Code: 08902
Country: United States of America
Voice Phone: (1) 7,32 545 7828
Fax Phone: (1) 732 545 9542
Email: sales@noriandprod.com
Web: http://www.norlandprod.coml
Description: Optical adhesives (which cure with UV light). used to adhere HOEs and for splicing fiber optic cables.
****

North Light Holograms Ltd.
PO Box 40
City, Province: Tai-an, Shandong
Postal Code: 271039
Country: China
Voice Phone: (86) 538 6511001
Fax Phone: (86) 538 6511001
Contact #1: Wang Jizhang
Description: Holography sales and distribution.
****

Northern Illinois University
Department Of Physics
City: Dekalb
State: IL
Postal Code: 60115
Country: United States of America
Voice Phone: (1) 815 753 1000
Email: rossing@niu.edu
Web: http://www.niu.edu/
Description: Scientific holography research. Projects vary in nature. Holographic interferometry for studying vibration modes such as in musical instruments.
****

NovaVision
419 Gould Street, Suite #3
City: Bowling Green
State: OH
Postal Code: 43402
Country: United States of America
Voice Phone: (1) 419 354 1427
Voice Phone: (1) 800 990 6682
Fax Phone: (1) 419 353 7908
Email: nova@wcnet.org
Web: http://www.wcnet.org/-nova!
Contact #1: Albert J. Caperna
Description: NovaVision (TM) direct holographic embossing technology allows printers to produce embossed holograms and other OVDs in-line, on-press at production speeds on a variety of substrate materials. Whereas conventional holograms are transferred from pre-image foil, the patented NovaVision process allows holograms to be created as an integral part of the printing process ... no inventory, lower costs, higher application speed.
****

Numazu College Of Technology
Dept Of Mechanical Engineering
3600 Ooka
City: Numazu-City, Shizuoka
Postal Code: 410-0022
Country: Japan
Voice Phone: (8 1) 559 20 3714
Contact #1: Dr. Koji Lkegami
Description: Holographic research.
****

Odhner Holographies
33 Juniper Drive
City: Amherst
State: NH
Postal Code: 03031
Country: United States of America
Voice Phone: (1) 603 673 8651
Fax Phone: (1) 603 673 8685
Email: info@stabilock.com
Web: www.stabilock.com
Contact #1: Jefferson E. Odhner
Description: Exclusive distributor of the Stabilock II inch fringe stabilizer (used to make brighter holograms), manufacture of custom holograms (trans/refl.). Specializing in HOE arrays (to 8" x 10") on silver halide.
SEE OUR ADVERTISEMENT
****

Oeserwerk Ernst Oeser & Sohne KG
Rigistrasse 20
City: Goppingen-Holzheim
Postal Code: D-73037
Country: Germany
Voice Phone: (49) (0)7 161 8009-0
Fax Phone: (49) (0)7161 8009-10
Contact #1: Ernst Dr. Qeser
Description: embossed hologram foils-holographic labels
****

Ojasmit Holographics
409 Vardhman Market Sector 17, VASHI
City: New Bombay
Postal Code: 400703
Country: India
Voice Phone: (91) 22 768 3526
Voice Phone: (91) 22 763 0373
Fax Phone: (91) 22 763 2509
Contact #1: Kailesh Shah
Description: Manufacturing and marketing of embossed holograms , photopolymers and dichromates for varied applications.
****

OMIKRON GmbH
Abteilung Feinchemikalien
Marktplatz 5
City: Neckarwestheim
Postal Code: D-74382
Country: Germany
Voice Phone: (49) (0)7133 17081
Fax Phone: (49) (0)7 133 17465
Web: www.omikron-online.de/cyberchernl
Description: Distributor of chemical holographic equipment.
****

Ontario College of Art/Holography
Art Division
100 McCaul St.
City, Province: Toronto, Ontario
Country: Canada
Voice Phone: (1) 416 977 6000 X263
Email: mpage@ocad.on.ca
Web: http://www.ocad.on.ca!
Contact #1: Michael Page
Description: Education & Seminars. Regular holography courses.
****

Ontario Science Centre
770 Don Mills Road
City, Province: Don Mills, Ontario
Postal Code: M3C I T3
Country: Canada
Voice Phone: (1) 416 429 4100 x 2820
Fax Phone: (1) 416 696 3197
Email: alena_kottova@fcgate1.osc.on.ea
Web: http://www.osc.on.ca!
Description: We have gallery of 15 holograms on permanent display and laser demonstration area. Holography workshops cover theory and practical uses of holography. Participants make their own reflection hologram.
****

Op-Graphics (Holography) Ltd.
Unit 4 - Technorth
7 Harrogate Road
City: Leeds
State: England
Postal Code: LS7 3NB
Country: United Kingdom
Voice Phone: (44) 113 262 8687
Fax Phone: (44) 113 237 4182
Email: n.hardy@ukonline.co.uk
Contact #1: Valerie Love
Description: Manufacturer of display holograms. Large selection of stock images in variety of formats and sizes. Commissioned work undertaken. Copying work for holographers undertaken .
SEE OUR ADVERTISEMENT
****

OpSec - USA
PO Box 700
City: Parkton
State: MD
Postal Code: 21120
Country: United States of America
Voice Phone: (1) 410 357 4491
Fax Phone: (1) 410 357 4485
Email: 103320.340@compuserve.com
Web: www.opticalsecurity.com
Contact #1: Dean Hill
Description: Embossed hologram manufacturing facility specializing in security and authentication applications. Custom work accepted. Stock items available, including 38 patterns of foil (16 colors). Company pioneered mass replication of embossed holograms (formerly Difco.)
****

OPTIC
Office. 35, 10 Plahotnogo St.
City: Novosibirsk
Postal Code: 630108
Country: Russia
Voice Phone: (7) 3832 43 29 22
Fax Phone: (7) 3832 43 29 22
Email: shoydin@ssga.ru
Web: http://www.ssga.ru:8080IFRIENDS/OPTIClindex.htm
Contact #1: Sergey Shoydin
Description: We were founded in 1992 as a

company for developing and producing holograms and holotechnology. During the years we have successfully introduced many new products into the Russian market. We have manufactured over 10 million holograms. We also seek collaboration with others to deliver their hologram products to Russia.
****

Optical Research Associates
3280 East Foothill Blvd. , Suite 300
City: Pasadena
State: CA
Postal Code: 91107-3103
Country: United States of America
Voice Phone: (1) 626 795 9101
Fax Phone: (1) 626 795 9102
Email: service@opticalres.com
Web: http :\\www.opticalres.com
Contact #1: Lia Titizian
Description: Optical Design Software. We sell the programs that allow you to create Holographic Optical Elements.
****

Optical Security Group
Corporate Headquarters - Suite 920
535 16th St.
City: Denver
State: CO
Postal Code: 80202
Country: United States of America
Voice Phone: (1) 303 534 4500
Voice Phone: (1) 773 665 8932
Fax Phone: (1) 303 534 1010
Email: staff@opticalsecurity.com
Web: http: //www.opticalsecurity.comJ
Contact #1 : Richard Bard
Description: Produce custom holographic labels as optical security devices for authentication applications. Full range of production services offered. Also produce lenticulars.
****

Optical Security Group - England
4EIF Gelders Hall Road
City, Province: Shepshed, Leicestershire
State: England
Postal Code: LE12 9NH
Country: United Kingdom
Voice Phone: (44) 1509 600 220
Voice Phone: (44) 173023 1144
Fax Phone: (44) 1509 508 795
Email: 100745.1342@compuserve.com
Description: A total secure service from concept design artwork to finished product-specializing in customer service and deli vering quality embossed holograp hic security and non-security work on time.
****

Optical Society of America (OSA)
2010 Mass Avenue NW
City: Washington
State: DC
Postal Code: 20036-1023
Country: United States of America
Voice Phone: (1) 202 223 8130
Voice Phone: (1) 800 762 6960
Fax Phone: (1) 202 223 1096
Email: osamem@osa.org
Web: http://www.osa.org
Description: Organization devoted to promoting optics and photonics research and applications. Publications include: Applied Optics, Optics Letters, Optics and Photonics News, Journal of Optical Society of America.
****

Optical Test Equipment
(a division of J.D. Moeller Optische Werke GmbH)
Rosengarten 10
City: Wedel

Postal Code: D-22880
Country: Germany
Voice Phone: (49) (0)4103 709 345
Fax Phone: (49) (0)4 103 709 375
Email: mail@moeller-wedel.com
Web: http://www.moeller-wedel.com
Contact #1: Carsten Schlewitt
Description: Manufactures custom optical components and optical test equipment including auto collimators, testing telescopes, focometers, goniometers, goniometer, spectrometers, and Fizeau-type interferometers.
****

Optical Works Ltd.
Ealing Science Centre
Treloggan Lane
City, Province: Newquay, Cornwall
State: England
Postal Code: TR7 IHX
Country: United Kingdom
Voice Phone: (44) 1637 87 7222
Fax Phone: (44) 1637 87 7211
Contact #1: E.O. Frisk
Description: Make optical components, lenses and scientific instruments.
****

Optics Plus Inc.
1369 East Edinger Avenue
City: Santa Ana
State: CA
Postal Code: 92705
Country: United States of America
Voice Phone: (1) 714 972 1948
Fax Phone: (1) 714 835 6510
Description: Manufacture optics; precision tool mounts (including lens and mechanical mounts).
****

Optilas GmbH
Boschstrasse 12
City: Puchheim
Postal Code: D-82178
Country: Germany
Description: Manufacturer and Designer of HOEs
****

Optimation
6765 South 400 West
City: Midvelle
State: UT •
Postal Code: 84047
Country: United States of America
Voice Phone: (1) 801 263 6575
Fax Phone: (1) 801 263 6576
Email: optimat@info.net
Web: http://www.edmmicromachining.com
Contact #1: Dean Jorgensen
Description: Specialize in the manufacture of excellent quality, burr-free pinholes for holographic and related optical applications. Call for catalog .
****

Optimation Holographics
3200 South Haskell, Suite 160
City: Lawrence
State: KS
Postal Code: 66046
Country: United States of America
Voice Phone: (1) 785 841 1642
Fax Phone: (1) 785 841 0439
Contact #1: Terry Faddis
Description: Large format holographic embossing facility. Complete in-house system including resist master and shim making (up to 16 inches wide).
****

Optineering
2247 E. La Mirada St.
City: Tucson
State: AZ
Postal Code: 18719

Country: United States of America
Voice Phone: (1) 520 882 2950
Fax Phone: (1) 520 882 6976
Email: kcreath@primenet.com
Web: http://www.primenet.comJ-kcreath
Contact #1: Kathy Creath
Description: Optical engineering consulting services specializing in optical testing, metrology, nondestructive evaluation, and optical design. Application areas include holography, NDT, speckle interferometry, microscopy, photography, and process control and monitoring .
****

Optlectra
Leisssstrasse 8
City: Feldkirchen - Westerham
Postal Code: D-83620
Country: Germany
Voice Phone: (49) (0)8063 888-8
Fax Phone: (49) (0)8063 888-9
Email: info@optlectra.com
Web: www.optlectra.com
Contact #1 : Werner Bleckwendt
Description: Distributor of optical holographic equipment .
****

Optopol Panoramic Metrology Consulting
Csiksomlyo u. 4.
City: Budapest
Postal Code: H -1025
Count ry: Hungary
Voice Phone: (36) I 463 2518
Voice Phone: (36) 1 335 5 139
Fax Phone: (36) 1 463 3178
Email: gregyss@next-lb.manuf.bme.hu
Contact #1: Pal Greguss
Description: Nonmultiplexed single-shot 360 degree panoramic holograms, based on the Panoramic Annular Lens (Pal-optic) invented by Dr. Greguss, are produced for metrological and other applications in science, technology and arts.
****

Orazem, Vito
Max Reger Strasse 17-19
City: Essen
Country: Germany
Voice Phone: (49) (0)201 2376 16
Fax Phone: (49) (0)201 230054
Email: 101727.1366@compuserve.com
Description: artist
****

Oregon Institute of Technology
Laser Optical Engineering Technology
3201 Campus Drive
City: Klamath Falls
State: OR
Postal Code: 97601-8801
Country: United States of America
Voice Phone: (1) 541 885 1698
Fax Phone: (1) 541 885 1666
Email: piecer@oit.osshe.edu
Web: http://www.oit.osshe.edu/
Contact #1: Robert Pierce
Description: The LOET program provides state of the art education by combining an applied laboratory approach to optical engineering technology together with theoretical classroom discussion. For program and course information, contact Dr. Robert Pierce.
****

Oregon Laser Consultants
455 Hillside Ave.
City: Klamath Falls
State: OR
Postal Code: 97601-2337
Country: United States of America
Voice Phone: (1) 541 882 3295
Email: olcbi1l@aip.org

Contact #1: Bill Deutschman
Description: Specialists in laser safety consulting and laser safety training. ANSI services for laser users and CDRH services for laser manufacturers. Also laser safety audits, employee training and electronic consulting.
****

Oriel Instruments
150 Long Beach Boulevard
City: Stratford
State: CT
Postal Code: 06615
Country: United States of America
Voice Phone: (1) 203 377 8282
Fax Phone: (1) 203 378 2457
Email: res_sales@oriel.com
Web: http://www.oriel.com
Description: A full line of optical components for holographic and related laboratory applications. Call for our catalog.
****

Ose Holografie-Design
LenneperstraBe 13-15
City: Wuppertal
Postal Code: 0-42289
Country: Germany
Voice Phone: (49) 202 825636
Contact #1 : Christian Ose-Wiese
Description: Manufacturing of master and display holograms.
****

OWIS GmbH
1m Gaisgraben 7
City: Staufen
Postal Code: D-79219
Country: Germany
Voice Phone: (49) (0)7633 9504 0
Fax Phone: (49) (0)7633 9504 44
Contact #1: Hubert Munzer
Description: Optical benches in different sizes. Mirror mounts (gimbal and kenematic)-mirrors and lenses.
****

Oxford Holographies
71 High Street
City: Oxford
State: England
Postal Code: OXI 4BA
Country: United Kingdom
Voice Phone: (44) 1865 250 505
Voice Phone: (44) 1865 250 506
Fax Phone: (44) 1865 250 505
Email: oxfordhologrphics@zetnet.co.uk
Contact #1: Nick Cooper
Description: Oxford Holographics has both a very well established retail and an expanding distribution operation, focusing on unusual and unique giftware, including holograms.
****

Pacific Holographics Inc.
503 Caledonia Street
City: Santa Cruz
State: CA
Postal Code: 95062
Country: United States of America
Voice Phone: (1) 408 425 4739
Fax Phone: (1) 408 425 4739
Email: pacholo@ix.netcom.com
Contact #1: Randy James
Description: Photo-resist mastering for embossed holography. Origination, design and consulting services offered. Extensive commercial portfolio.
****

Page , Michael
22 Bayview Avenue
City, Province: Toronto, Wards Island
Postal Code: M5J IZI
Country: Canada
Voice Phone: (1) 416 203 7243

Fax Phone: (1) 416 203 7243
Contact #1: Michael Page
Description: Holographic artist and technician.
****

Panatron Inc.
P.O. Box 2687
City: Pomona
State: CA
Postal Code: 91769-2687
Country: United States of America
Voice Phone: (1) 909 629 0748
Voice Phone: (1) 800 669 7945
Fax Phone: (1) 909 620 0378
Email: panatron@aol.com
Web: http://www.panatron.coml
Description: Supplies complete support, parts and service on all lasers. Also manufactures mirrors, lenses, rods and other parts for lasers. Laser repair and used lasers.
****

Pangaea Design
PO Box 2028
City: New York
State: NY
Postal Code: 10009
Country: United States of America
Voice Phone: (1) 888 772 6423
Description: Model maker for holography.
****

Parallax Gallery
Shop R-l, Harbor Rocks Hotel
Nurses Walk, The Rocks
City: Sydney
State: NSW
Postal Code: 2000
Country: Australia
Voice Phone: (61) 29247 6382
Fax Phone: (61) 29 247 6382
Contact #1: Tony Butteriss
Description: Hologram Gallery. Large vari ety of holograms for sale to the public.
****

Parker Foils, Inc.
Building 3W, P.O. Box 949
Chimney Rock Road
City: Bound Brook
State: NJ
Postal Code: 08805
Country: United States of America
Voice Phone: (1) 732 271 5770
Fax Phone: (1) 732 271 9393
Description: Manufacturer of hot stamping foils . Holographic patterns and imaging.
****

Peacock Laboratories, Inc.
1901 S. 54th St.
City: Philadelphia
State: PA
Postal Code: 19143
Country: United States of America
Voice Phone: (1) 215 729 4400
Fax Phone: (1) 215 729 1380
Email: plabs@bellatlantic.net
Description: Established 1930. Dedicated to developmental research in mirror manufacturing, silver metalizing, and protective coatings. Innovators of silver spray processes and dual-nozzle spray guns. Consultants and suppliers of silvering solutions and chemicals for electroconductive, decorative and reflective applications.
****

Pepper, Andrew
46 Crosby Road
City, Province: West Bridgford, Nottingham
State: England
Postal Code: NG2 5GH
Country: United Kingdom
Voice Phone: (44) 7050 133 624
Fax Phone: (44) 7050 133 625

Email: pepper@monand.demon.co .uk
Web: http://www.holo.comlpeper/search.html
Contact #1 : Andrew Pepper
Description: Fine art holography. Limited editions, unique pieces, collaborations. Main work in reflection holography.
****

Phantastica
Suchtener Strasse 4a
City: Arnsberg
Postal Code: D-59757
Country: Germany
Voice Phone: (49) (0)293 81917
Fax Phone: (49) (0)293 29441
Contact #1: Gerd M. Albrecht
Description: Makers and distributors of articles related to embossed holograms and diffraction foil, including earrings and other jewelry, badges, pens, mobiles. Main focus is street-crafts and Christmas markets.
****

Photo Research, Inc.
9731 Topanga Canyon Place
City: Chatsworth
State: CA
Postal Code: 91311-4926
Country: United States of America
Voice Phone: (1) 818 341 5151
Fax Phone: (1) 818 341 7070
Email: sales@photoresearch.com
Web: http://www.photoresearch.com
Description: Photo Research is the world leader in manufacturing precision instruments that measure light and color. We serve the following markets: CRTIFPD, automotive, aerospace (commercial and military), motion picture, R & D, and many other related industries. For over 50 years, our leadership has delivered world-class light measurement solutions.
****

Photo Sciences
2542 W. 237th Street
City: Torrance
State: CA
Postal Code: 90505
Country: United States of America
Voice Phone: (1) 310 784 7460
Fax Phone: (1) 310 539 6740
Email: scsales@photo-sciences.com
Web: http://www.photo-sciences .coml
Contact #1: John Stogsdill
Description: PSI was among the earliest producers of photo masks for the computer and electronics industry. PSI's reputation as a high quality photomask manufacturer has become a well accepted fact within the industry.
****

Photon League Of Holographers Ontario
401 Richmond Street West Suite B03
City, Province: Toronto, Ontario
Postal Code: M5U 3A8
Country: Canada
Voice Phone: (1) 416 599 9332
Voice Phone: (1) 416 203 7243
Contact #1: Claudette Abrams
Description: Artist run non-profit holography studio. Technical workshops throughout the year. 2 tables 50mw HeNe. Copy Lab. Stereogram LCD HOP. Associate Membership $151 year Lab Users Program $60/year.
****

Photonics Spectra
Laurin Publishing Co. Inc.
2 South Street
City: Pittsfield
State: MA
Postal Code: 01202
Country: United States of America
Voice Phone: (1) 413 499 0514
Fax Phone: (1) 413 442 3180

Email: photonics@laurin.com
Web: http://www.laurin.com
Description: Trade publication covering the field of optics, lasers, electro-optics, and related imaging research, as well as commercial applications.
****

Physik Instrumente (PI) GmbH & Co.
Polytecplatz 5-7
City: Waldbronn
Postal Code: D-76337
Country: Germany
Voice Phone: (49) (0)724 604-100
Fax Phone: (49) (0)724 604-145
Contact #1: Karl Dr. Spanner
Description: Holography, laser, optical components, miscellaneous, oscillation insulators, vibrating dampers, PZT actuators, sensors, PZT ceramics.
****

PHYWE Systeme GmbH
Robert Bosch Breite 10
City: Gottingen
Postal Code: D-37079
Country: Germany
Voice Phone: (49) (0)551 604-0
Fax Phone: (49) (0)551 604-107
Contact #1: Mr. Platz
Description: Distributor of optical/mechanical and chemical equipment.
****

Pilkington Optronics
Glascoed Road
St. Asaph
City, Province: Clwyd, North Wales
State: England
Postal Code: LL17 OLL
Country: United Kingdom
Voice Phone: (44) 1745 583 301
Fax Phone: (44) 1745 584 258
Web: http://www.thejob.com/pilkington/
Description: Manufacturer of DCG and photopolymer HOEs and related optical components for Heads Up Displays, etc.
****

Planet 3-D
201 Silver Fox Lane
City: Downingtown
State: PA
Postal Code: 19335
Country: United States of America
Voice Phone: (1) 610 873 6192
Fax Phone: (1) 610 873 6194
Email: rcossa@aol.com
Contact #1: Rich Cossa
Description: Marketing of holo. products.
****

Point Source Productions
P.O. Box 55
City: Boulder Creek
State: CA
Postal Code: 95006
Country: United States of America
Voice Phone: (1) 831 457 1426
Fax Phone: (1) 831 338 3438
Email: hololab@cruzio.com
Contact #1: Bob Hess
Description: We are an independent recording stndio offering product design and technical imaging consultations, mastering and ganging services (specializing in silver-halide and photopolymer), and "short-run" or limited ed. transfer services.
****

Polaris Research Group
24400 Highland Road
City: Richmond Heights
State: OH
Postal Code: 44143
Country: United States of America
Voice Phone: (1) 216 383 9480

Fax Phone: (1) 216 383 9488
Email: polarisrg@aol.com
Web: w3.gwis.com/-polaris/whatis.html
Contact #1: Howard Fein
Description: Polaris Research Group offers NDT to the commercial market. We use Holography- based technology to measure vibration, stress, and structural characteristics in a wide range of applications. Modal and vibration analysis; bonded and composite structure test; and flaw, defect, and delamination identification are just a few of the primary applications for these techniques.
****

Polaroid Corporation
Optical Films Division
I Upland Road, N2-IL
City: Norwood
State: MA
Postal Code: 02062
Country: United States of America
Voice Phone: (1) 781 3864516
Voice Phone: (1) 800 225 2770
Fax Phone: (1) 781 386 5093
Email: holography.info@polaroid.com
Web: http://www.holoroid .com
Description: Fully integrated supplier of highest quality, mass produced optical films including polarizers, flexible anti-rflection films, plastic transparent conductors, retarders and holographic optical elements for a wide range of applications including enhanced LCD.
****

Polytec GmbH
Polytec Platz 5-7
City: Waldbronn
Postal Code: D-76337
Country: Germany
Voice Phone: (49) (0)7243 604 0
Fax Phone: (49) (0)7243 69944
Contact #1: H. G. Lossau
Description: Distributor of holographic laser equipment.
****

Potomac Photonics, Inc.
4445 Nicole Drive
City: Lanham
State: MD
Postal Code: 20706
Country: United States of America
Voice Phone: (1) 301 459-3033
Fax Phone: (1) 301 459-3034
Email: gbehrmann@potomac-Iaser.com
Web: http://www.potomac-Iaser.com
Contact #1: Greg Behrmann
Description: Manufactures compact UV lasers and tabletop micro machining workstations. Potomac offers rapid prototyping of computer generated holograms and diffractive optical elements.
****

PPM Promotion Products Munich GmbH
Hohenzollernstrasse 10'
Ambassadepassage-I. Stock
City: Munich
Postal Code: D-80801
Country: Germany
Voice Phone: (49) (0)89 338616
Contact #1: Thomas Kubeile
Description: Promotion and distribution of holograms and holographic jewelry, toys, gifts. Custom made holograms.
****

Print-M-Boss
5124 Kirti Nagar Indl. Area
City: New Delhi
Postal Code: 110017
Country: India
Voice Phone: (91) 11 530586
Fax Phone: (91) 11 544 1144
Contact #1: Ravinder Singh

Description: Manufacturers of embossed holograms with in-house facility for shim making. Would be interested in buying copyrights for various images and patterns. Like to make contacts for mastering.
****

Process Technologies
436 West Rawson Ave.
City: Oak Creek
State: WI
Postal Code: 53154
Country: United States of America
Voice Phone: (1) 414 571 9200
Fax Phone: (1) 414 571 9202
Email: pti@execpc.com
Web: www.execpc.com/-pti
Contact #1: Manfred Stelter
Description: Provides photoresist coated plates, ronchi rulings, reticles and masks.
****

PullTime 3-D Laboratories
29 West 26 Street
City: New York
State: NY
Postal Code: 10010
Country: United States of America
Voice Phone: (1) 212 889 5994
Fax Phone: (1) 212 889 5926
Email: pulltime3d @aol.com
Web: http://www.vision3d.com/pulltime3d1
Contact #1 : Gerald Marks
Description: PullTime 3-D Laboratories is responsible for 3D broadcast television and home video for clients including CBS Records, Fox Television, The Rolling Stones, AT&T, Howard Stern and Atlantic Records. Over 25 million PullTime 3-D glasses produced to date.
****

Pyramid Foils, Inc.800 Bond St.
City: Elizabeth
State: NJ
Postal Code: 07201
Country: United States of America
Voice Phone: (1) 908 527 OliO
Fax Phone: (1) 908 354 1461
Description: Manufacturer of quality hot stamping foils and diffraction foils.
****

Qingdao Gaoguang Holography Tech. Co
Feng-Xian-Lu #8
City, Province: Qingdao, Shangdong
Postal Code: 266071
Country: China
Contact #1: Yuan Baoqing
Description: Holography class.
****

Qingdao Qimei Images, Inc.
Qingdao Economic Development Zone
City: Qingdao
Postal Code: 266555
Country: China
Voice Phone: (86) 532 6898751
Fax Phone: (86) 532 6898751
Contact #1: Yang Caizhi
Description: Sales and distribution.
****

Quan Zhou Pacific Laser Images
Bei-Men Huan-Cheng-Lu Gongsi-Da-Xia
City, Province: Quanzhou, Fujian
Postal Code: 362000
Country: China
Voice Phone: (86) 595 2783945
Fax Phone: (86) 595 2784626
Contact #1: Wu Rongkun
Description: Hologram distribution.
****

Quantel
17, av de l' Atlantique

ZA de Courtaboeuf, BP 23
City: Les Ulis Orsay Cedex
Postal Code: F-91941
Country: France
Voice Phone: (33) I 6929 1700
Fax Phone: (33) I 6929 1729
Description: Manufacturer of lasers and other light sources, laser accessories.
****

Radiant dyes accessories GmbH
Friedrichstrasse 58
City: Wermelskirchen
Postal Code: D-42929
Country: Germany
Voice Phone: (49) (0)2196 81061
Fax Phone: (49) (0)2196 3422
Contact #1: Dr. P. Jauemik
Description: Distributor of optical/mechanical holographic equipment.
****

Rainbow Symphony Inc.
6860 Canby Ave. #120
City: Reseda
State: CA
Postal Code: 91335
Country: United States of America
Voice Phone: (1) 818 708 8400
Fax Phone: (1) 818 708 8470
Email: 3dglasses@rainbowsymphony.com
Web: http://www.rainbowsymphony.com/
Contact #1: Mark Margolis
Description: Manufacturers of uniquely designed holographic and diffraction products for the gift, novelty, advertising, specialty, premium incentive, souvenir and museum markets.
SEE OUR ADVERTISEMENT
****

Ralcon
Box 142
8501 South 400 West
City: Paradise
State: UT
Postal Code: 84328
Country: United States of America
Voice Phone: (I) 435 245 4623
Fax Phone: (I) 435 245 6672
Email: rdr@ralcon.com
Web: http://www.xmission.com/-ralcon
Contact #1: Richard Rallison
Description: Design, development and fabrication of volume holographic optical elements, (HOEs) including gratings, scanners, multifocus devices, heads up and down di splays and notch filters formed in dichromated gelatin or photopolymer.
SEE OUR ADVERTISEMENT
****

Real Image
PO Box 566
City: Pacifica
State: CA
Postal Code: 94044
Country: United States of America
Voice Phone: (1) 650 355 8897
Fax Phone: (1) 650 355 5427
Contact #1: Roy Bradshaw
Description: Incorporation of patented holographic designs into fishing tackle and fishing lures.
****

Reconnaissance International Ltd. (UK)
Runnymede Malthouse
Runnymede Road
City, Province: Egham, Surrey
State: England
Postal Code: TW20 9BD
Country: United Kingdom
Voice Phone: (44) 1784497008
Fax Phone: (44) 1784 497001

Email: 100142.1164@compuserve.com
Web: http://www.hmt.com/holography/hnews/index.html
Contact #1: Ian Lancaster
Description: We are an international consultancy for market and industry information and analysis. Publisher of "Holography News", "Holo-Pack/Holo-print Guidebook" and "Authentication News". Clients studies are fully confidential.
****

Reconnaissance International Ltd. (USA)
3003 Arapahoe St., Suite 213
City: Denver
State: CO
Postal Code: 80205
Country: United States of America
Voice Phone: (1) 3Q3 293 3000
Fax Phone: (1) 303 293 8661
Email: ReconnUSA@aol.com
Web: http://www.reconnaissance-intl.com
Contact #1: Lewis Kontnik
Description: North American office. We are an international consultancy for market and industry information and analysis. Publisher of "Holography News", "Holo-Pack/Holoprint Guidebook" and "Authentication News". Clients studies are fully confidential.
****

Red Beam, Inc.
9011 Skyline Blvd.
City: Oakland
State: CA
Postal Code: 94611
Country: United States of America
Voice Phone: (1) 510 482 3309
Fax Phone: (1) 510 482 1214
Contact #1: Lon Moore
Description: Mastering facility specializing in the design and production of masters suitable for high volume corporate applications, especially on photopolymer films and embossed materials. Clients include Activision, AT&T, NFL (Superbowl) and Polaroid. Also produces a line of trademarked giftware holograms distributed by "Lightrix" .
****

Regal Press Inc.
Holographics Division
129 Guild Street
City: Norwood
State: MA
Postal Code: 02062
Country: United States of America
Voice Phone: (1) 781 769 3900
Voice Phone: (1) 800 447 3425
Fax Phone: (1) 781 551 0466
Web: www.regalpress .com
Contact #1: William Duffey
Description: The Regal Press, Inc. has expertise in all areas of print production, including engraving, lithography, thermography, embossing, foil-stamping, and holography. We hold a worldwide patent for REGAL MARQUE, a simulated private watermarking process, and we provide our customers with REGAL EXPRESS guaranteed overnight delivery and 24-hour rush Business Cards.
****

Repro 17 SA
Av. Marillac Les Minimes B.P.36
City, Province: La Rochelle, Cedex
Postal Code: 17002
Country: FrancelMonaco
Voice Phone: (33) 5 46451900
Fax Phone: (33) I 46454550
Email: jb_lachaud@yahoo.com
****

Reva's Holographic Illusions
446 South Main Street

City: Frankenmuth
State: MI
Postal Code: 48734
Country: United States of America
Voice Phone: (1) 517 652 3922
Voice Phone: (1) 800 390 7547
Fax Phone: (1) 517 652 6503
Email: holoshop@revas-holograms.com
Web: http://www.revas-holograms.com
Contact #1: Reva Krick
Description: GallerylRetail store, with over 250 holograms on display. We feature a full line of holographic jewelry, gifts, toys, etc. Established in 1992.
****

Reynolds Metals Co.
Flexible Packaging Division
6603 West Broad St.
City: Richmond
State: VA
Postal Code: 23230
Country: United States of America
Voice Phone: (1) 804 281 2082
Voice Phone: (1) 804 281 3969
Fax Phone: (1) 804 281 2238
Email: C.L.Baynes@rmc.com
Web: http://www.rmc.com
Contact #1: Rich Patterson
Description: Holographic specialty cartons and printed paper materials for distilled spirits and wine, pharmaceuticals, confections, personal care, and other consumer goods. Holographic flexible light web materials for pouches, lidding, and overwraps. Full service from design to finishing.
****

RGB Holographics Inc.
10 Shell meetinghouse Road
City: Canterbury
State: NH
Postal Code: 03224
Country: United States of America
Voice Phone: (1) 603 783 9238
Fax Phone: (1) 603 783 9255
Email: lasernene@aol.com
Contact #1: Dr. Paul Kruzel
Description: MarketinglManufacturer of mass produced and custom photopolymer holograms produced by proprietary technology and utilizing DuPont Omnidex photopolymer films. Large and small format holograms available for commercial applications .
****

Rheinisches Landesmuseum Bonn
Colmantstrasse 14-16
City: Bonn
Postal Code: D-53115
Country: Germany
Voice Phone: (49) (0)228 7294216
Contact #1: Prof. Klaus Honnef
Description: former Matthias Lauk collection.
****

Rice Systems, Inc.
1150 Main Street, Suite C
City: Irvine
State: CA
Postal Code: 92614
Country: United States of America
Voice Phone: (1) 949 553 8768
Fax Phone: (1) 949 553 0307
Email: ricesys@aol.com
Web: http://www.ricesystems .com/
Contact #1: Colleen Fitzpatrick
Description: Laser metrology and diagnostic measurements, HNDT fluid measurements. Combustion diagnostics. Integrated optics and non linear optical material (R&D and product development). Very successful SBIR company.
****

Richardson Grating Laboratory
(a division of Spectronic Instruments)
820 Linden Ave.
City: Rochester
State: NY
Postal Code: 14625
Country: United States of America
Voice Phone: (1) 716 248 4000
Voice Phone: (1) 800 654 9955
Fax Phone: (1) 716 454 1568
Email: gratings@spectronic.com
Web: htlp:llwww.spectronic.com/
Contact #1: Susan Willard
Description: The Richardson Grating Laboratory has been a world leader in the design and manufacture of ruled and holographic diffraction grating for fifty years.
****

Richmond Holographic Studios Ltd
6 Yorkton St.
City: London
State: England
Country: United Kingdom
Voice Phone: (44) 171 739 9700
Fax Phone: (44) 171 739 9707
Email: rhs@augustin.demon.co.uk
Contact #1: Edwina Orr
Description: Pulsed laser display holography. R&D using HOE's for autostereoscopic displays.
****

Rita Wittig Fachbuchverlag
Chemnitzer Strasse 10
City: Htickelhoven
Postal Code: D-41836
Country: Germany
Voice Phone: (49) (0)2433 84412
Fax Phone: (49) (0)2433 86356
Contact #1: Dr. Siegmar Wittig
Description: Publisher of books for the practical use of holography.
****

Robert Sherwood Holographic Design
1380 Wendover Dr.
City: Charlottesville
State: VA
Postal Code: 22901
Country: United States of America
Voice Phone: (1) 804 971 2910
Fax Phone: (1) 804 971 2998
Email: evoke@aol.com
Web: http://www.holographicdesign.com/
Contact #1 : Robert Sherwood
Description: RSHD, Inc. provides custom commercial holographic products and services. Specialized management of complex material constructions and conversion of holographic materials. Products include: PS labels, Heat applied labels, films, foils , laminates and photopolymer products.
****

Rochester Inst. Of Technology
Center for Imaging Science
One Lomb Memorial Drive
City: Rochester
State: NY
Postal Code: 14623-5604
Country: United States of America
Voice Phone: (1) 716 475 2411
Fax Phone: (1) 716 475 5988
Email: info@rit.edu
Web: http://www.rit.edu/
Description: Research on HOEs, holographic materials, CGHs. Instruction in holography and related topics in the Department of Imaging and Photographic Technology, and the Center for Imaging Science.
****

Rochester Photonics Corporation
330 Clay Road
City: Rochester
State: NY
Postal Code: 14623
Country: United States of America
Voice Phone: (1) 716 272 3010
Fax Phone: (1) 716 272 9374
Email: sales @rphotonics.com
Web: http://www.rphotonics .com
Description: RPC specializes in the design and manufacturing of diffractive optical components and subsystems. Precision diffractive mastering, molding, replication, and tes ting services are provided. Products include: hybrid refractive/diffractive lenses and subassemblies, microlens arrays, diffractive phase plates, engineered diffusers, and holographic gratings.
****

Rofin-Sinar Laser GmbH
Berzeliusstrasse 85
City: Hamburg
Postal Code: D-22113
Country: Germany
Voice Phone: (49) (0)40 733 630
Fax Phone: (49) (0)40 733 63 100
Description: C02 and Nd:YAG lasers for materials processing-Laser components-Laser processing devices and machines.
****

Rolyn Optics
706 Arrow Grand Circle
City: Covina
State: CA
Postal Code: 91722-2199
Country: United States of America
Voice Phone: (1) 626 915 5707
Fax Phone: (1) 626 915 1379
Email: sales@rolyn.com
Web: http:\\www.rolyn.com
Description: General selection of optical items. Catalogue available.
****

ROSS BOOKS
P.O. Box 4340
City: Berkeley
State: CA
Postal Code: 94704
Country: United States of America
Voice Pl)one: (1) 800 367 0930
Voice Phone: (1) 510 841 2474
Fax Phone: (1) 510 841 2695
Email: staff@rossbooks.com
Web: http:\\www.rossbooks.com
Contact #1: Alan Rhody
Description: Publisher of the "HOLOGRAPHY MARKETPLACE" Editions 1-8 (a worldwide database and sourcebook), the "HOLOGRAPHY HANDBOOK - Making Holograms the Easy Way" (world's best selling laboratory manual), and other related titles. Educational, research and information services also provided.
****

Roth Ulrich
Schmale Strasse 5
City: Reutlingen
Country: Germany
Voice Phone: (49) (0)172 7300581
Fax Phone: (49) (0)7121 44687
Description: Distributor for all types of holograms.
****

Rottenkolber Holo-System GmbH Betrieb
Kirchheim
Bergweg 47
City: Amerang
Postal Code: D-83123
Country: Germany
Voice Phone: (49) 89 9030021
Fax Phone: (49) 89 904 39 83

Contact #1: Hans Dr. Rottenkolber
Description: holographic interferometry
****

Rowland Institute For Science
100 Edwin H. Land Blvd.
City: Cambridge
State: MA
Postal Code: 02142
Country: United States of America
Voice Phone: (1) 617 497 4657
Contact #1: lean-Marc Fournier
Description: Scientific holography research. NDT, Lippman photography
****

Royal Holographic Art Gallery
122 Market Square
560 Johnson Street
City, Province: Victoria, British Colombia
Postal Code: V8W 3C6
Country: Canada
Voice Phone: (1) 250 384 0123
Fax Phone: (1) 250 384 0123
Email: office@holograms.bc.ca
Web: www.holograms.bc.ca
Contact #1: Derek Galon
Description: Gallery offers full range of holographic art, holograms (including fine art Russian holograms), and holo-gifts. Retail, wholesale and low-cost custom work on our new RED STAR film. Also holography equipment from Russia. We ship worldwide.
SEE OUR ADVERTISEMENT
****

Royal Institute of Technology
Dept. of Materials Processing
Industrial Metrology
City: Stockholm
Postal Code: S-10044
Country: Sweden
Voice Phone: (46) 8 790 7832
Voice Phone: (46) 8 796 6899
Email: nilsa@matpr.kth. se
Contact #1: Nils Abramson
Description: Industrial Metrology comprises conventional engineering metrology and laser- based metrology, especially industrial applications of display holography, holographic interferometry and Light-in-Flight recording by holography, which is largely the result of the research and development at the Department. The principal objective of the group is research and education; to develop new measurement principals for applying lasers in the industry and to disseminate knowledge of known laser-based methods of measurement.
****

Ruey-Tung, Miss. Hung
A 202
Chigasati-Coat Nango 6-7-12
City: Chigasaki-Shi, Kanagawa
Postal Code: 253
Country: Japan
Voice Phone: (81) 467 857 750
Contact #1: Hung Ruey-Tung
Description: Holographic Fine Artist.
****

S. C. Ahlemeyer
Heerstrasse 178
City: Duisburg
Postal Code: D-47053
Country: Germany
Voice Phone: (49) (0)203 60966 -0
Contact #1: Frank Ahlemeyer
Description: Distributor of holographic material
****

S.O.P.R.A.
Societe de Production et de Recherche
Appliquee
26 & 28 rue Pierre Joigneaux

City: Bois-Colombes
Postal Code: F-92270
Country: France
Voice Phone: (33) 1 47 81 0949
Fax Phone: (33) 1 42 42 29 34
Contact #1: Robert Stehle
Description: Laser equipment, holographic kit and camera for interferometry.
****

Saginaw Valley State University
7400 Bay Road
City: University Center
State: MI
Postal Code: 48710-0001
Country: United States of America
Voice Phone: (1) 517 790 4000
Fax Phone: (1) 517 790 2717
Web: http://www.svsu.edu/
Description: Course instruction on holography; research includes HOEs, multiplex and rainbow holography.
****

Saint Mary's College
Art Department
City: Notre Dame
State: IN
Postal Code: 46556
Country: United States of America
Voice Phone: (I) 219 284 4000
Web: http://www.saintmarys.edu/
Description: Holographic fine artist. Extensive portfolio. Holography Instructor. Call for class schedule.
****

SAM Museum
3-27-3 Isoji, Minato-ku
1117 Kitakaname
City: Osaka
Postal Code: 552
Country: Japan
Voice Phone: «8 1)) 6 572 0036
Fax Phone: «81)) 6 574 8136
Contact #1: Akinobu Fukuda
Description: Museum that exhibits holograms.
****

San Jose State University
Physics Dept. and Inst. for Modern Optics
One Washington Square
City: San Jose
State: CA
Postal Code: 95192-0106
Country: United States of America
Voice Phone: (I) 408 924 5245
Fax Phone: (1) 408 924 2917
Web: http://www.sjsu.edu/
Description: Research and development work on 1) holographic fingerprint sensor, 2) holographic fingerprint verification, 3) display holography on DCG, and 4) holographic optical elements.
****

Sandia National Laboratories
1515 Eubank SE
City: Albuquerque
State: NM
Postal Code: 87123
Country: United States of America
Voice Phone: (1) 505 845 0011
Web: http://www.sandia.gov/
Description: Sandia National Laboratories is able to do research in all phases of holography.
****

Saxby, Graham
3 Honor Ave.
Goldthorn Park
City, Province: Wolverhampton, West Midlands
State: England
Postal Code: WV4 5HF

Country: United Kingdom
Voice Phone: (441) 902 341 291
Contact #1: Graham Saxby
Description: Research scientist; author of "Practical Holography".
****

Schall Messen GmbH - Optatec
Gustav Werner Strasse 6
City: Frickenhausen
Postal Code: D-72633
Country: Germany
Voice Phone: (49) (0)7025 9206 -0
Fax Phone: (49) (0)7025 9206-20
Email: info@schall-messen.de
Web: www.schall-messen .de
Contact #1: Bettina Knauer
Description: Organizer of technical fair in Frankfurt .
****

Scharr Industries
40 East Newberry Road
City: Bloomfield
State: CT
Postal Code: 06002
Country: United States of America
Voice Phone: (1) 860 243 0343
Voice Phone: (1) 800 284 7286
Fax Phone: (1) 860 242 7499
Contact #1 : Peg Horne
Description: Scharr holographic ·embossing: wide web, on polyester, polypropylene, polyethylene, PVC, and nylon. We "coat, laminate, metalize standard and custom patterns. Products include film to paper, board, transfer film, PSA and static cling.
****

School Of Holography
Museum Of Holography/Chicago
11 34 W. Washington Blvd.
City: Chicago
State: IL
Postal Code: 60607
Country: United States of America
Voice Phone: (1) 312 226 1007
Fax Phone: (1) 312 829 9636
Email: museum@concentric.net
Web: http://www.cris.coml-museumh
Contact #1: Loren Billings
Description: Founded in 1978, the oldest continuous school of holographic instruction in the world . Basic courses in holography have been taught to thousands of students. In addition there are special workshops and tutorials for advanced study.
****

Science Kit & Boreal Labs
777 East Park Drive
City: Tonawanda
State: NY
Postal Code: 14150-6784
Country: United States of America
Voice Phone: (1) 716 874 6020
Fax Phone: (1) 716 874 9572
Description: Suppliers of educational science materials es pecially suitable for junior and senior high school coursework. Comprehensivemail order catalog includes holography kits, holography books, related optical components and more.
****

SEAREACH, Pic
Seareach House
Wantz Road
City, Province: Dagenham, Essex
State: England
Postal Code: RMIO 8PS
Country: United Kingdom
Voice Phone: (44) 181 595 3212
Fax Phone: (44) 181 5934615
Email: info@seareach.plc.uk
Web: http://www.seareach.plc.ukl

Description: SEAREACH plc offer a range of 'off-the-shelf' adhesive holograms for you to apply to your products, letters, envelopes, etc. These stickers offer the security of the hologram but without the cost of origination and artwork. These holograms are tamper evident and are a cheap alternative to having your own personal design produced. Prices start from as little as £99.99 for 1,000. Please view some of our prices at our web site.
****

Selltexx
Rosenkoppel I
City: Monege
Postal Code: D-25436
Country: Germany
Voice Phone: (49) (0)4 122 83804
Contact #1: Martin Buttler
Description: Distributor for all types of holograms.
****

Shandong Academy of Sciences
Keyuan Road
City: Jinan Shandong
Postal Code: 250014
Country: China
Voice Phone: (86) 615 615102 316
Contact #1 : Zhu De Shun
Description: Laser & holography exhibit.
****

Shanghai Dahua Printing Factory
Pu-Dong-Xin-Qu Wang-Gang Xin-Hong-Cun
Tang-Lu-Gong-Lu #2498
City: Shanghai
Postal Code: 201201
Country: China
Contact #1: Gong Yuanzhong
Description: Finishing work for holography.
****

Shanghai Kanlian S & T Development Co.
Zhongshan-Xi -LU #1521
City: Shanghai
Postal Code: 200233
Country: China
Voice Phone: (86) 21 64813107
Fax Phone: (86) 21 64647030
Contact #1: Zhou Bingda
Description: Hologram distribution.
****

Sharon McCormack Holography
P.O. Box 38
City: White Salmon
State: WA
Postal Code: 98672
Country: United States of America
Voice Phone: (1) 509 493 4850
Fax Phone: (1) 509 493 4830
Email : sharon@gorge.net
Web: www.gorge. net/business/holography/
Contact #1: Sharon McCormack
Description: Holographic fine artist. Complete stereogram production, including 360 degree viewable. Filming, animation and computer graphics services offered. Also exhibit, design, and consultation services. Extensive commercial portfolio for major corporate clients. Recent work includes life sized commissions and installations.
****

Sharp Corp.
1-9-2 Nakase, Mihama-ku,
Chiba-shi,
City: Chiba
Postal Code: 261-0033
Country: Japan
Voice Phone: (81) 43 299-8711
Fax Phone: (81) 43 299-8709
Contact #1: Shunichi Sato
Description: Research
****

Sheng Ding Holography Development Co.
Lian Tong Garden #1, Rm. 704, Lou Hu Qu
518004 Shen Zhen
City: Guang Dong
Country: China
Voice Phone: (86) 755 5701420
Fax Phone: (86) 755 5701420
Email: holohot@nenpub.szptt.net.cn
****

Shenzhen Reflective Materials Factory
Shenzhen University
Rm 117-119 Lab Building
City, Province: Shenzhen, Guangdong
Postal Code: 518060
Country: China
Voice Phone: (86) 755 6660277-2236
Voice Phone: (86) 755 6660970
Fax Phone: (86) 755 755 6660462
Contact #1: Ye lingde
Description: Holography classes.
****

Shipley Chemical Co.
455 Forrest Street
City: Marlboro
State: MA
Postal Code: 01752
Country: United States of America
Voice Phone: (1) 800 837 2515
Voice Phone: (1) 508 481 7950
Fax Phone: (1) 508 485 9113
Web: www.rohmhaas.com
Contact #1: Stu Price
Description: Primary manufacturer of photoresist.
Sold wholesale by quarts and gallons as liquid. For
precoated plates, see listings for Towne and Process
Technology.
****

Shonan Institute of Technology
1-1-25 Nishi-kaigan, Tsujido,
Fujisawa-shi,
City: kanagawa
Postal Code: 251-0046
Country: Japan
Voice Phone: (81) 466 34 4111
Fax Phone: (81) 46 35 8897
Email: satok@elec.shonan-it.ac.jp
Contact #1: Koki Sato
Description: Electro Holography
****

Shriram Holographics
104, Kirti Deep
Nanagal Raya
City: New Delhi
Postal Code: 110046
Country: India
Voice Phone: (91) 55 96697
Fax Phone: (91) 11 5552986
Contact #1: Rajeev Jain
Description: Embossed holography
****

Silhouette Technology Inc.
10 Wilmot Street
City: Morristown
State: NJ
Postal Code: 07962-1479
Country: United States of America
Voice Phone: (1) 973 539 2110
Fax Phone: (1) 973 539 5797
Contact #1 : Toicia Murphay
Description: Produces custom HOEs under contract.
HOP maker. Heads-up display. DOE & DOE printers.
****

Sillcocks Plastics International
128 Main St
City: Hudson
State: MA
Postal Code: 01749
Country: United States of America

Voice Phone: (1) 978 562 6200
Voice Phone: (1) 800 526 4919
Fax Phone: (1) 978 562 7128
Email: spisales@silicocks.com
Web: http://www.silex.com!
Description: Producer of flat plastic products, printed
or unprinted, which can feature hotstamped holograms
and laminated holograms. Products include credit
cards , promotional cards and other custom specialties
and POP products.
****

Silverbridge Group
Box 489
City, Province: Powassan, Ontario
Postal Code: POH IZO
Country: Canada
Voice Phone: (1) 705 724 6164
Fax Phone: (1) 705 724 6249
Contact #1: lames Hepburn
Description: Limited edition DCG holograms in large
format size.
****

Simian Co.
298 Harvey West
City: Santa Cruz
State: CA
Postal Code: 95060
Country: United States of America
Voice Phone: (1) 408 457 9052
Fax Phone: (1) 408 457 9051
Email: simian@cruzio.com
Description: Manufacturer of high quality masters for
embossed holography. Originations can be 2D/3D,
3D, animation and motion, or any combination. High
production capacity with quick turnaround. Owned by
"Bridgestone Technologies".
****

Sinclair Optics, Inc.
6780 Palmyra Road
City: Fairport
State: NY
Postal Code: 14450
Country: United States of America
Voice Phone: (1) 716425 4380
Fax Phone: (1) 716425 4382
Email: sales@sinopt.com
Web: http://w.ww.sinopt.com!
Contact #1: Douglas P. Sinclair
Description: Software for Computer Generated Holograms.
****

Siros Technologies, Inc.
101 Daggett Drive
City: San Jose
State: CA
Postal Code: 95134
Country: United States of America
Voice Phone: (1) 408 944 9300
Fax Phone: (1) 650 938 3896
Email: jforese@sirostech.com
Web: http://www.sirostech.com!
Contact #1: Burt Hesselink
Description: Holographic data storage R&D
****

Siavich - International Wholesale Office
Uab Geola
P.O. Box 343
City: Vilnius
Postal Code: LT-2006
Country: Lithuania
Voice Phone: (370) 2 232 737
Fax Phone: (370) 2 232 838
Email: sales@slavich.com
Web: www.slavich.com
Description: International wholesale office for
Siavich holographic emulsions. Alternate
Emails: geola@post.omnitel.net and
technical@geola.com
SEE OUR ADVERTISEMENT
****

Siavich - Joint Stock Company
2 Mendeleeva Square
City: Pereslavl-Zalessky
Postal Code: 152140
Country: Russia
Description: Makers of ultra fine grain silver halide
emulsion for use in holography.
SEE OUR ADVERTISEMENT
****

Smith & McKay Printing Co. Inc.
96 North Almaden Boulevard
City: San Jose
State: CA
Postal Code: 95110-2490
Country: United States of America
Voice Phone: (1) 408 292 8901
Fax Phone: (1) 408 292 0417
Email: smithmckay@aol.com
Contact #1: Dave McKay
Description: Expert hot-stampers of foil holograms
onto paper products. Dimensional printing and fine
lithography. Parent company of "Holographic Impres-
sions".
****

Smith, Steven L.
City: Medford
State: MA
Country: United States of America
Voice Phone: (1) 617 253 0626
Voice Phone: (1) 781 391 7302
Fax Phone: (1) 617 253 8823
Email: sls@media.mit.edu
Web: http://www.media.mit.edul-sls/
Contact #1: Steven Smith
Description: Professional Holographer since 1979.
Extensive background in all Holographic imaging
technologies. Photoresist, Dichromate, Photo-polymer,
and related imaging systems. Image replication tech-
nologies, Embossed, Photo-polymer. Published author,
Lecturer. Active researching in new imaging technolo-
gies . Media Lab, MIT.
****

Sommers Plastic Products
81 Kulier Road
City: Clifton
State: NJ
Postal Code: 07015
Country: United States of America
Voice Phone: (1) 973 777 7888
Voice Phone: (1) 800 225 7677
Fax Phone: (1) 973 345 1586
Email: sales@sommers.com
Web: http://www.sommers.com
Description: Sommers turns holography into profitable
fashions. Epoxy and polyurethane domings, PVC and
rubber labels, stretch fabrics and vinyls are a few of the
value added ingredients used to transform holography
into profitable products for Nike, Adidas, Warner Bros,
Calvin Klein and many more.
****

Sonoma State University
Physics Dept.
1801 E. Cotati Ave.
City: Rohnert Park
State: CA
Postal Code: 94928
Country: United States of America
Voice Phone: (1) 707 664 2119
Fax Phone: (1) 707 664 3012
Web: http://www.sonoma.edu/
Description: Holography workshops offered . Call for
class schedule.
****

Sophia University
Faculty Of Science & Technology
7-1, Kioi-Cho Chiyoda-Ku
City: Tokyo

Country: Japan
Fax Phone: (81) 3 3238 3341
Contact #1: Kazue Ishikawa
Description: Holography research.
****

Southern Indiana Holographics
6841 Newburgh Rd
City: Evansville
State: IN
Postal Code: 47715
Country: United States of America
Voice Phone: (1) 812 474 0604
Fax Phone: (1) 812473 0981
Email: l-.iohann@msn.com
Contact #1: Larry Johann
Description: Holographic Fine Artist.
****

Spatial Holodynamics (India) Pvt. Ltd.
1041105 Shah & Nahar Estate
Off. Dr. E. Moses Road
City: Worli, Bombay
Postal Code: 400 018
Country: India
Voice Phone: (91) 22 493 0975
Voice Phone: (91) 22 492 1069
Fax Phone: (91) 22 495 0585
Contact #1: Yogesh Desai
Description: Holographic embossing using the latest DI-HO System. Total service from designing of holograms up to holographic shims. Or, if desired, glass Photo Resist Masters.
****

Spatial Imaging Limited
6 Marlborough Rd.
City, Province: Richmond, Surrey
State: England
Postal Code: TWIO 6JR
Country: United Kingdom
Voice Phone: (44) 181 332 1948
Voice Phone: (44) 1932 564899
Fax Phone: (44) 1932 564 899
Email: jeff@holograms.co.uk
Web: http://www.ho~ograms .co.ukl
Contact #1: Jeffrey Robb
Description: Holographic Origination for all media including embossed, photopolymer and silver halide. Holographic origination systems including digital stereogram mastering and dot matrix. Pulsed portraiture. Tech. transfer.
****

Spectra-Physics GmbH
Siemensstrasse 20
City: Darmstadt
Postal Code: D-64289
Country: Germany
Voice Phone: (49) 6151 7 08-0
Fax Phone: (49) 6151 79102
Email: 100137.304@compuserve.com
Web: www.splasers.com
Description: Manufacturer and distributor of lasers-German office.
****

Spectra-Physics Lasers Inc.
1330 Terra Bella Ave.
City: Mountain View
State: CA
Postal Code: 94039-7013
Country: United States of America
Voice Phone: (1) 800 775 5273
Voice Phone: (1) 650 966 5546
Fax Phone: (I) 650 964 3584
Email: splaser@ix.netcom.com
Web: http://www.splasers.com
Contact #1: Curt Cavoon
Description: World 's largest supplier of CW and pulsed gas and solid state laser systems, including a comprehensive optical accessories line and a worldwide customer service network.

SEE OUR ADVERTISEMENT
****

Spectratek Inc.
5405 J andy Place
City: Los Angeles
State: CA
Postal Code: 90066
Country: United States of America
Voice Phone: (1) 310 822 2400
Fax Phone: (1) 310 822 2660
Web: www.spectratektech.com
Contact #1: Randy Bouverat
Description : Spectratek manufacturers the highest quality diffraction patterns which are the only ones available without seams or pattern breaks. These patterned films are available in a variety of formats, including adhesive backed for labels, laminated to card stock, or films for packaging and other applications.
****

SpectroLas GmbH
Freisinger Strasse [9a
City: Langenbach
Postal Code: D-854 [6
Country: Germany
Voice Phone: (49) (0)8761 61030
Fax Phone: (49) (0)8761 70798
Email: SpectroLas@t-online.de
Description: Distributor of optical/mechanical holographic equipment and HOEs.
****

SPIE
The International Society for Opt!cal Engineering
P.O. Box 10
City: Bellingham
State: WA
Postal Code: 98227
Country: United States of America
Voice Phone: (1) 360 676 3290
Fax Phone: (1) 360 647 1445
Email: spie@spie.org
Web: http://www.spie.org/
Description: SPIE - The international Society for Optical Engineering is a nonprofit educational society dedicated to advancing engineering and scientific applications of optical, electro-optical, and optoelectronic instrumentation, systems, and technologies.

SEE OUR ADVERTISEMENT
****

SPIE's Holography Working Group Newsletter
Society of Photo-Optical Instrumentation Engineers
P.O. Box 10
City: Bellingham
State: WA
Postal Code: 98227-0010
Country: United States of America
Voice Phone: (1) 360 676 3290
Fax Phone: (1) 360 647 1445
Email: info-holo-request@spie.org
Web: http://www.spie.org/
Description : The ""Holography Working Group .... newsletter is published semiannually by SPIE - The International Society for Optical Engineering, for its International Technical Working Groups on Holography.
****

Spindler & Hoyer GmbH & Co.
Konigsallee 23
City: Goettingen
Postal Code: D-37081
Country: Germany
Voice Phone: (49) (0)551 6935-0
Voice Phone: (49) (0)551 6935-971
Fax Phone: (49) (0)551 6935-166
Email: sales@spindlerhoyer.de
Web: www.spindlerhoyer.de
Contact #1: Mr. Keilholz
Description: Manufacturer of precision optics- mechanics and laser technology.

****

Spot Agentur flir Holographie und Werbung
An St. Katharinen 2
City: Cologne
Postal Code: D-50678
Country : Germany
Voice Phone: (49) (0)221 315500
Fax Phone: (49) 221 322426
Contact #1: Walter Trebst
Description: Distribution of holograms, exhibition.
****

Springer Verlag Berlin
Postfach 311340
City: Berlin
Postal Code: D-I0643
Country: Germany
Voice Phone: (49) (0)30 82787-0
Fax Phone: (49) 0 82787-301
Email: orders@springer.de
Description: Publi sher of books for the technical use of holography.
****

Springer-Verlag New York
175 Fifth Ave.
City: NY
State: NY
Postal Code: 10010
Country: United States of America
Voice Phone: (1) 212 460 1500
Voice Phone: (1) 800 777 4643
Fax Phone: (1) 201 348 4505
Email: orders@spri nger-ny.com
Web: http://www.springer-ny.com
Description : Publishers of an "Optical Sciences" series of books, including "Silver Halide Recording Materials for Holography", by H. Bjelkhagen.
****

Standard Paper Box Machine, Co. , Inc.
347 Coster Street
City: Bronx
State: NY
Postal Code: 10474
Country: United States of America
Voice Phone: (1) 718 328 3300
Voice Phone: (1) 800 367 8755
Fax Phone: (1) 718 842 7772
Description: Manufacturer of hot stamping equipment with holographic and die cutting capabilities.
****

Stanford University
Mechanical Engineering Dept.
Mail Code 4021
City: Stanford
State: CA
Postal Code: 94305-4021
Country: United States of America
Voice Phone: (1) 650 723 2123
Fax Phone: (1) 650 723 3521
Email: dnelson@leland.stanford.edu
Web: http://www.stanford.edu/
Contact #1: Drew Nelson
Description: Use of holographic interferometry (with rapid thermoplastic recording of holograms) for measurements of small deformations and for residual stresses in materials via stress release technique.
****

Star Foil Technology Ltd.
Goddard Rd., Units A - B,
Whitehouse Industrial Estate
City, Province: Ipswich, Suffolk
State: England
Postal Code: IPI 5NP
Country: United Kingdom
Voice Phone: (44) (0)1 473 462315
Fax Phone: (44) (0)1 473 462357
Description: Manufacturer of hot stamping equipment. USA rep is Foilmark.
****

Star Magic
745 Broadway (below 8th St.)
City: New York
State: NY
Postal Code: 10003
Country: United States of America
Voice Phone: (1) 212 228 7770
Email: staff@starmagic.com
Web: http:\\www.starmagic.com
Description: Retail store featuring Space Age gifts, holograms, novelties, etc.
****

Star Magic
1256 Lexington Ave. (85th St.)
City: New York
State: NY
Postal Code: 10028
Country: United States of America
Voice Phone: (1) 212 988 0300
Email: staff@starmagic.com
Web: http://www.starmagic.com
Description: Retail store featuring Space Age gifts, holograms, novelties, etc.
****

Star Magic
275 Amsterdam St.
City: NY
State: NY
Postal Code: 10023
Country: United States of America
Voice Phone: (1) 212 769 2020
Email: sales@starmagic.com
Web: http:\\www.starmagic.com
Description: Retail store featuring Space Age gifts, holograms, novelties, etc.
****

Starcke, Ky.
Ratastie 6
City: Kokemaki
Postal Code: FIN-32800
Country: Finland
Voice Phone: (358) 39 5460 700
Fax Phone: (358) 39 5467 230
Contact #1: Ari-Veli Starcke
Description: Starcke KY is the leading company selling holograms in Scandinavia. Hologram Hot Stamping.
****

Steeg & Reuter GmbH
Philosophiestrasse 116
City: Giessen
Postal Code: D-35396
Country: Germany
Voice Phone: (49) (0)641 4007 0
Fax Phone: (49) (0)641 4007 84
Email: sales@steegreuter.de
Web: www.steegreuter.de
Description: Manufacturer of optics and crystaloptics. Main activities: Polarization, laser interference optics, surface mirrors, lenses for UVIVISIIR- X-ray monochromators. The company was founded in 1855 ..
****

Steinbichler Optical Technologies U.S.A.
40000-T Grand River, Suite 101
City: Novi
State: MI
Postal Code: 48375
Country: United States of America
Voice Phone: (1) 888 349 5641
Fax Phone: (1) 248 426 0643
Web: http://www.steinbichler.coml
Description: Mfrs. Of 3D Optical Test & Measurement Systems. Holography For Non-Destructive Testing & 3D Visualization. Service Bureau ..
****

Steinbichler Optotechnik GmbH
Am Bauhof 4
City: Neubeuern
Postal Code: D-83115

Country: Germany
Voice Phone: (49) (0)8035 87040
Fax Phone: (49) (0)8035 1010
Email: sales@steinbichler.com
Web: http://www.steinbichler.com
Contact #1: H. Steinbichler
Description: Development and sales of optical measuring and test systems -e .g. holographic interferometer-ESPI (electronic speckle Pattern Interferometer)-contour measurement systems-nondestructive inspection (shearography)-image analysis software.
****

Steiner GmbH
Talbachstrasse 14-14a
City: Siegen Eiserfeld
Postal Code: D-57080
Country: Germany
Voice Phone: (49) (0)271 382035
Fax Phone: (49) (0)271 385265
Description: Distributor of chemical lab equipment.
****

Stensborg Inc.
Center for Advanced Technology/RIS0
Frederiksborgvej 399
City: Roskilde
Country: Denmark
Email: jan.stensborg@cat.risoe.dk
Web: http://www.catscience.dk/compani/
descrip .html#stensborg
Contact #1: Mr. Jan Stensborg
Description: We offers advi sory and consultancy services to industries wishing to use holography as a visual means of communication.
****

Steuer KG GmbH & Co.
Ernst-Mey-Strasse 7
City: Leinfelden-Echterdingen
Postal Code: D-70771
Country: Germany
Voice Phone: (49) (0)711 16068 0
Fax Phone: (49) (0)711 16068 63
Contact #1: Mr. Seitz
Description: Manufacturer of holographic hotstamping mach,ines.
****

STI - Europe
Huttons Yard, Mapledurwell
City, Province: Basingstoke, Hampshire
State: England
Postal Code: RG25 2LP
Country: United Kingdom
Voice Phone: (44) 1256 346208
Fax Phone: (44) 1256 329238
Description: STI is a manufacturer of complex holographic products. We offer a wide variety of products for security, promotional, collectable, and packaging applications. The company specializes in the development of security devices for ant! counterfeiting and identification applications.
****

Stiletto Studios
Freinwalder Str. 13a
City: Berlin
Postal Code: D-13359
Country: Germany
Voice Phone: (49) 30 4936829
Description: holography in art- furniture with holographic elements.
****

Superbin Co. Ltd
3F-339
Section 2 Ho Ping E Road
City: Taipei
Postal Code: 10662
Country: Taiwan
Voice Phone: (886) 02 701 3626
Fax Phone: (886) 02 70 I 3531

Contact #1: Edward Hwang
Description: Exclusive Chinese representative of Coherent (Argon, Krypton Laser, Dye Laser); Continuum (ruby Laser, Nd: YAG laser); Newport (optical components).Also supply embossed hologram-manufacturing equipment/ material and consulting service.
****

Suzhou University
Laser Research Lab
Research Lab
City, Province: Suzhou, Jiangsu
Postal Code: 215006
Country: China
Voice Phone: (86) 512 5215257
Fax Phone: (86) 512 5215257
Contact #1: Chen Linsen
Description: Holography classes
****

Swift Instruments
1190 North 4th St.
City: San Jose
State: CA
Postal Code: 95112
Country: United States of America
Voice Phone: (1) 408 293 2380
Description: Microscope objectives and related optics.
****

Synchron Pty Ltd.
Chempet
P.O. Box 36921
City: Capetown
Postal Code: 7442
Country: South Africa
Voice Phone: (27) 21 55 1' 1790
Fax Phone: (27) 21 52 5291
Contact #1: Sean Kritzinger
Description: Agent and di stributor of holographic foils and labels. Also in house hot stamping of holographic images.
****

Syracuse University
Department of Chemistry
111 College Place, 1-014C
City: Syracuse
State: NY
Postal Code: 13244-4100
Country: United States of America
Voice Phone: (1) 315 443 2925
Fax Phone: (1) 315 443 4070
Email: ijbarani@mailbox.syr.edu
Web: http://www-che.syr.edu/Sponsler.html
Contact #1 : Michael B. Sponsler
Description: Research in Liquid Crystals as Holographic Recording Media.
****

Tair Hologram Company
Behterevsky, 8
City: Kiev
Postal Code: 252053
Country: Ukraine
Voice Phone: (38) 044 269 13 77
Email: feofan@megamed.kiev.ua
Contact #1: Alexander Monchak
Description: Fine art holograms.
****

Taiyuan Shiji Holography Ltd.
Shi-Fan-Jie #23
City, Province: Taiyuan, Shanxi
Postal Code: 030006
Country: China
Voice Phone: (86) 351 9001996
Fax Phone: (86) 351 7060457
Contact #1: An Shouzhong
****

Tama Art University
Department Of Physics

INTERNATIONAL BUSINESS DIRECTORY Ta - Th

1723 Yarimizu Hachiouji-Shi
City: Tokyo
Postal Code: 102-0095
Country: Japan
Voice Phone: (81) 426 768 611
Fax Phone: (81) 426 762 935
Contact #1: Hidetoshi Katsuma
Description: Research on Holographic TV, Holography Movie.
****

Tavex, Ltd.
45 prospect Nauki
City: Kiev
Country: Ukraine
Voice Phone: (380) 44 265-6178
Voice Phone: (380) 44 265 6178
Fax Phone: (380) 44 265-5871
Email: sav@tav.kiev.ua
Web: http://www.tav.kiev.ua
Contact #1: Alexander Stolyarenko
Description: Develop and produce photothermoplastic cameras - reversible photosensitive medium and devices on its basis for hologram registration. Optics.
****

Technical University Zvolen
Faculty of Wood Technology
Dept. of Physics and Applied Mechanics
City: Zvolen
Postal Code: SK-960 53
Country: Slovakia
Voice Phone: (42) 855 635
Fax Phone: (42) 855 321 811
Email: stano@tuzvo.sk
Contact #1 : Stanislav Urgela
Description: Holographic interferometry. Measurement of temperature fields, deformations and vibrations applied to wood technology, wooden plates, musical instruments and material quality control.
****

Technische Fachhochschule Berlin
FB 2 / Labor flir Laseranwendungen
Seestrasse 64 '
City: Berlin
Postal Code: D-13347
Country: Germany
Voice Phone: (49) (0)30 4504 3917
Voice Phone: (49) (0)30 4504 3918
Fax Phone: (49) (0)30 4504 3959
Email: eichler@tfh-berlin.de
Contact #1: Jiirgen Prof. Dr. Eichler
Description: Holographic Interferometry-Display Holography-Medical Applications.
****

Technische Universitat Berlin
Optisches Institut Pl1
Strasse des 17. Juni 135
City: Berlin
Postal Code: D-10623
Country: Germany
Voice Phone: (49) (0)30 314 22498
Fax Phone: (49) (0)30 314 26888
Email: eichler@physik.tu-berlin.de
Web: http://www.physik.tu-berlin.de
Contact #1: Hans Joachim Prof. Dr. Eichler
Description: Scientific holography research: Optical holographic data storage-real-time holography- material research-semiconductors- liquid crystals-new lasers.
****

Technische Wien a Institut flir Raurnliche ...
Karlsplatz 13/2561
City: Wien
Postal Code: A- 1040
Country: Austria
Voice Phone: (43) (0)1 58801-3382
Fax Phone: (43) (0)1 5041147
Email: bmartens@Email.tuwien.ac.at
Contact #1 : Dr. Bob Martens
****

Technoexan Ltd
Polytechnicheskaya, 26
City: St. Petersburg
Postal Code: 194021
Country: Russia
Voice Phone: (7) 812 247 9383
Voice Phone: (7) 812 247 5273
Fax Phone: (7) 812 247 5333
Contact #1: Igor Lovygin
Description: Power semiconductors, Lasers, opto-electronics devices (IR range), many channel 1- and p- diodes, equipment for high format art and picture hologram, holographic registers, school packages for showing optic effects.
****

Technolas Laser Technik Gmbh
Lochhamer Schlag 19
City: Grafelfing
Postal Code: D-82166
Country: Germany
Voice Phone: (49) (0) 89 854 5040
Fax Phone: (49) (0) 89 854 561
Description: Lasers-medical
****

Teo Hong Silom Co., Ltd.
Banga Towers B.- 17th FI.
2/3 Moo 14 Banga-Trad K.M.6.5. Bangaew
City, Province: Samutprakarn, Bangplee
Postal Code: 10540
Country: Thailand
Voice Phone: (662) 3120045 to 69
Fax Phone: (662) 312 0700 to 'I
Email: teohung@asiaaccess.neUh
****

Tesa
Beiersdorf AG
Unnastrasse 48
City: Hamburg
Postal Code: D-20245
Country: Germany
Voice Phone: (49) (0) 40 4909-0
Fax Phone: (49) (0) 40 4909-3434
Description: Holography adhesive tapes
****

Textile Graphics, Inc.
(also Holography Presses On)
201 North Fruitport Road
City: Spring Lake
State: MI
Postal Code: 49456
Country: United States of America
Voice Phone: (1) 616 842 5626
Fax Phone: (1) 616 842 5653
Contact #1 : Jan Bussard
Description: Holographic stock or custom images (all shapes and sizes) applied with heat or pressure for adhesion to all substrates. Specialize in stickers and textile applications. Sealed edges prevent delamination in all weather; washable. Worldwide distributors sought!
SEE OUR ADVERTISEMENT
****

The Foreign Dimension
(see Foreign Dimension)
SEE OUR ADVERTISEMENT
****

The Hologram Store
8770 170 Street - Unit 2673
City, Province: Edmonton, Alberta
Postal Code: T5T T4J2
Country: Canada
Voice Phone: (1) 403 444 3333
Description: Located in the world's largest mall, the Hologram Store offers all types of holograms for sale to the public.
****

The Hologram Store, Ltd.
#101 , llOO Robson Street
City, Province: Vancouver, BC

Postal Code: V6E 1B2
Country: Canada
Voice Phone: (1) 403 438 2537
Fax Phone: (1) 403 431 2914
Email: mckern@teusvelocity.net
Contact #1: John Sandra
Description: Independent retail store selling holograms. Specializing in holographic and unique products. Wholesale and mail order.
****

Thorlabs Inc.
Laubacher Weg 27
City: Griinberg
Country: Germany
Voice Phone: (49) (0)6401 2209-36
Fax Phone: (49) (0)640 I 2209-46
Email: ikabert@thorlabs.com
Contact #1: Inge Kabert
Description: Distributor of optical equipment.
****

Thorlabs Inc.
435 Route 206
City: Newton
State: NJ
Postal Code: 07860
Country: United States of America
Voice Phone: (1) 973 579 7227
Fax Phone: (1) 973 383 8406
Email: sales@thorlabs.com
Web: http://www.thorlabs.com
Description: High quality equipment for optics and photonics research including first surface mirrors, optical component mounts, power meters, etc.
****

Three Deep Co.
(see 3Deep Co.)
540 Massachusetts Ave.
City: Boston
State: Massachusetts
Country: United States of America
Voice Phone: (1) 617 912-1040
Fax Phone: (1) 617 912-1040
Email: acheimets@3deepco.com
Web: www.3deepco.com
Contact #1: Alex Cheimets
Description : Supplies Russian silver halide and DCG emulsions on plates. Green sensitive, red sensitive, and new full color silver halide emulsions available. Can also recommend appropriate developing and processing procedures.
SEE OUR ADVERTISEMENT
****

Three Dimensional Imagery
P.O. Box 858
City: Vienna
State: VA
Postal Code: 22183-0858
Country: United States of America
Voice Phone: (1) 703 573 0935
Email: smichael@3dimagery.com
Web: http://www.3dimagery.com
Contact #1: Steve Michael
Description: Production of white-light reflection display holograms, animated computergenerated holograms, and custom holograms.
****

Three-D Light Gallery
109-A The Commons
City: Ithaca
State: NY
Postal Code: 14850
Country: United States of America
Voice Phone: (1) 607 273 1187
Fax Phone: (1) 607 347 6454
Contact #1: Jonathan Pargh
Description: Artistic holography; holography gallery/shop.

Tianjin Holdor Optics Inc. China
11 Tianjin Binguan Nandao
City: Tianjin
Postal Code: 300061
Country: China
Voice Phone: (86) 22 28359338
Voice Phone: (1) 402 466-7468
Fax Phone: (86) 22 28359338
Email: yuanwbtj@public.tpt.tj.cn
Contact #1: Weiben Yuan
Description: Manufacturer of silver-halide emulsions; monochromatic and panchromatic; for reflection and transmission holography; resolution 3 -10,000 l/mm; various sized precoated plates available. Also manufacture optical isolation tables, holographic cameras and true color art reflection holograms. USA distributor - Control Optics.

Tianjin Water Laser Holography Image Co.
Bin-Guan-Nan-Dao #5
City: Tianjin
Postal Code: 300061
Country: China
Contact #1: Song Qihong
Description: Holography sales and distribution.

Tokai University
Department of Electro Photo Optics
3F 1-1-8, Shinsenri-Nishimachi,
City: Hiratsuka City
Postal Code: 259-1207
Country: Japan
Voice Phone: (81) 463 50 2157
Fax Phone: (81) 463 59 2594
Contact #1: Hideshi Yokota
Description: Holography research - artistic holography.

Tokyo Institute Of Technology
Imaging Science And Engineering
4259 Nagatsuda Midori-Ku
City: Yokohama
Postal Code: 226-0026
Country: Japan
Voice Phone: (81) 45 924 5190
Fax Phone: (81) 45 924 5175
Contact #1: Masahiro Yamaguchi
Description: Holographic Display, 3-D Imaging Science.

topac GmbH
Department Holography
Carl - Miele Strasse 202-204
City: Guetersloh
Postal Code: D-33311
Country: Germany
Voice Phone: (49) (0)5241 803302
Fax Phone: (49) (0)5241 8060870
Email: tophol@aol.com
Web: http://members .aol.com/tophol/ hhome_e.htm
Contact #1: Irrnhild Hoffstadt-Braeutigam
Description: topac GmbH is a full service hologram producer with complete production line for all hologram processes. Specialty is embossed hologram production for security and authenticity devices.

Topag GmbH
Kiesstrasse 58
City: Darmstadt
Postal Code: D-64283
Country: Germany
Voice Phone: (49) (0)6151 4259-78
Fax Phone: (49) (0)6151 4259-88
Contact #1: Dr. E. Jager
Description: Manufacturer of lasers

Topcon Inc.
75-1 Hasunuma-Machi Ttabasi-Ku
City: Tokyo
Postal Code: 174-0052
Country: Japan
Voice Phone: (81) 3 3558 2549
Fax Phone: (81) 3 3966 5011
Contact #1: Reiji Hashimoto
Description: Hologram manufacturer.

Toppan Printing Co. , Ltd.
Tech. Research Inst. Advance Products Research Lab
4-2-3 Takanodai - Minami
City: Saitama, Sugitomachi Kit.- gun
Postal Code: 345-8508
Country: Japan
Voice Phone: (81) 480 33 9079
Fax Phone: (81) 480 33 9022
Email: fiwata@tri.toppan.co.jp
Web: http://www.toppan.com/
Contact #1: Fujio Iwata
Description: All types of holograms

Toshihiro Kubota, dept of electronics
Faculty of Engineering and Design,
Kyoto Institute of Technology
City: Kyoto, Matsugasaki, Sakyo-ku,
Postal Code: 606-0962
Country: Japan
Voice Phone: (81) 75 7247443
Fax Phone: (81) 75 724 7400
Email: kubota@dj.kit.ac.jp
Contact #1: Toshihiro Kubota
Description: Works on holography include the imaging characteristics, recording material, color holography for fundamental study, and their applications to holographic display and optical devices.

Total Register Inc.
71 Commerce Drive
City: Brookfield
State: CT
Postal Code: 06804
Country: United States of America
Voice Phone:- (1) 203 7400199
Fax Phone: (1) 203 740 0177
Email: sall!s@totalregister.com
Web: http://www.totalregister.com/
Contact #1: John Gallagher
Description: Manufacturer of registration devices for hot-stamping presses. Manufacturer of registered rotary die cutting equipment and hot-stamping machines. Registered hologram sheeting and die cutting services.

Towne Technologies
6-10 Bell Ave.
City: Somerville
State: NJ
Postal Code: 08876-0460
Country: United States of America
Voice Phone: (1) 908 722 9500
Fax Phone: (1) 908 722 8394
Email: sales@townetech.com
Web: http://www.townetech.com
Contact #1: Sal LoSardo
Description : Towne Technologies is a producer of fine quality holographic photoresist plates with or without a sub-layer of IronOxide. These plates are spin-coated with striation free photoresist in sizes up to 15"" x 15"".
SEE OUR ADVERTISEMENT

Toyama National College Of Marit
1-2 Ebie-Neriya
City: Tokyo
Postal Code: 933 0235

Country: Japan
Voice Phone: (81) 766 86 5100
Contact #1: Dr. Kenji Kinoshita
Description: Holographic Stereogram

Trace Holographic Art & Design, Inc.
107 Inglewood Court
City: Charlottesville
State: VA
Postal Code: 22901-2619
Country: United States of America
Voice Phone: (1) 804 984 4239
Fax Phone: (1) 804 984 5490
Email: fcattapr@traceholo.com
Web: http://www.traceholo.com/
Contact #1: Fernando Catta-Preta
Description: Trace Holographic Art & Design is a company dedicated to the art of embossed holography, from the initial design phase all the way through to its final application. Come and visit our web site.

Transfer Print Foils, Inc.
Holographic Division
(a division of Holopak Technologies, Inc.)
21B Cotters Lane - P.O. Box 538
City: East Brunswick
State: NJ
Postal Code: 08816
Country: United States of America
Voice Phone: (1) 732 238 1800 x 4415
Voice Phone: (1) 800 235 3645 x4415
Fax Phone: (1) 732 651 1660
Web: www.tpfholo.com
Contact #1: Marc O. Woontner
Description: T.P.F. provides holographic products for a variety of end. uses. From security images (as transfer coatings or laminated patches) which enhance the security of credit cards, licenses, bank documents. tickets and gate passes; to value added decorative finishes for greeting cards, book jackets, trophies and picture frames. T.P.F holographic security operations currently provides security solutions for 431 government agencies worldwide.
SEE OUR ADVERTISEMENT

Traumlaboratorium
Traubenstrasse 41
City: ApenlAperberg
Postal Code: D-26689
Country: Germany
Voice Phone: (49) (0)4489 5198
Web: Elmar
Description: High quality art holograms

Triple-D Laser Imaging
Bergselaan 13-B
City: Rotterdam
Postal Code: NL-3037 BA
Country: Netherlands
Voice Phone: (31) 10465 6331
Fax Phone: (31) 10 465 6331
Email: aca@wirehub.nl
Web: http://www.home.wirehub.nl/-aca/
Contact #1: A.C. Akveld
Description: New business specializing in the sale of holograms and related optics.

Turing Institute
Boyd Orr Bldg.
University Ave.
City: Glasgow
State: Scotland
Postal Code: G12 8NN
Country: United Kingdom
Voice Phone: (44) 141 337 6410
Fax Phone: (44) 141 339 0796
Email : tim@turing.gla.ac.uk

Web: http:\\\www.turing.gla.ac.uk
Contact #1 : Tim Niblett
Description: 3D imaging systems based on multiple cameras, digital processing and proprietary software. Designed especially for medical and clinical applications where precise 3D modeling and accurate measurement are required. The Turing Institute is the trading name of Greenagate Ltd.
****

Tyler Group
218 Linden Avenue
City: Moorestown
State: NJ
Postal Code: 08057
Country: United States of America
Voice Phone: (1) 609 234 1800
Fax Phone: (1) 609 8660351
Contact #1: Peyton Old
Description: Holographic image security consulting and application service. Extensive experience with high production document and packaging authentication. Specialists in Novavision system.
****

U.K. Gold Purchasers, Inc.
(DBA Holograms Unlimited)
11 0 Central Park Mall
City: San Antonio
State: TX
Postal Code: 78216
Country: United States of America
Voice Phone: (1) 210 530 0045
Voice Phone: (1) 800 722 7590
Fax Phone: (1) 210 530 0048
Email: marvinuram@hotmail.com
Web: . www.eden.comJ-mainlinkltx/satiartiraii index.htm
Contact #1 : Marvin Uram
Description: Full line distributor of hologram and related products for specialty retailers - representing more than 80 firms. One stop shopping at competitive prices. SEE OUR ADVERTISEMENT
****

Uk Optical Supplies
84 Wimborne Road West
City, Province: Wimborne, Dorset
State: England
Postal Code: BH2l 2DP
Country: United Kingdom
Voice Phone: (44) 1202 886 831
Fax Phone: (44) 1202 886 831
Contact #1 : Ralph Cullen
Description: Supplying probably the world's largest selection of Holographic/Optical components which are best quality, best value. Designed by experienced holographers. Component selection and laboratory/studio set-up advice freely available. Also buys and sells a large selection of used and surplus equipment.
****

Ultra-Res Corporation
1395 Greg St. - Suite 107
City: Sparks
State: NY
Postal Code: 89431
Country: United States of America
Voice Phone: (1) 702 355 1177
Fax Phone: (1) 702 359 6273
Email: alex@acds.com
Web: hup://www.acds.com
Contact #1: Alex Chaihorsky
Descrip tion: Manufa cturers of the Ultra-Res Instant Holographic camera system. Capable of making small transmission and reflection holograms "instantly", without darkroom processing, on slides or film rolls. Useful for interferometry, setup verification, HOEs for phase filters and proofing for embossed runs.
****

Uni Innsbruck
Institut fUr Experimentelle Physik
City: Innsbruck
Postal Code: A- 6020
Country: Austria
Voice Phone: (43) (0)512 507-6331
Fax Phone: (43) (0)512 507-2921
Email: Alois.Mair@uibk.ac.at
Contact #1: Alois Mair
Description: R&D in CGH
****

Unifoil Corporation
217 Brook Avenue
City: Passaic
State: NJ
Postal Code: 07055
Country: United States of America
Voice Phone: (1) 973-365 2000 Ext.243
Fax Phone: (1) 973 365 0924
Email: unifoil@unifoil.com
Web: http:\\\www.unifoil.com
Contact #1: Joseph Funicelli
Description: Manufacturer of holographic paper and board, both laminated and nonlaminated products. Stock or custom holography available. Paper or board can be optically sheeted or cut.
****

Uniphase Lasers
(a division of Uniphase Corp.)
163 Baypoint Parkway
City: San Jose
State: CA
Postal Code: 95134
Country: United States of America
Voice Phone: (1) 408 434 1800
Voice Phone: (I) 800 644 8674
Fax Phone: (I) 408 433 3838
Email: sales@uniphase.com
Web: http:\\\www.uniphase.com
Description: Manufacturer of lasers suitable for holography.
****

Uniphase Vertriebs-GmbH
Arbeonstrasse 5
City: Eching
Postal Code: D-85386
Country: Germany
Voice Phone: (49) (0)89 319 60 26
Voice Phone: (49) (0)89 319 30 02
Description: Manufacturer and distributor of lasers. ion-laser-he-ne laser-diode pumped solid state laser.
****

Univ. de Liege
Sart Tilman
Hololab, Physique B5
City: Liege
Postal Code: B-4000
Country: Belgium
Voice Phone: (32) 4 166 3626
Fax Phone: (32) 4 166 2355
Email: lion@gw.unipc.ulg.ac.be
Contact #1: Yves F. Lion
Description: Holography Courses
****

Universidade Do Porto
Laboratorio De Fisica
Praca Gomes Teixeira
City: Porto
Postal Code: P-4000
Country: Portugal
Voice Phone: (351) 2 557 0700
Description: Holographic non-destructive testing; Academic holography research .
****

Universita Di Roma
La Sapienza Dipt Di Fisica
Piazzale Aldo Moro 2
City: Rome
Postal Code: 1-00185
Country: Italy
Voice Phone: (39) 6 559 9776
Description: Scientific research.
****

Universite De Neuchatel
Institut De Microtechnique
2, Rue A L Breguet
City: Neuchatel
Postal Code: CH-2000
Country: Switzerland
Voice Phone: (41) 32 720 09 20
Fax Phone: (41) 32 720 0990
Email: info@fsrm.ch
Web: http://www.fsrm.ch
Contact #1: Rene Dandliker
Description: Industrial research.
****

Universite De Neuchatel
Institut De Microtechnique
2-Rue A L Breguet
City: Neuchatel
Postal Code: CH-2000
Country: Switzerland
Voice Phone: (41) (0)32 720 0920
Fax Phone: (41) (0)32 720 0990
Email: info@fsrm.ch
Web: http://www.fsrm.ch
Contact #1: Rene Dandliker
Description: Industrial research.
****

Universite Laval
Dept Physique - C.O.P.L
Pavilion Vachon
City, Province: University City, Quebec
Postal Code: GIK 7P4
Country: Canada
Voice Phone: (1) 418 656 2131
Voice Phone: (1) 418 656 3436
Web: http://www.fsg.ulaval.ca/
Contact #1 : Roger A. Lessard
Description: Holography education. Research in holographic recording materials (photopolymer) for optical data storage, CGH and diffractive optics.
****

University Of Alabama at Huntsville
Center For Applied Optics
City: Huntsville
State: AL
Postal Code: 35899
Country: United States of America
Voice Phone: (1) 256 890 6030
Fax Phone: (1) 256 895-6618
Web: http://www.uah.edu/
Description: Scientific holography research, NDT& HOE.
****

University Of Alicante
Applied Physics/Cent De Holograf
Facultad De Ciencias
City: Alicante Apdo
Postal Code: 99
Country: Spain
Voice Phone: (34) 566 1200
Contact #1: A. Fimia.
Description: Artistic holography; HOEs; workshops.
****

University Of Arizona
Optical Sciences Center
City: Tucson
State: AZ
Postal Code: 85721
Country: United States of America
Voice Phone: (1) 520 621 6997
Fax Phone: (1) 520 621 9613
Web: www.arizona.edu
Description: Industrial and scientific holog

Holographic non-destructive testing. Primarily graduate. Some undergraduate courses available.
****

University of Bundeswehr
LRT 7.1
Werner Heisenberg Weg 39
City: Neubiberg
Postal Code: D-85579
Country: Germany
Voice Phone: (49) (0)89 6004-2566
Fax Phone: (49) (0)89 6004-4092
Contact #1: Dip!. Ing. Jelena Wahler
Description: R&D in fluid holographic interferometry.
****

University Of Dayton
Research Institute
300 College Park
City: Dayton
State: OH
Postal Code: 45469-0102
Country: United States of America
Voice Phone: (1) 937 229 3515
Fax Phone: (1) 937 229 3873
Email: info@udri.udayton.edu
Web: http://www.udri.udayton.edu/
Description: Scientific research, industrial research; NDT - courses.
****

University of Erlangen
Physics Institute-Dept. of Optics
Staudtstrasse 7
City: Erlangen
Postal Code: D-91023
Country: Germany
Voice Phone: (49) (0)9131 858382
Fax Phone: (49) (0)9131 858423
Contact #1: G. Prof. Dr. Hausler
Description: We offer design and manufacturing of microlenses and diffractive optical elements (lenses-beamsplitters-beam shaping elements). We can test micro optical elements by means of interferometers.
****

University of Jena
Institut fUr Angewandte Optik
Friibelstieg 1
City: Jena
Postal Code: D-07743
Country: Germany
Voice Phone: (49) (0)3641 947663
Email: p6flbu@rz.uni-j ena.de
Contact #1: Dr. Burkhard Fleck
Description: R&D in realtime holographic memories.
****

University of Karlsruhe
Elektrotechnik
Kaiserstrasse 12
City: Karlsruhe
Postal Code: D-76131
Country: Germany
Voice Phone: (49) (0)721 608 2480
Voice Phone: (49) (0)721 608 2492
Contact #1: Gerhard Grau Prof. Dr. techno
Description: Research on synthetic (computer generated) holograms
****

University of Kassel
FB for Engineering
Miinchebergstrasse 7
City: Kassel
Postal Code: D-34109
Country: Germany
Voice Phone: (49) (0)561 804-2771
Fax Phone: (49) (0)561 804-2330
Contact #1: Prof. Dr.-Ing. E. Steinmetz
Description: Seminars for engineers and scientists- focus on holography in engineering, holography in the field of printing and graphics.
****

University of Latvia
Institute of Solid State Physics
8 Kengaraga Str.
City: Riga
Postal Code: LV-I063
Country: Latvia
Email: teteris@acad.latnet.lv
Web: http://www.cfi.lu.lv/
Description: R&D and production of photoresists for holography suitable up to 650 nm. Design & production of embossed 2D/3D holograms. Ni shim electroforming.
****

University of Magdeburg
Informatik - Computergrafik
Postfach 4120
City: Magdeburg
Postal Code: D-39016
Country: Germany
Web: simsrv.cs. uni-magdeburg.de/-alfl
english/maerz/maerz.html
Contact #1: Thomas Benziger
Description: R&D in CGH
****

University Of Michigan
Dept. of Electrical Engineering
Room 1108 EECS Building
City: Ann Arbor
State: MI
Postal Code: 48109-2122
Country: United States of America
Voice Phone: (1) 734 764 1817
Fax Phone: (1) 734 763 1503
Web: www.umich.edu
Description: Scientific holography research . Design HOEs. Courses on holography.
****

University of Munich
Institute Of Medical Optics
Barbarastrasse 16
City: Munich
Postal Code: D-80797
Country: Germany
Voice Phone: (49) (0)89 2105 3000
Fax Phone: (49) (0)89 1240 6301
Contact #1: Mr. Zurek
Description-: Medical Holography-Scientific holographic research.
****

University of MUnster
Laboratory of Biophysics
Robert-Koch-Str. 45
City: MUnster
Postal Code: D-48129
Country: Germany
Voice Phone: (49) (0)251 83 6888
Fax Phone: (49) (0)251 83 8536
Email: biophys@gabor.uni-muenster.de
Contact #1: Gert von Bally
Description: holography and Interferometry in medicine. Environmental research and cultural heritage protection.
****

University of Oldenburg
FB Physik
Postfach 2503
City: Oldenburg
Postal Code: D-26111
Country: Germany
Voice Phone: (49) (0)441 798-3512
Fax Phone: (49) (0)441 789-3201
Email: helmers@uwa.physik.unioldenburg.de
Contact #1: Heinz Helmers
Description: Holographic interferometry.
****

University Of Rochester
The Institute Of Optics

City: Rochester
State: NY
Postal Code: 14627
Country: United States of America
Voice Phone: (1) 716 275 2322
Fax Phone: (1) 716 273 1072
Email: info@optics.rochester.edu
Web: http://www.optics.rochester.edu
Description: Scientific and industrial holography research; interferometry; particle testing & measurement. Primarily graduate. Some undergrad courses include holography related studies.
****

University Of Southern California
Department Of Physics
University Park
City: Los Angeles
State: CA
Postal Code: 90089-0484
Country: United States of America
Voice Phone: (1) 213 740 2311
Web: www. usc.edu
Description: R&D program. Utilizes holographic gratings for for passive and active optical switches. primarily for fiber opitc communications.
****

University of Stuttgart
Institute Of Applied Optics
Pfaffenwaldring 9
City: Stuttgart
Postal Code: D-70569
Country: Germany
Voice Phone: (49) (0)711 685 6075
Fax Phone: (49) (0)711 685 6586
Contact #1: Hans Tiziani
Description: Scientific holography research; interferometry, NDT, HOE's.
****

University Of Tokyo
Faculty Of Engineering
1-1, Tennodai
City: Toyko
Country: Japan
Voice Phone: (81) 3 3812 2111
Fax Phone: (81) 3 3818 5706
Email: info@t.u-tokyo.ac.jp
Web: http://www.t.u-tokyo.ac.jp/
Contact #1: T. Uyemura
Description: Scientific and Medical holography research; Interferometry.
****

University Of Tsukuba
Institute Of Art & Design
Shinminato
City: Tsukuba
Postal Code: 305
Country: Japan
Voice Phone: (81) 298 53 2833
Fax Phone: (81) 298 53 2833
Contact #1: Shunsuke Mitamura
Description: Artistic holography, courses.
****

University Of Wisconsin/Madison
Dept. Of Engineering - Professional Deve!.
432 North Lake Street
City: Madison
State: WI
Postal Code: 53706
Country: United States of America
Voice Phone: (1) 608 262 8708
Fax Phone: (1) 608 263 3160
Email: custserv@epd.engr.wisc.edu
Web: http://www.engr.wisc.edu/
Description: Continuing Education courses on laser system design and application.
****

Uen erseher & Associates
-09 112 West Glen Oaks Blvd.
City: Glendale
State: CA
Postal Code: 91202
Country: United States of America
Voice Phone: (1) 818 549 0534
Contact #1: Fred Unterseher
Description: Artistic holography and holography education. Originates masters for mass production holograms in embossed, DCG and photopolymer materials. Both pulsed and CW lasers available.
****

Uvex Safety Inc.
10 Thurber Blvd.
City: Smithfield
State: RI
Postal Code: 02917
Country: United States of America
Voice Phone: (1) 401 232 1200
Voice Phone: (1) 800 343 3411
Fax Phone: (1) 401 232 1830
Email: sales@uvex.com
Web: http ://www.uvex.com
Description: Manufacturer of laser safety eyewear, as well as various respirators.
****

Van Leer Metallized Products
24 Forge Park
City: Franklin
State: MA
Postal Code: 02038
Country: United States of America
Voice Phone: (1) 508 541 7700
Voice Phone: (1) 800 343 6977
Fax Phone: (1) 508 541 7777
Email: vlmpinfo@vlmpusa.com
Web: http://www.vlmpusa.com
Description: Producer of unique holographic metallized papers:HoloPRISM™ holographic papers, Valvac™ metallized papers and HoloSECURE™ security papers. Increase product awareness in a variety of applications including labels, advertisements, laminated constructions and specialty promotions. Can be easily converted to meet your needs.
****

Veeco Instruments Inc.
2650 East Elvira Road
City: Tucson
State: AZ
Postal Code: 85706-7123
Country: United States of America
Voice Phone: (1) 520 741 1044
Fax Phone: (1) 520 294 1799
Email: sales@wyko.com
Web: http://www.wyko.com
Contact #1: Don McNeil
Description: Interferometry and analysis.
****

Verlag 3D Magazin
Tannenbergstrasse 36
Ci ty : Stuttgart
Postal Code: D-70374
Country: Germany
Voice Phone: (49) (0)711 524026
Email: 3D-Magazin@stereo.s.bawue.de
Web: www.tisco.com/3d-web/3dmag/3dmag.htm
Contact #1: Alexander Klein
Description: Publisher of 3D Magazine.
****

Ve rwal tungsgeselJschaft Holographische Technik Gmb
Schlossgartenallee 3
City: Schwerin
Postal Code: D-19061
Country: Germany

Web: http://www.igd.fhg.de/wwwligd-a1/holografielindex_d.html
Contact #1 : Roman Wojcicik
Description: Distrib. of holography equip.
****

Vincennes University
1002 North First Street
City: Vincennes
State: IN
Postal Code: 47591
Country: United States of America
Voice Phone: (1) 812 888 8888
Email: info@vinu.edu
Web: www.vinu.edu
Description: Offering holography workshops for high school teachers, & college level courses in holography. We have 4 researchgrade optical tables, as well as argon and krypton lasers. Call for details.
****

VinTeq, Ltd.
611 November Lane / Autumn Woods
City: Willow Springs
State: NC
Postal Code: 27592-7738
Country: United States of America
Voice Phone: (1) 919 639 9424
Fax Phone: (1) 919 639 7523
Email: vinson@vinteq.com
Web: http://www.vinteq.com
Contact #1: Joachim Vinson
Description: Distributor for HRT silver halide plates. Red, Blue, Green 'and panchromatic silver halide emulsions. (HRT GmbH in Steinau/Germany).
SEE OUR ADVERTISEMENT
****

Virtual Image
(a division of Printpack, Inc .)
PO box 1198
City: Litchfield
State: CT
Postal Code: 06759
Country: United States of America
Voice Phone: (1) 860 567 2022
Fax Phone: (I) 860 567 8699
Email: tvinstadt@printpack.com
Web: http://www.virtimage.com
Contact #1: Tom Vinstadt
Description: Manufacturer of high quality, embossed BOPP holographic film, primarily for the packaging industry. Specialty is large runs in wide web.
SEE OUR ADVERTISEMENT
****

Visions Unlimited
637 NW 12th Ave
City: Deerfield Beach
State: FL
Postal Code: 33442
Country: United States of America
Voice Phone: (1) 954 429 1017
Fax Phone: (1) 954421-2391
Web: http ://www.spectore.com
Description: Retail gift store offering Titanium jewelry and holograms.
****

Visual Visionaries
2011 Clement St., Suite 4
City: San Francisco
State: CA
Postal Code: 94121
Country: United States of America
Voice Phone: (1) 415 8764307
Description: Consulting and marketing firm. Exhibitions and educational services. Over 14 years experience with holographic production,
display and sales. Professional and reliable.
****

VISUELL

Dalbkerstrasse 92A
City: Oerlinghausen
Postal Code: D-33813
Country: Germany
Voice Phone: (49) (0)5202 71879
Fax Phone: (49) (0)52027 1825
Contact #1: Karsten Habighorst
Description: artist
****

Volkswagen AG
Forschung und Entwicklung
Postfach 1785
City: Wolfs burg
Postal Code: D-38436
Country: Germany
Voice Phone: (49) (0)536 925 824
Fax Phone: (49) (0)536 972 444
Contact #1: M.-A. Dr. Beeck
Description: Industrial research-Interferometry; Holographic non-destructive testingLaser combustion diagnostics.
****

Volvo-Flygmotor
S-461
City: Trollhattan
Postal Code: S-81
Country: Sweden
Voice Phone: (46) 0520 94471
Contact #1: Robert Frankmark.
Description: Holographic NDT.
****

Vornhusen Mark
W orthstrasse 89
City: Osnabruck
Country: Germany
Voice Phone: (49) (0)541 572897
Email: mavornhusen@metronet.de
Web: www.shop.de/hpIl123/a
Contact #1: Mark Vornhusen
Description: artist
****

Voxel
26081 Merit Circle - Suite 117
City: Laguna Hills
State: CA
Postal Code: 92653
Country: United States of America
Voice Phone: (1) 949 348 3200
Fax Phone: (1) 949 348 8665
Email: sales@voxel.com
Web: http://www.voxel.com
Description: Medical imaging research. Company is developing a 3D visual display for non-invasive imaging techniques to be used as a diagnostic tool. Looking for financing to complete business plan.
****

W. Cordes GmbH + Co.
Offsetdruckerei und Kartonagenfabrik
Wasserbreite 71
City: Bunde
Postal Code: D-32257
Country: Germany
Voice Phone: (49) (0)5223 15901
Fax Phone: (49) (0)5223 15900
Contact #1 : H. E. Dipl.-Ing.Cordes
Description: Hotstamping holograms-stickerholograms-packaging using paper and carton for advertising-exhibitions
****

Waldner Laboreinrichtungen GmbH
Georgenstasse 35 PB 5 I
City: Berlin
Country: Germany
Voice Phone: (49) (0)30 201741-0
Fax Phone: (49) (0)30 201741-21
Email: waldneclabor@t-online.de
Web: www.waldner.de
Description: Chemical lab equipment.

Waseda University
Dept Of Applied Physics
Hongo 7-3-1 Bunkyo-Ku
City: Tokyo
Postal Code: 160
Country: Japan
Voice Phone: ((81)) 3 209 321
Description: Medical holography research.
****

Wave Mechanics
450 North Leavitt
City: Chicago
State: IL
Postal Code: 60612
Country: United States of America
Voice Phone: (1) 312 829 9283
Fax Phone: (1) 312 829 8557
Contact #1: Deni Drinkwater
Description: Artistic holographer; silver halide transmission & reflection ; consultant.
****

Wavefront Research, Inc.
616 West Broad Street
City: Bethlehem
State: PA
Postal Code: 18018-5221
Country: United States of America
Voice Phone: (1) 610 974 8977
Fax Phone: (1) 610 974 9896
Email: tws@wavres.com
Contact #1: Thomas Stone
Description: Basic and applied research and development in holographic optical elements, applications of holography, non linear optics, and novel optical systems.
****

Wavefront Technology
15149 Garfield Ave.
City: Paramount
State: CA
Postal Code: 90723
Country: United States of America
Voice Phone: (1) 562 634 0434
Fax Phone: (1) 562 634 7758
Email: wavfrnt@idt.net
Contact #1: Joel Petersen
Description: Specialists primarily serving the embossing industry. Embossed hologram mastering, recombining and ganging. Prototype, short run embossing. Rigid sheet embossing up to 4 x 8 foot in transmission or reflection.
****

Werner Jiilich
Optische & Elektronische Gerate
Rheingasse 8-10
City: Bonn
Postal Code: D-53113
Country: Germany
Voice Phone: (49) (0)228 692212
Fax Phone: (49) (0)228 631339
Description: Distributor of optical holographic equipment.
****

Werner, Markus
c/o Christel Kuhl
AJbert Kindle Strasse 18
City: Cologne Wei den
Country: Germany
Voice Phone: (49) (0)2234 7170 I
Email: mwerner@ph-cip.Uni-koeln.de
Description: Artist
****

Wesley, Ed
2124 West Irving Park Road
City: Chicago
State: IL
Postal Code: 60618-3924

Country: United States of America
Voice Phone: (1) 312 345 3998
Contact #1: Ed Wesly
Description: Holographic fine artist, author, and researcher.
****

Westmead Technology Ltd
Unit 7, Wilton Road
St. Georges Industrial Estate
City, Province: Camberley, Surrey
State: England
Postal Code: GU15 2QW
Country: United Kingdom
Voice Phone: (44) 127 668 5455
Fax Phone: (44) 127 662 810
Email: westmead@globalnet.co.uk
Web: http://www.westmead-technology.coml
Contact #1: Stephen Kyle
Description: Westmead Technology offers purpose built systems for the embossing and security holographic industry. This comprises of rotary embossing, WT 360 electroforming, WT Microshim high-resolution dot matrix origination systems, Lightgate-mechanical recombination systems, Impress X -Y Compliant embossing rollers, WT SEI & WT SEM. We also carry our full installations of holographic facilities to ISO standards. All our equipment is UK manufactured and to CE standard.
****

Whiley Foils Limited
(see API Foils)
****

Witchcraft Tape Products, Inc.
P.O. Box 937
City: Coloma
State: MI
Postal Code: 49038
Country: United States of America
Voice Phone: (1) 616 468 3399
Voice Phone: (1) 800 521 0731
Fax Phone: (1) 616 468 3391
Email: ronw@wtp-inc.com
Contact #1: Ronald Warczynski
Description: Full service manufacturer of quality embossed holograms, precision registered die cutting hot stamping, specialty lamination, product assembly and packaging. This is our 25th year of furnishing quality products to the industry.
****

Wonders of Holography Gallery
PO Box 1244
City: Jeddah
Postal Code: 21431
Country: Saudi Arabia
Voice Phone: (966) 2 652 0052
Voice Phone: (966) 2 653 4004
Fax Phone: (966) 2 651 1325
Contact #1: A.M. Baghdadi
Description: Retail holograms of all kinds. We are the first and only holography gallery in the Gulf countries. We resell all types of holograms and we produce laser shows.
****

Worcester Polytechnic Institute
Mechanical Engineering Department
100 Institute Road
City: Worcester
State: MA
Postal Code: 01609-2280
Country: United States of America
Voice Phone: (1) 508 831 5000
Fax Phone: (1) 508 831 5713
Email: rjp@wpi.edu
Web: https://www.wpi.edu
Description: Center for Holographic Studies and Laser Micro-Mechatronics - Photonics studies for undergrad, graduate and post graduate students. Scientific, medical & industrial holography research; Interferometry;

Holographic non-destructive testing.
****

World Holographics
2934 Beverly Glen Circle #400
City: Los Angeles
State: CA
Postal Code: 90077
Country: United States of America
Voice Phone: (1) 310 474 1935
Fax Phone: (1) 310 446 9194
Email: world3D@aol.com
Contact #1: Greg Schuman
Description: We are a full service producer of both holographic and lenticular product. We specialize in the production of promotional premiums for our clients who include many of the nations largest marketers.
****

Wuhan Packaging and Printing United Co.
Han-Kou Da-Shuang-Jie #335
City, Province: Wuhan, Hubei
Postal Code: 430022
Country: China
Contact # 1: Luo Wenjun
Description : Hologram finisher.
****

Wuxi Light Impressions Inc.
Mei Cun
City, Province: Wuxi, Jiangsu
Postal Code: 214000
Country: China
Voice Phone: (86) 510 8150292
Fax Phone: (86) 510 8150292
Contact #1: Wang Guoping
Description: Holography distribution.
****

Xiamen Grand World Laser Label Products
Si-Ming-Nan-Lu 412-#12
2nd Floor
City, Province: Xiamen, Fujian
Postal Code: 361005
Country: China
Voice Phone: (86) 592 2083178
Fax Phone: (86) 592 2083179
Contact #1: Huang Lishan
****

Yu Feng Laser Images Co.
Shantou Economic Zone
Long-Hu-Qu Zhu-Chi-Lu Zhong-Duan
City, Province: Shantou, Guangdong
Postal Code: 515041
Country: China
Voice Phone: (86) 754 8893697
Voice Phone: (86) 754 8891015
Fax Phone: (86) 754 8893677
Contact # 1 : Xue Minqin
Description: Holography distribution.
****

Zanders Feinpapiere AG
Veldener Str. 121-131
City: Diiren
Postal Code: D-52349
Country: Germany
Voice Phone: (49) (0)2202 15-0
Fax Phone: (49) (0)2202 15-2806
Description: Manufacturer of paper-carrier material for hot stamped holograms.
****

Zebra Imaging, Inc
PO Box 81247
City: Austin
State: TX
Postal Code: 78708-1247
Country: United States of America
Voice Phone: (1) 512 251 5100
Fax Phone: (1) 512251 5123
Email: ferdi@zebraimaging.com
Web: www.zebraimaging.com
Contact #1: Alex Ferdman

Description: Large scale reflection holographv. Holographic billboards. Automated holographic output peripherals.
****

a c, Peter
Am Vogelsherd 20
City: Essen
Country: Germany
Voice Phone: (49) (0)201 402185
Fax Phone: (49) (0)201 408231
Description: Consulting, organization of hologram exhibitions, publications on holography.
****

Zero Gravity
9101 International Drive
City: Orlando
State: FL
Postal Code: 32819
Country: United States of America
Voice Phone: (1) 407 903 0824
Fax Phone: (1) 407 903 0834
Web: http://www.spectore.com
Description: Retail gift store offering Titanium jewelry, holograms and other unique products.
****

Zhuhai Xiangzhou Great Wall Laser
Anticounterfeiting Label Co.
Zi-Jing-Lu #40
City, Province: Zhuhai, Guang-Dong
Postal Code: 519000
Country: China
Contact #1: Wu Ziping
****

ZKM Zentrum flir
Medientechnologie
Museum fiir neue Medien
Lorenzstrasse 19
City: Karlsruhe
Postal Code: D-76133
Country: Germany
Kunst
Voice Phone: (49) (0)721 8100-1301
Email: ufrohne@zkm.de
Web: www.zkm.de
Contact #1: Ursula Frohne und
Description: Education for artists and other interested parties on holography and computer generated holograms.
****

Zone Holografix
5338 B Vineland Ave.
City: North Hollywood
State: CA
Postal Code: 91601
Country: United States of America
Voice Phone: (1) 818 985 8477
Fax Phone: (1) 818 549 0534
Email: Cunterseher@yahoo.com
Contact #1: Fred Unterseher
Description: Originates masters for mass production holograms in embossed, DCG and photopolymer materials. Both pulsed and CW lasers available.
SEE OUR ADVERTISEMENT
****

A Listing of Individuals
Last name CA-Z), First name, Business affiliation, Country

A

Abendroth, Detlev, AKS Holographie-Galerie GmbH, Germany
Abouchar, Natalalie, Foreign Dimension (The), China
Abrams, Claudette, Abrams, Claudette, Canada
Abrams, Claudette, Photon League Of Holographers Ontario, Canada
Abramson, Nils, Royal Institute of Technology, Sweden
Ahlemeyer, Frank, S. C. Ahlemeyer, Germany
Akhmedjanov, Dipl Phys. I., Gesellschaft fiir Holografie mbH - GfH, Germany
Akveld, A.c., Triple-D Laser Imaging, Netherlands
Albrecht, Gerd M., Phantastica, Germany
Albright, Steve, O"ptimation Holographics, USA
Alschuler, William R., California Institute of the Arts, USA
Alten, Susanne, Spindler & Hoyer GmbH & Co., Germany
Armmelung, Hr., Micos GmbH & Co KG, Gennany
Anders, Ulrich, Holographie Konzept GmbH, Germany
Andersen, Chad, Meredith Instruments, USA
Anderson, Dick, Anderson Lasers, Inc. , USA
Anderson, Mike, Holomex, UK
Ando, Hiroshi, Nippondenso Co. , Ltd. , Japan
Andrade, Ana Alexandra, INETI - Institute of Information Technologies, Portugal
Andrews, Mathew, 3D-4D Holographics, UK
Anton, Theodor J., Deutscher Drucker Verlagsgesellschaft mbH & Co. KG, Germany
Arkin, Bill, Holo-Spectra, U.SA
Amott, Bob, Holographyx Inc., USA
Arnott, Bob, Holographyx Inc. , USA
Aymar, Bill, Laser Las Vegas, USA

B

Baghdadi, A.M. , Wonders of Holography Gallery, Saudi Arabia
Baghdadi, Abdul Wahab, Hololaser Gallery, United Arab Emerates
Baker, Marion, Krystal Holographics International Inc., USA
Balogh, Tibor, Artplay Holographika Studio, Hungary
Baoqing, Yuan, Qingdao Gaoguang Holography Tech. Co, China
Bard, Richard, Optical Security Group, USA
Barefoot, Paul D., Holophile, Inc., USA
Barre, Pascal, Holos Art Gallery, Switzerland
Bassin, Barry, Infrared Optical Products, Inc., USA
Battin, Dave, Holo Art, USA
Bauer, Josef, Bauer-Josef, Germany
Bautista, Ramon, Empaques y Envolturas Holograficas, S.A. de c.Y. , Mexico
Baynes, Cheryl, Reynolds Metals Co., USA
Bazargan, Kaveh, Focal Image Ltd, UK
Bear, Sol, Hologram World, Inc., USA
Beauregard, Alain, Lasiris Inc., Canada
Beck, Mrs., Coherent Deutschland GmbH, Germany
Beeching, Dave, CFC Applied Holographics, USA
Beeck, M.-A., Volkswagen AG, Germany
Begleiter, Erich, Dimensional Foods Co., USA
Behrmann, Greg, Potomac Photonics, Inc., USA
Beichert, GUnter, Holografie IIlusion mit Licht, Germany
Bellini, Victor, Bellini, Victor, USA
Benito, Ramon, Karas Studios Holografia, Spain
Bentley, John, Dimuken (GB), UK
Benton, Stephen A., M.I.T. (Massachusetts Institute of Technology), USA
Benyon, Margaret, Benyon, Margaret - Holography Studio, UK
Benziger, Thomas, University of Magdeburg, Germany
Berkhout, Rudie, Berkhout, Rudie, USA
Berrie, T, Art Foil Graphic Machinery, UK
Bianchi, Herman-Josef, Arbeitskreis Holografie B.Y., Germany
Billings, Loren, Holographic Design Systems, USA
Billings, Loren, Museum Of Holography/Chicago, USA
Billings, Loren, School Of Holography, USA
Billo, Andreas, Institute for Applied Physics, Germany
Bingda, Zhou, Shanghai Kanlian S & T Development Co. Ltd. , China
Birch, Clive, Kendall Hyde Ltd., UK

Last name (A-Z). First name. Business affiliation. Country

Birenheide, Dr. Richard, HRT GmbH, Germany

Bjelkhagen, Hans, Dr. Hans Bjelkhagen, UK

Bleckwendt, Werner, Optlectra, Germany

Blosvern, Moss, Kinetic Systems, Inc., USA

Bobeck, Paula, DuPont (E.!. DuPont De Nemours & Co.), USA

Bohan, Brian, Cambridge Laser Labs, USA

Bohlinger, Harald, HB Laserkomponenten, Germany

Boone, Prof. Pierre, Ctr for Applied Research in Art & ... , Belgium

Booth, Roberta, Booth, Roberta, USA

Bosco, Eric, Centre d'Art Holographique et Photonique, Canada

Bosco, Eric, Holostar, Canada

Bose, Hans J., Highlite, Germany

Bourgault, Debbie, Laser Focus World, USA

Bouverat, Randy, Spectratek Inc., USA

Boyd, Patrick, Boyd, Patrick, UK

Bradshaw, Roy, Real Image, USA

Brill, Louis, Illuminations, USA

Broadbent, Donald, Broadbent Consulting, USA

Brooker, Dennis, D. Brooker & Associates, USA

Brown, David, Optical Research Associates, USA

Brown, John, Light Impressions International, Ltd., UK

Brown, Kevin, K.C. Brown Holographics, UK

Brown, Kevin, Holographic Dimensions, USA

Brown, William, INNOLAS (UK) Ltd., UK

Bruck, Richard, Bruck, Richard Holography, USA

Bruegmann, Machteld, Hologram Co. RAKO GmbH, Germany

Bruhn, Prof. Ines, Design + Kunst e.V Chernnitz, Germany

Burder, David, 3D Images Ltd., UK

Burgmer, Brigitte, Burgmer Brigitte, Germany

Burke, Ed, Hologram Development Corp., Canada

Burney, Michael, Chronomotion, USA

Burns, Joseph, Hologram Research, Inc., USA

Bussard, Jan, Textile Graphics, Inc., USA

Bussard, Jan, Holography Presses On (HPO), USA

Bussaut, Laurent, Art, Science & Tech. Institute (ASTI), USA

Butteriss, Tony, Parallax Gallery, Australia

Butteriss, Tony, New Dimension Holographics, Australia

Buttler, Martin, Selltexx, Germany

C

Caizhi, Yang, Qingdao Qimei Images, Inc., China

Caperna, Albert J., NovaVision, USA

Carlsson, Torgny, Royal Institute of Technology, Swede?

Carmichael, Lisa, Checkpoint Security Services Limited, UK

Carvino, Cheryl, Foilmark Holographic Images, USA

Casasent, David, Laboratory for Optical Data Processing, USA

Casdin-Silver, Harriet, Casdin-Silver Holography, USA

Cason, Thad, Infinity Laser Laboratories, USA

Catta-Preta, Fernando, Trace Holographic Art & Design Inc. USA

Cavoon, Curt, Spectra-Physics Lasers Inc., USA

Chaihorsky, Alex, Ultra-Res Corporation, USA

Chang, Milton, New Focus, Inc, USA

Chapman, Diane, FlexSystems USA, USA

Cheimets, Alex, 3Deep Company, USA

Cheirnets, Alex, Three Deep Co., USA

Chen, Alex C.T., Infox Corporation, Taiwan

Cheng, Fan, Guangdong Dongguan South Holoprint Co., China

Cherry, Greg, Cherry Optical Holography, USA

Chiang, Mark, Fong Teng Technology, Taiwan

Chiou, Craig, Holo Images Tech Co., Ltd., Taiwan

Chmielewski, Gabriele, Holographie Konzept GmbH, Germany

Chou, Billy, Hiat Image Technology Group, Inc. , Taiwan

Christie, Paul, Liti Holographics, USA

Chuntian, Xu, Guangzhou Chuntian Industrial Tech. Inc., China

Cifelli, Dan, Cifelli, Dan, USA

Clarke, Walter, Global Images, UK

Clafien, Walter, LichtBlicke Classen & Voss, Germany

Claudius, Peter, Multiplex Moving Holograms, USA

Claytor, Linda H. , Fresnel Technologies Inc., USA

Conklin, Don, Glass Mountain Optics, USA

Last name (A-Z). First name. Business affiliation. Country

Connors, Betsy, Acme Holography, USA

Cooper, Nick, Oxford Holographics, UK

Cordner-Guled, Valeska, HOL 3-Galerie fur Holography, Germany

Cos sa, Rich, Planet 3-D, USA

Cossette, Marie-Andree, Holography & Media Institute ... , Canada

Coursen, Dan, G.M. Vacuum Coating Lab, Inc., USA

Cowen, Rick, Advanced Deposition Technologies, Inc., USA

Cox, David, American Laser Corporation, USA

Cox, Dr. 1. Allen, Honeywell Technology Center, USA

Creath, Kathy, Optineering, USA

Crenshaw, Melissa, Melissa Crenshaw Holography Studio, Canada

Crenshaw, Milessa, Capilano College, Canada

Cross, Lloyd, 3-D Systems, USA

Cubberly, George, Excitek Inc., USA

Cullen, Karoline, Holocrafts, Canada

Cullen, Ralph, Uk Optical Supplies, UK

Cvetkovich, Thomas J., Chroma gem Inc., USA

D'Entremont, Joseph P., Lenox Laser, USA

D

Da-hsiung, Hsu, Beijing Univ. of Posts & Tele. ... China

Dahn, Wolfgang G. 0., Gigahertz-Optik, Germany

Darner, Cynthia, AD 2000, Inc., USA

Dandliker, Rene, Universite De Neuchatel, Switzerland

Dandliker, Rene, Universite De Neuchatel, Switzerland

Dann, David, David Dann Modelmaking Studios, USA

Dausmann, Gunther, HSM, Germany

De Odorico, B., Dilas Diodenlaser GmbH, Germany

de Roos, Marc, Deep Space Holographics, Canada

Deem, Rebecca, Zone Holografix, USA

Deem, Rebecca, Deem, Rebecca, USA

DeFreitas, Frank, Holoworld.com (The Holography Studio), USA

DeFreitas, Frank, DeFreitas Holography Studio; USA

del-Prete, Sandro, Galerie Illusoria, Switzerland

del-Prete, Sandra, Galerie Illusoria, Switzerland

Denisyuk, Yuri N., Denisyuk, Yuri N., Russia

Desai, Yogesh, Spatial Holodynamics (India) Pvt. Ltd. , India

Detrich, Ed, OpSec - USA, USA

Deutschman, Bill, Oregon Laser Consultants, USA

Deutschmann, Gunther, AHT 3D-Medien, Germany

Deutschmann, Gunther, Holographie Fachstudio ... , Germany

Devens, Scott, Holographyx Inc., USA

Diamond, Mark, Diamond Images, Inc., USA

Dicker, Mark, Holographic Consulting Agency, UK

Dietrich, Edward, Optical Security Group, USA

Dirksen, Alfred, Alfred Dirksen + Sohn Mod .. , Germany

Dittrich, Ronald, KomSa Kornmunikationstechnik, Germany

Dixon, John, Holman Technology, Inc. , USA

Dixon, Ken, Holman Technology, Inc., USA

Dolgoff, Gene, Floating Images, Inc., USA

Dreelaw, Dawn, International Data Ltd. , UK

Drinkwater, Deni, Wave Mechanics, USA

Duarte, Brad, Polaroid Corporation, USA

Duffey, William, Regal Press Inc., USA

. Duignan, Michael, Potomac Photonics, Inc., USA

Dutton, Keith, Laser International, UK

E

Edgar, John, Brandtjen & Kluge, Inc." USA

Edhouse, Simon, 3DIMAGE, Australia

Edhouse, Simon, Australian Holographics, Australia

Eimers, Tilmann, Holographic Laserdesign, Germany

Engelmann, Heiko, Modern Marketing, Germany

Erickson, Ronald R. , Media Interface, Ltd., USA

Evans, Ronald, Collector's Castle, USA

F

Faddis, Terry, Optimation Holographics, USA

Fattal, Isaac, Krystal Holographics International Inc., USA

INDIVIDUALS

Last name (A-Zl. First name. Business affiliation. Country

Fein, Howard, Polaris Research Group, USA
Felix, Patricia, Holograms and Lasers International, USA
Felix, Patricia, Holograms and Lasers International, USA
Felix, Perry, Holograms and Lasers International, USA
Felix, Perry, Holograms and Lasers International, USA
Fendel, Achim, AF Elektronik, Germany
Ferdman, Alex, Zebra Imaging, Inc, USA
Feroe, James, Feroe Holographic Consulting, USA
Ferri, Lucilla Croce, FhG Fraunhofer Gesellschaft, Germany
Fiedler, Mr., Art Ig Wohndesign - Holografie, Germany
Fimia., A., University Of Alicante, Spain
Fink, Detief, Fink Feinoptik, Germany
Fischler, Ben, Imagination Plantation, USA
Fisher, Gary, Man/Environment, Inc., USA
Fitzpatrick, Colleen, Rice Systems, Inc., USA
Fleck, Dr. Burkhard, University of Jena, Germany
Florence, Mike, James River Products, USA
Foerster, Thomas, Holostudio Beate Krengel, Germany
Forese, John, Siros Technologies, Inc., USA
Fornari, Arthur David, Fornari, Arthur David, USA
Forsberg, Mona, HoloMedia Ab/Hologram Museum, Sweden
Fournier, Jean-Marc, Rowland Institute For Science, USA
Franck, Gerhard, G. Franck OptroniK GmbH - GFO, Germany
Frank, A., Bernhard Halle Nachf. GmbH & Co., Germany
Frankmark., Robert, Volvo-Flygmotor, Sweden
Frieb, M.T, Holographie Anubis, Germany
Frisk, E.O., Optical Works Ltd., UK
Frohne, Ursula, ZKM Zentrum fur Kunst und Medien ... , Germany
Fuechtenbusch, Annette, Holodesign, Germany
Fukuda, Akinobu, SAM Museum, Japan
Fukuda, Shinichi, Kimmon Electric Co., Ltd., Japan
Fukurna, Mineko, Japan Communication Arts Co., Japan
Funicelli, Joseph, Unifoil Corporation, USA

G

Gabrielson, Dan, International Holographic Paper, USA
Gallagher, Dan" Total Register Inc., USA
Gallagher, Dawn, Cfc Northern Bank Note Co., USA
Gallagher, John, Total Register Inc., USA
Galon, Derek, Royal Holographic Art Gallery, Canada
Garcia, Diego, M.LT Museum, USA
Garrett, Steve, Midwest Laser Products, USA
Gericke, H. G., Holo Time Gericke, Germany
Gericke, W. K., Holo Time Gericke, Germany
Gibb, Jim, Automated Holographic Systems (AHS), USA
Gibson, lA., Ag Electro-Optics Ltd., UK
Gillespie, Don, El Don Engineering, USA
Gillespie, Mike, Jodon Inc., USA
Ginouves, Paul, Coherent, Inc. - Laser Group, USA
Glaser, Shelly, Glaser· Technical Consulting, Israel
Glazer, Stewart, Crown Roll Leaf, Inc." USA
Goldstein, Robert, Lasermetrics, Inc., USA
Golen, VP, Grace, Holographic Dimensions, Poland S.A., Poland
Gorglione, Nancy, Cherry Optical Holography, USA
Gorglione, Nancy, Gorglione, Nancy, USA
Gorglione, Nancy, Laser Affiliates, USA
Gougeon, Pierre, Dimension 3, Canada
Graf, Mike, Letterhead Press, Inc., USA
Graffeo, Gus, Digital Matrix Corp., USA
Graham, Ben, Lexel Laser, Inc., USA
Grau Prof. Dr. techn. , Gerhard, University of Karlsruhe, Germany
Greenspan, Alex, Digital Matrix Corp., USA
Greguss, Pal, Optopol Panoramic Metrology Consulting, Hungary
Greiner, Gotz, Holografie Manufaktur, Germany
Grindel, Mark, Continental Optical, USA
Grohmann, Mr., b+g Banse und Grohmann GmbH, Germany
Groote, Manfred, HMS-Elektronik, Germany
Grueneberg, Joachim, Decolux GmbH, Germany
Gugg-Helminger, Anton E. A., Gigahertz-Optik, Germany
Guoping, Wang, Wuxi Light Impressions Inc., China

Last name (A-Zl. First name. Business affiliation. Country

Gustafsson., Jonny, Holovision AB, Sweden
Gutekunst, H. & R., Linea - Bregaglia, Switzerland

H

Habighorst, Karsten, VISUELL, Germany
Haidong, Liu, Beijing Dongfang Laser Printing Tech. Co., China
Halkes, Adrian 1, Far East Holographics, China
Halotek, John, Foilmark Holographic Images, USA
Haney, Lorri, Nimbus Manufacturing, Inc., USA
Hangjia, Sun, Jiangsu Sida Images, Inc., China
Hankin, Alan, Bridgestone Technologies, Inc., USA
Hardy, Nick, Op-Graphics (Holography) Ltd., UK
Harris, Ken, Dimensional Arts, USA
Harris, Ken, HoloCom, USA
Hart, John, Holotek, USA
Harttnan, John, Batelle Pacific Northwest National Labs, USA
Harttnann, Vera, Laser Solution Management, Germany
Hashimoto, Chikara, Central Glass Co., Ltd., Japan
Hashimoto, Reiji, Topcon Inc., Japan
Hassen, Chuck, Holo Sciences, LLC, USA
Hausler, G. Prof. Dr., University of Erlangen, Germany
Heck, David, Spectra-Physics Lasers Inc., USA
Hedley, David, Whiley Foils Limited (API Foils), UK
Heil, Wendy, Advanced Optics, Inc., USA
Hein, Elke, Die Dritte Dimension, Germany
Helmers, Heinz, University of Oldenburg, Germany
Hepburn, James, Silverbridge Group, Canada
Hermann, Christian, Leonhard Kurz GmbH, Germany
Herr, Doug, Bobst Group Inc., USA
Hess, Bob, Point Source Productions, USA
Hesselink, Burt, Siros Technologies, Inc., USA
Hicklin, Schryl, API Foild, USA
Hill, Dean, OpSec - USA, USA
Hill, Zohra, HOL 3-Galerie fur Holographie GmbH, Germany
Hillard, Bill, NeoVision Productions, USA
Hocke, WolfPeter, AD HOC Public Relations GmbH, Germany
Hoffman, John, Holographic Design Systems, USA
Hoffstadt-Braeutigam, Irrnhild, topac GmbH, Germany
Hofiing, Dr. R., GESA - Arbeitskreis fur optische ... , Germany
Hofmann, Christina, Holotopia II, Germany
Hofmann, Christina, Galerie WesteriandiSylt, Germany
Hofmann, Christina, Holografie Hofmann, Germany
Hofmann, Martin, Galerie WesteriandiSylt, Germany
Hofmann, Martin, Holotopia II, Germany
Hofmann, Martin, Holografie Hofmann, Germany
Hofmann-Lange, Dr. rer. nat. Brigitte, Germany
Holden, Lawrence, API Foils, UK
Hollinsworth, TR., Expanded Optics Limited, UK
Hollmann-Langecker, Liesel, Art Agentur Koln, Germany
Honda, Miss Prof. Toshio, Chiba University, Japan
Honnef, Prof. Klaus, Rheinisches Landesmuseum Bonn, Germany
Hoose, John, Richardson Grating Laboratory, USA
Hoppe, Guido, Buntstift, Germany
Horn, Jiirgen, Dt. Gesellschaft fur Stereografie - DGS, Germany
Horn, Rolf, HoloMedia Ab/Hologram Museum, Sweden
Horne, Peg, Scharr Industries, USA
Hotta, Tsuyoshi, Dai Nippon Printing Co., Ltd., Japan
HSU, Dr. Fu Kuo, Electro-Optics Lab, NECTEC, Thailand
Hsu, Jonathan, Holo Impressions Inc, Taiwan
Huajian, Gu, Suzhou University, China
Hufmann, Michael, Holodesign, Germany
Hui, Chen, Beijing Xi Ji Wo Computer Graphic Working Co.,China
Hultzsch, Rainer, Carls Zeiss GmbH, Germany
Hurwitz, Noah, Imagination Plantation, USA
Hwang, Edward, Superbin Co. Ltd, Taiwan

I

Ibbs, Ken, LiCONiX, A Melles Griot Company, USA
ide, Makoto, Nippon Polaroid K.K., Japan
Infantes, Victoria E., Karas Studios Holografia, Spain

Last name (A-Z). First name. Business affiliation. Country

Inoue, Yutaka, Brainet Corporation - International Division, Japan
Iovine, John, Art Lab, USA
Ishihara, Dr. Satoshi, Ministry Of International Trade, Japan
Ishii, Setsuko, Ishii, Ms. Setsuko, Japan
Ishikawa, Kazue, Sophia University, Japan
I verson, Mark, Datacard Corporation, USA
Iwata, Fujio, Toppan Printing Co., Ltd. , Japan

J

Jager, Dr. E., Topag GmbH, Germany
Jain, Rajeev, Shriram Holographics, India
James, Randy, Pacific Holographics Inc., USA
Jauernik, Dr. P, Radiant dyes accessories GmbH, Germany
Jelic, Nikola, 3D Technologies & Arts, Slovenia
Jeong, T.H., Lake Forest College, USA
Jeong, T.H., Integraf, USA
Jerit, John, American Paper Optics Inc., USA
Jiang, Yaguang, China Ann Arbor Holographical Institute, China
Jingde, Ye, Shenzhen Reflective Materials Factory, China
Jizhang, Wang, North Light Holograms Ltd., China
Johann, Larry, Southern Indiana Holographics, USA
John, Pearl, John, Pearl, USA
Jorgensen, Dean, Optimation, USA
Joyce, Chris, Oxford Holographics, UK
Jung, Dieter, Academy of Media Arts Cologne, Germany
Junger, Hr., CHIRON Technolas GmbH, Germany

K

Kabert, Inge, Thorlabs Inc., Germany
Kac, E, Art Institute Of Chicago (The School of the ...), USA
Kalka, Jutta, PPM Promotion Products Munich GmbH, Germany
Kallenbach, Heinz, H. Kallenbach - H.M.V, Germany
Kane, Brian, General Design, USA
Karaganova, Svetlana, Australian Holographics, Australia
Karlovac, Nevin, Accuwave Corp., USA
Kassover, Kathy, Crown Roll Leaf, Inc." USA
Katsuma, Hidetoshi, Tama Art University, Japan
Katz, Jordan, Grafix Plastics, USA
Kauffman, John, Kauffman, John, USA
Keilholz, Mr., Spindler & Hoyer GmbH & Co., Germany
Keller, Manuela, Steuer KG GmbH & Co., Germany
Kendall, Christen, Metrologic Instruments, Inc., USA
Kenny, Mike, MWK Industries, USA
Kettel, Klaus, Krystal Holographics Vertriebs-GmbH, Germany
Keyaniyam, Behnam, Holographic Identity, Germany
Kiely, Annette, De La Rue Holographics, UK
King, R. Eric, Laser Innovations, USA
Kinoshita, Dr. Kenji, Toyama National College Of Marit, Japan
Kirk, Ronald L., Holovision Systems Inc., USA
Kita, Ms. Nonoko, Holographic Display Artists & Eng. Club (HODIC), Japan
Klein, Alexander, Verlag 3D Magazin, Germany
Kleinhenz, G., Holotec, Germany
Kleinherne, Klaus, Holo Service/Service-Druck, Germany
Kleinherne, Reiner, Holo Service/Service-Druck, Germany
Kleinherne, Reiner, Holo-Idee Reiner Kleinherne, Germany
Klimasewski, Tim, Burleigh Instruments, Inc., USA
Klix, Armin, Armin Klix Holographie, Germany
Knauer, Bettina, Schall Messen GmbH - Optatec, Germany
Knocke, Frank, F & E Labor, Germany
Knuth, Eckard, Laserfilm, Germany
Kodera, Mitsuo, Ardee International, Japan
Koizumi, Fumihiko, Asahi Glass Co., Japan
Kollin, Joel, Holonix, USA
Kontnik, Lewis, Reconnaissance International Ltd. , USA
Koril, Gary, Creative Label, USA
Koril, Jerry, Creative Label, USA
Komienko, Sergey, Feofaniya Ltd. , Ukraine
Kraak, J.O. , 3-D Hologrammen, Netherlands
Krautter, Thomas, HOLOGRAPHIA, Germany

Last name (A-Z). First name. Business affiliation. Country

Krengel, Beate, Holostudio Beate Krengel, Germany
Kretschmer, Karl, Kremo, Germany
Krick, Reva, Reva's Holographic Illusions, USA
Krick, Robert, -Reva's Holographic Illusions, USA
Kritzinger, Sean, Synchron Pty Ltd., South Africa
Krosel, Hans, HAKRO, Germany
Krueger, Dave, Holograms International, USA
Krueger, Jean, Holograms International, USA
Kruzel, Dr. Paul, RGB Holographics Inc., USA
Kubeile, Thomas, PPM Promotion Products Munich ... , Germany
Kubota, Toshihiro, Toshihiro Kubota, dept of electronics, Japan
Kukhtarev, Nicholai, Alabama A&M University, USA
Kukhtarev, Tanya, Alabama A&M University, USA
Kiirsten, Dieter, Melles Griot GmbH, Germany
Kuwayama, Tetsuro, Canon Inc. R&D Headquarters, Japan
Kwok, Joyce, Laser Art Studio Ltd, China
Kyle, Stephen, Westmead Technology Ltd, UK
Kyle, Stephen, Laza Holograms Ltd., UK

L

Labelle, Scott, Holographic Label Converting (HLC), USA
Lacey, Lee, Holo-Source Corporation, USA
Lamb, Mark, Cfc Northern Bank Note Co., USA
Lancaster, Ian, Reconnaissance International Ltd., UK
Langer, w., Dornier Medizintechnik GmbH, Germany
Larim, Jim, Laser Optics, Inc., USA
Lau, Prf. Dr. B., Fachhochschule Ulm a IITA, Germany
Lauk, Matthias, Lauk & Partner GmbH, Germany
Law, Linda, Linda Law Holographics, USA
Lee, Julie, Ahead Optoelectronics, Inc., Taiwan
Lembessis, Alkis, Hellenic Institute Of Holography, Greece
Lembessus, Alkis, Cavomit, Greece ,
Leseberg, Detlef, Leseberg Dr. Dellef, Germany
Lessard, Roger A. , Universite Laval, Canada
Lev, Steven, Chromagem Inc., USA
Levine, Chris, iC Holographics, UK
Levy, Rob, Holo-Source Corporation, USA
Levy, Uri, Holo-Or Ltd, Israel
Lieberman, Dan, HoloWebs, LLC, USA
Lieberman, Larry, Larry Lieberman Holography, USA
Liedlbauer, Eric, Highlite, Germany
Liedtke, Mario, mario liedtke pro design, Germany
Lifshen, Alan, Lone Star Illusions, USA
Lin, Yow-Snin, Holoart Studio, Taiwan
Lind, Michael, Batelle Pacific Northwest National Labs, USA
Linsen, Chen, Suzhou University, China
Lion, Yves F. , Univ. de Liege, Belgium
Lishan, Huang, Xiamen Grand World Laser Label Products, China
Lissack, Selwyn, Laserworks, USA
Liu, Wai-Min, Control Optics, USA
Lkegami, Dr. Koji, Numazu College Of Technology, Japan
Lloyd, Glen, 3D Ltd., Switzerland
Long, Mike, Pacific Holographics Inc., USA
LoSardo, Sal, Towne Technologies, USA
Lossau, H. G., Polytec GmbH, Germany
Love, Valerie, Op-Graphics (Holography) Ltd. , UK
Lovygin, Igor, Technoexan Ltd, Russia
Ludwig, Peter, Holographie & Design, Germany
Liipfert, Eckhard, DLR e.V , Germany
Luton, Chris, Holocrafts Europe Limited., UK

M

MacArthur, Ana, Ana MacArthur, USA
MacShane, Jim, MacShane Holography, USA
Magarinos, Jose R., Holographic Optics Inc., USA
Maier, Andreas, HAM Kristall Technologie, Germany
Mair, Alois, Uni Innsbruck, Austria
Mairiedl, Horst, Decolux GmbH, Germany
Makansi, Munzer, Fiber Engineering, USA
Malott, Michael, Laser Light Designs, USA

INDIVIDUALS

INDIVIDUALS

Last name (A-Z). First name. Business affiliation. Country

Richardson . Martin, London Holographic Image Studio, UK
Richter. Harald, MeiLing - Lasertechnik, Germany
Rickert. Sue, Hologram Land, USA
Ritter. Alf, University of Magdeburg, Germany
Robb. Jeffrey, Spatial Imaging Limited, UK
Robiette, Nigel, Colour Holographics, UK
Robinson, Deborah, Lightrix, Inc., USA
Robinson, George, Hologram Land, USA
Robur, Lubomir, Feofaniya Ltd., Ukraine
Rodia, Carl M., Carl M. Rodia And Associates, USA
Rongkun, Wu, Quan Zhou Pacific Laser Images, China
Rosowski, Ralf, Holographic Studios, Germany
Ross, Franz, Ross Books, USA
Ross, Franz, Holography Marketplace, USA
Ross, Jonathan, Holograms 3D, UK
Ross, Michael, IBM Almaden Research Center, USA
Rost, Thomas, K. Thielker & T. Rost-Holographie ... , Germany
Roth, Ulrich G., Holographie Roth, Germany
Rottenkolber, Hans, Rottenkolber Holo-System GmbH Betrieb
Kirchheim, Germany
Roule, Richard, Label Systems, USA
Rueck, A. B., AB Riick Holoart, Germany
Ruey-Tung, Hung, Ruey-Tung, Miss. Hung, Japan
Ruiz-Rosales, Jorge, Corp. Mexicana De Impresion ... , Mexico
Rumo, Wang, Minjian Laser Holography, China

S

Saarinen, Jyrki, Heptagon Oy, Finland
Salter, Dr. Phillip, British Aerospace Pic., UK
Sandra, John, The Hologram Store, Ltd., Canada
Sapan, Jason, Holographic Studios, USA
Sato, Koki, Shonan Institute of Technology, Japan
Sato, Shunichi, Sharp Corp., Japan
Saxby, Graham, Saxby, Graham, UK
Schaper, Stefan, Directa GmbH, Germany
Scheir, Peter, AD 2000, Inc., USA
Schenker, Frank M. , Aquarius-Vertrieb, Germany
Schipper, Wilfried, Hologram Company RAKO GmbH, Germany
Schlewitt, Carsten, Optical Test Equipment, Germany
Schmidt, Marina, Coherent Liibeck GmbH, Germany
Schnitzer"Chemnitzer Werstoffmechanik GmbH, Germany
Schrieber, Matthew, Holographic Images Inc., USA
Schomer, Jiirgen, Rottenkolber Holo-System ... , Germany
Schulze, Ute, Directa GmbH, Germany
Schulze Brockhausen, Eva, EPA - Elektro-Physik. .. , Germany
Schuman, Greg, World Holographics, USA
Schvartzman, Frederic, Foreign Dimension (The), China
Schweer, Jorg, Holoptics, Germany
Schweitzer, Dan, Center For The Holographic Arts, USA
Schweitzer, Dan, New York Holographic Laboratories, USA
Sciammarella, Cesar, Illinois Institute Of Technology, USA
Scott, G.H., Customer Service Instrumentation, UK
Seitz, Mr., Steuer KG GmbH & Co., Germany
Selbach, H., Polytec GmbH, Germany
Shafer, Brad, Engineering Animation, Inc., USA
Shah, Kailesh, Ojasmit Holographics, India
Shahjahan, Mr., Dimensions, Pakistan
Sharma, Govind, Holostik India Pvt. Ltd., India
Sharpe, Frank, Datasights Ltd., UK
Sherwood, Robert, Robert Sherwood Holographic Design, USA
Shimon, Hameiri, Holography Israel, Israel
Shindell, David, Ann Arbor Optical Co., USA
Shindell, David, Data Optics, Inc., USA
Shouzhong, An, Taiyuan Shiji Holography Ltd., China
Shoydin, Sergey, OPTIC, Russia
Shun, Zhu De, Shandong Academy of Sciences, China
Simson, Bernd, Capilano College, Canada
Simson, Bernd, General Holographics, Inc., Canada
Simson, Paula, General Holographics, Inc., Canada
Sinclair, Douglas P., Sinclair Optics, Inc., USA

Last name (A-Z). First name. Business affiliation. Country

Singer, Kenneth, Capitol Converting Equipment, Inc., USA
Singh, Ravinder, Print-M-Boss, India
Siveriver, Leonid, Avant-Garde Studio, USA
Sivy, George, Image Engine, USA
Skipnes, Olav, Interferens Holografi D.A. , Norway
Smith, Cheryl, Evergreen Laser Corp., USA
Smith, S.D., Art Foil Graphic Machinery, UK
Smith, Steven, Smith, Steven L., USA
Soales, Bob, CVI Laser Corporation, USA
Sobotka, Prof. Dr. Werner, Fachhochschule St. Poiten, Austria
Song, Chung, Dan Han Optics, Korea
Song, Li, Inspeck, Inc. , Canada
Sosa, Laura, Hologramas, S.A. DE C.V, Mexico
Sott, Gudrun, AKS Holographie-Galerie GmbH, Germany
Souparis, Hughes, Hologram Industries, France
Sowdon, Michael, Fringe Research Holographics, Canada
Spanner, Karl, Physik Instrumente (PI) GmbH & Co., Germany
Speer, Erik, Holarium, Germany
Spiegel, Gary, Newport Corporation, USA
Spierings, Waiter, Dutch Holographic Laboratory BY, Netherlands
Sponsler, Michael B., Syracuse University, USA
St. Cyr, Suzanne, Holographic Applications, Inc, USA
Stack, John, Edmund Scientific Company, USA
Staiger, Brigitta, Holosta Holographie-Galerie, Germany
Starcke, Ari-Veli, Starcke, Ky., Finland
Stehle, Robert, S.O.P.R.A., France
Steinbichler, H., Steinbichler Optotechnik GmbH, Germany
Steinfeld, Belle, Dell Optics Company, Inc., USA
Steinmetz, Prof. Dr.-Ing. E., University of Kassel, Germany
Steiter, Manfred, Process Technologies, USA
Stensborg, Mr. Jan, Stensborg Inc., Denmark
Stepien, Pawel, Hololand S.C., Poland
Stich, Boguslaw, Holografia Polska, Poland '
Stief, Gerhard, Museum 3. Dimension, Germany
Stogsdill, John, Photo Sciences, USA
Stoller, Carol, Another Dimension Inc. (Spectore/ADI), USA
Stolyarenko, Alexander, TAVEX, Ltd., Ukraine
Stone, Thomas, Wavefront Research, Inc., USA
Stooss, Richard, Krystal Holographics Vertriebs-GmbH, Germany
Strassner, Hans M., HMS-Elektronik, Germany
Styns, Erik, Free University Of Brussels., Belgium
Su, Dr. 1.1., Industrial Technology Research Inst. , Taiwan
Sugarman, Stephen, Holographic Products, USA
Summer, Carol, Art Lab, USA
Surana, Rajendra, Ojasmit Holographics, India
Swetter, Erik, 3-D Hologrammen, Netherlands
Swinehart, Patricia, CFC Applied Holographics, USA
Synowiec, George, Lumonics Ltd., UK

T

Taylor, Rob, Forth Dimension Holographics, USA
Tesche, Dr. Bernd, Dt. Gesellschaft fur Elektro ... , Germany
Teteris, Janis, University of Latvia, Latvia
Thielker, Klaus, K. Thielker & T. Rost-Holographie ... , Germany
Thiemon, Ms., Daimler Benz Aerospace, Germany
Tholen, Maureen, 3M - Safety and Security Systems, USA
Thoma, John, API Foild, USA
Thomas, Jackie, Laser Institute Of America, USA
Thompson, Bridget, iC Holographics, UK
Thuston - Lighty, Cathy, M.LT. Museum, USA
Tianji, Wang, Guangzhou Inst.of Electronics, China
Tidmarsh, David, Applied Holographics, Pic., UK
Tiemon, M., Dornier Medizintechnik GmbH, Germany
Titizian, Lia, Optical Research Associates, USA
Tiziani, Hans, University of Stuttgart, Germany
Tobin, John, Macquarie Holographics, Australia
Tobin, John, Moonbeamers, Australia
Toland, Lee, Meredith Instruments, USA
Tolia, Dr. Arun, Spatial Holodynamics (India) Pvt. Ltd., India
Tomking, Donald, Kurz Foils, USA

INDIVIDUALS

Tong, Chen Guo, Morning Light Holograms, China
Trayner, David, Richmond Holographic Studios Ltd, UK
Trebst, Walter, Spot Agentur fur Holographie und .. , Germany
Tribillon, Dr.Jean Louis, Holo-Laser, France
Tschudi, Prof. Dr. Theo, Dt. Gesellschaft fur . . , Germany
Tsuj iuchi., Jumpei, Chiba University, Japan
Tuffy, Francis, API Foils, UK
Tunnadine, Graham, 3D-4D Holographics, UK
Tzong, Tang Yaw, Institute Of Optical Science, Taiwan

U

Unbehaun, Klaus, Deutscher Drucker Verlagsgesellschaft mbH & Co. KG, Germany
Unbehaun, Klaus, Holopublic Unbehaun, Germany
Unterseher, Fred, Zone Holografix, USA
Unterseher, Fred, Unterseher & Associates, USA
Upatnieks, Juris, Applied Optics, USA
Uram, Marvin, Holograms Unlimited, USA
Uram, Marvin, U.K. Gold Purchasers, Inc., USA
Urgela, Stanislav, Technical University Zvolen, Slovakia
Uwe, Saurda, Holographie Labor, Germany
Uyemura, T., University Of Tokyo, Japan

V

Van Duyne, Tom, Laser Resale Inc., USA
Vancurova, Jana, Lightgate, Ltd, Czech Republic
Varga, Miklos, Hologram Varga Miklos, Hungary
Varney, Chris, Electro Optics Developments Ltd., UK
Vergnes, Florian, Geola, Lithuania
Vertongen, Emmanuel, Krystal Holographics SARL, France
Vila, Doris, Doris Vila Holographics, USA
Vince, Bill, Anaspec Ltd., Holography and Image Systems, UK
Vinson, Joachim, VinTeq, Ltd., USA
Vinstadt, Tom, Virtual Image (a division of Printpack, Inc.), USA
Vogel, Jon, Holographics (Uk) Ltd., UK
von Bally, Gert, University of Miinster, Germany
Vornhusen, Mark, Vornhusen Mark, Germany
Voss, Katharina, LichtBlicke Classen & Voss, Germany
Vulcano, Charles, Holographic Finishing, Inc ., USA
Vulcano, Michael, Holographic Finishing, Inc., USA

W

Wade, Heather, Foil Stamping and Embossing Association, USA
Wagensonner, M., Holotec, Germany
Wahl, Mike, Coburn Europe GmbH, Germany
Wahler, Dip!. Ing. Jelena, University of Bundeswehr, Germany
Wale, R. D., Galvoptics Ltd. , UK
Walters, Glenn 1., Advanced Deposition Technologies, Inc., USA
Wang, Hunter, Holography Development Co.Ltd., China
Wanlass, Mike, Spectratek Inc., USA
Wappelt, Andreas, Andreas Wappelt - Photonics Direct, Germany
Warczynski, Ronald, Witchcraft Tape Products, Inc., USA
Wassel, Manfred, AD HOC Public Relations GmbH, Germany
Wegeler, Marc, Lauk & Partner GmbH, Germany
Weil, Jeffrey, Holographic Dimensions, USA
Weili, Zhu, Institute for Holographic Tech., China
Weinstein, Beth, New York Hall Of Science, USA
Weitzel, Tilo, ETH - Eidgenossische Technische ... , Switzerland
Wen, Pei, Beijing Sanyou Laser Images Co., China
Wenjun, Luo, Wuhan Packaging and Printing United Co., China
Wernenski, John, Jodon Inc., USA
Wernicke, Prof. Dr. Giinther, Humboldt Univ. Berlin, Germany
Wesly, Ed, Wesley, Ed, USA
Wheeler, Bill, American Banknote Holographics, Inc. , USA
White, John, Coburn Corporation, USA
White, Steve, Electro Optical Industries, Inc., USA
Wild, Urs. P., ETH - Eidgenossische Technische ... , Switzerland
Willard, Susan, Richardson Grating Laboratory, USA
Wilson, Brett, Lazart Holographics, Australia

Wilzner, Ulf, GTO Lasertechnik GmbH, Germany
Windeln, Wilbert, ETA-Optik Gmbh, Germany
Windsor, Robert K., American Society for NDT, USA
Winopal, Gerhard, Gerhard Winopal Forschungsbedarf, Germany
Withycombe, Nicola, Laza Holograms Ltd., UK
Witte, Bob, Dri-Print Foils, USA
Wittig, Dr. Siegmar, Rita Wittig Fachbuchverlag, Germany
Wober, Irmfried, Holography Center of Austria, Austria
Wojcicik, Roman, Verwaltungsgesellschaft Holo. .. , Germany
Wollenweber, Andreas, Magick Signs Holografie, Germany
Woolford, Jimmy, HDIPanama, Panama
Woontner, Marc 0., Transfer Print Foils, Inc. , USA

X

Xuzhang, Zeng, Chengdu Xinxing Inst. for Development.., China

Y

Yamaguchi, Masahiro, Tokyo Institute Of Technology, Japan
Yamazaki, Hitoshi, Hyogo Prefectual Museum Modem Art, Japan
Yao, Wang, Beijing Fantastic Hologram Products, China
Yijun, Qin, Foshan Holosun Packaging Co. Ltd, China
Yokota, Hideshi, Tokai University, Japan
Yoshikawa, Dr. Hiroshi, Nihon University, Japan
Yuan, Weiben, Tianjin Holdor Optics Inc. China, China
Yuanzhong, Gong, Shanghai Dahua Printing Factory, China

Z

Zacharovas, Dr. Stasys, Geola, Lithuania
Zeller, Mrs. M., Meisinger Verlag, Germany
Zhaoqun, Zhang, Image Technical Development Co., China
Zheng, Zhang, Chongqing Yinhe Laser Products Ltd., China
Zhongming, Lu, Nanjing Sanle Laser Technology R&D, China
Ziping, Fu, Beijing Hologram Printing Tech. Co, China
Ziping, Wu, Zhuhai Xiangzhou Great Wall Laser, China
Zucker, Richard, Bridgestone Technologies, Inc., USA
Zurek, Mr. , University of Munich, Germany

14

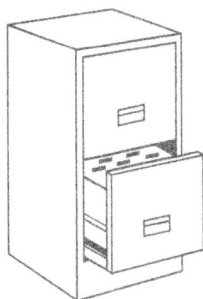

Cross-Index Tables

TABLE 1 - BUSINESSES THAT SELL OR EXHIBIT HOLOGRAMS

Country	Business Name	Distributor Wholesaler	Retail Shop	Mail Order Catalog	Artist-With Portfolio	Pulsed Portrait Studio	Broker-Sales Rep	Touring exhibit	Gallery/museum display
Australia	3D Optical Illusions	X	X	X			X		
Australia	3DIMAGE						X		
Australia	Australian Holographics	X	X	X	X		X		X
Australia	Holograms Fantastic & Optical Illusions	X	X	X			X		
Australia	HOPSec/dii	X					X		
Australia	Lazart Holographics	X	X		X		X		
Australia	Menning, Melinda				X		X		
Australia	Moonbeamers	X							
Australia	New Dimension Holographics	X	X		X		X		
Australia	Parallax Gallery	X	X		X		X		
Austria	Holography Center of Austria	X	X	X	X	X	X	X	
Austria	Korn Michael				X				
Austria	Lacher Peter				X				
Canada	Abrams, Claudette				X		X		
Canada	Capilano College				X				
Canada	Deep Space Holographics	X			X		X		
Canada	Dimension 3	X		X	X		X		
Canada	Fringe Research Holographics		X		X				X
Canada	General Holographics	X		X	X		X		
Canada	Holocrafts	X	X	X	X		X		
Canada	Hologram Development Corp.				X				
Canada	Holography & Media Inst. of Quebec				X				
Canada	Island Graphix				X		X		
Canada	Melissa Crenshaw Holography Studio				X		X		
Canada	Mu's Laser Works						X		
Canada	Ontario Science Centre		X		X				
Canada	Page, Michael				X		X		
Canada	Royal Holographic Art Gallery	X	X	X					
Canada	Silverbridge Group	X			X		X		
Canada	The Hologram Store		X	X					
China	Beijing Fantastic Hologram Products	X	X				X		
China	Beijing Hologram Printing Tech. Co	X							
China	Beijing Sanyou Laser Images Co.	X							
China	Far East Holographics	X							
China	Foreign Dimension (The)	X	X	X	X				
China	Foreign Dimension (The)	X	X		X		X		
China	Guangdong Dongguan South Holoprint	X							
China	Guangzhou Chuntian Industrial Tech.						X		
China	Holography Development Co.Ltd.	X					X		
China	Jiangsu Sida Images	X							
China	Minjian Laser Holography	X					X		
China	Morning Light Holograms	X							
China	North Light Holograms Ltd.	X							
China	Qingdao Gaoguang Holography Tech.	X							
China	Qingdao Qimei Images	X							
China	Quan Zhou Pacific Laser Images	X					X		
China	Shandong Academy of Sciences								X
China	Shanghai Kanlian S & T Development	X							
China	Taiyuan Shiji Holography Ltd.	X							
China	Tianjin Water Laser Holography Image	X							

TABLE 1 - BUSINESSES THAT SELL OR EXHIBIT HOLOGRAMS

Country	Business Name	Distributor Wholesaler	Retail Shop	Mail Order Catalog	Artist-With Portfolio	Pulsed Portrait Studio	Broker-Sales Rep	Touring exhibit	Gallery/museum display
China	Wuxi Light Impressions Inc.	X							
China	Xiamen Grand World Laser Label	X							
China	Yu Feng Laser Images Co.	X							
China	Zhuhai Xiangzhou Great Wall Laser	X							
Columbia	Imagenes Holograficas De Columbia	X					X		
Czech Rep.	Lightgate						X		
Denmark	Stensborg Inc.						X		
Finland	Starcke, Ky	X					X		
France	Atelier Holographique De Paris	X			X		X		
France	GEHOL sarl		X		X		X		
France	Hologram Industries	X		X	X		X		
France	Holomedia France	X	X				X		
France	Holomedia France	X	X				X		
France	Krystal Holographics France SARL	X		X					
France	Magic Laser	X					X	X	
France	Mulhem, Dominique				X				
Germany	3D Vision	X					X		
Germany	Abtei Brauweiler		X						X
Germany	AD HOC Public Relations GmbH						X		
Germany	AKS Holographie-Galerie GmbH	X	X						X
Germany	Alfred Dirksen + Sohn						X		
Germany	Andreas Wappelt - Photonics Direct	X	X	X	X		X		
Germany	Arbeitskreis Holografie B.V.				X				
Germany	Armin Klix Holographie	X		X			X		
Germany	Art Agentur Köln						X		
Germany	Art Ig Wohndesign - Holografie	X							
Germany	Artbridge Light Studios			X	X		X		
Germany	Buntstift				X				
Germany	Burgmer Brigitte				X				
Germany	Design + Kunst e.V. Chemnitz				X				
Germany	Die Dritte Dimension	X	X	X			X		
Germany	Dietmar Oehlmann				X				
Germany	Directa GmbH	X	X						
Germany	ETA-Optik Gmbh	X							
Germany	Fielmann-Verwaltung KG				X				
Germany	Frank M. Schenker's Aquarius-Vertrieb	X	X						
Germany	Galerie 3D		X						
Germany	Galerie Westerland/Sylt		X						
Germany	Gesellschaft für Holografie mbH - GfH	X	X						
Germany	H. Kallenbach - H.M.V.	X	X						
Germany	Heiss Peter Dr. Priv. Doz.				X				
Germany	HOL 3-Galerie fur Holographie GmbH	X			X				
Germany	Holar Seele KG					X			
Germany	Holarium		X						X
Germany	Holo Service/Service-Druck	X	X						
Germany	Holo-Idee Reiner Kleinherne						X		
Germany	Holografie Galerie		X						X
Germany	Holografie Hofmann		X		X				X
Germany	Holografie Manufaktur				X				
Germany	Holografie Studio Nürnberg				X				
Germany	Holografie Vertrieb	X							
Germany	HOLOGRAPHIA	X							
Germany	Holographic Identity	X							
Germany	Holographic Studios				X				
Germany	Holographie & Design		X		X			X	X
Germany	Holographie Anubis	X	X	X		X			
Germany	Holographie Fachstudio						X		
Germany	Holographie Konzept GmbH	X							
Germany	Holographie Labor	X					X		
Germany	Holographie Roth	X						X	
Germany	Holoptics	X						X	
Germany	Holopublic Unbehaun						X		
Germany	Holosta Holographie-Galerie	X					X		
Germany	Holotec				X	X			
Germany	Holotopia II		X						
Germany	Hotek Holografie Full service						X		
Germany	Introduct GmbH	X							
Germany	K. Thielker & T. Rost-Holographie	X						X	
Germany	Koch Herbert	X							
Germany	Krystal Holographics Vertriebs-GmbH	X		X					
Germany	Langen R.		X						

TABLE 1 - BUSINESSES THAT SELL OR EXHIBIT HOLOGRAMS

Country	Business Name	Distributor Wholesaler	Retail Shop	Mail Order Catalog	Artist-With Portfolio	Pulsed Portrait Studio	Broker-Sales Rep	Touring exhibit	Gallery museum display
Germany	Laser Trend	X							
Germany	Laserfilm	X							
Germany	Lauk & Partner GmbH	X	X				X		
Germany	Leonhard Kurz GmbH						X		
Germany	Lichtwelt Gratings	X							
Germany	Magick Signs Holografie	X	X		X		X		
Germany	Marita Schlemmer GmbH	X							
Germany	Meulien Odile				X		X		
Germany	Museum 3. Dimension		X						X
Germany	N.V.K. Holographie	X							
Germany	Orazem, Vito				X				
Germany	Phantastica	X							
Germany	PPM Promotion Products Munich	X							
Germany	Rheinisches Landesmuseum Bonn		X						
Germany	Roth Ulrich	X							
Germany	Schall Messen GmbH - Optatec						X		
Germany	Selltexx	X							
Germany	Spot Agentur für Holographie	X	X						
Germany	Stiletto Studios				X		X		
Germany	Tesa	X	X						
Germany	Traumlaboratorium	X			X				
Germany	VISUELL				X				
Germany	Vomhusen Mark				X				
Germany	Werner, Markus				X				
Germany	Zec, Peter						X		
Hungary	Artplay Holographika Studio	X					X		
Hungary	Hologram Varga Miklos	X			X		X		
Hungary	Optopol Panoramic Metrology Con.				X		X		
India	Holographic Security Marking Systems	X		X			X		
India	Ojasmit Holographics	X					X		
India	Shriram Holographics						X		
Israel	Glaser - Technical Consulting						X		
Israel	Holography Israel						X		
Japan	Ardee International	X					X		X
Japan	Brainet Corporation - International Div.	X		X			X		
Japan	Hyogo Prefectual Mus. of Modern Art		X						
Japan	Ishii, Setsuko				X				
Japan	Japan Communication Arts Co.	X		X					
Japan	Light Dimension	X	X				X		
Japan	Light Dimension	X	X				X		
Japan	Nippon Polaroid K.K.	X							
Japan	Ruey-Tung, Hung				X				
Japan	SAM Museum		X						
Japan	Tokai University				X				
Japan	Toppan Printing Co.	X							
Lithuania	Geola	X	X				X		
Mexico	HOLOGRAMAS	X					X		
Netherlands	3-D Hologrammen	X	X				X		
Netherlands	Dutch Holographic Laboratory BV	X		X					
Netherlands	HOLOTECH -Texel		X						
Netherlands	Triple-D Laser Imaging	X					X		
Norway	Interferens Holografi D.A.								X
Pakistan	Dimensions	X					X		
Poland	Holografia Polska	X					X		
Poland	Hololand S.C.	X					X		
Portugal	INETI - Inst. of Info. Technologies						X		
Russia	OPTIC	X			X		X		
Saudi A.	Wonders of Holography Gallery	X	X		X		X		
Slovenia	Likom								X
S. Africa	Synchron Pty Ltd.	X					X		
Spain	Holosco, Ernest Barnes						X		
Spain	Karas Studios Holografia		X		X				
Spain	Museu D' Holografia	X	X		X				X
Sweden	Holography Group TEM				X				
Sweden	HoloMedia Ab/Hologram Museum	X	X				X		X
Sweden	Holovision AB				X				
Switzerland	3D Ltd.	X							
Switzerland	Galerie Illusoria		X		X				X
Switzerland	Holos Art Gallery		X		X				X
Switzerland	Linea - Bregaglia		X						
Taiwan	Holo Images Tech Co.	X					X		

TABLE 1 - BUSINESSES THAT SELL OR EXHIBIT HOLOGRAMS

Country	Business Name	Distributor Wholesaler	Retail Shop	Mail Order Catalog	Artist-With Portfolio	Pulsed Portrait Studio	Broker-Sales Rep	Touring exhibit	Gallery/ museum display
Taiwan	Holo Impressions Inc	X					X		
Taiwan	Holoart Studio	X			X				
Thailand	Electro-Optics Lab						X		
Ukraine	Tair Hologram Company						X		
UAE	Hololaser Gallery		X						
UK	3D Images Ltd.	X	X	X			X		
UK	3D-4D Holographics	X	X		X		X		
UK	Anaspec Ltd.				X		X		
UK	Benyon, Margaret				X				
UK	Boyd, Patrick				X		X		
UK	Colour Holographics	X			X				
UK	Holocrafts Europe Limited.	X			X		X		
UK	Holograms 3D						X	X	
UK	Holographic Consulting Agency						X		
UK	Holographics (Uk) Ltd.						X		
UK	K.C. Brown Holographics				X		X		
UK	Laza Holograms Ltd.	X		X					
UK	Light Impressions International	X		X			X		
UK	London Holographic Image Studio	X		X					
UK	Op-Graphics (Holography) Ltd.	X							
UK	Oxford Holographics	X		X			X		
UK	Pepper, Andrew				X		X		
UK	SEAREACH	X		X					
UK	Spatial Imaging Limited	X			X				
USA	3-D Systems				X				
USA	Acme Holography				X		X		
USA	AD 2						X		
USA	Amagic Holographics	X		X					
USA	American Paper Optics Inc.	X							
USA	Ana MacArthur				X		X		
USA	Another Dimension Inc. (Spectore/ADI)	X	X						
USA	Art Institute Of Chicago								X
USA	Art, Science & Technology		X						X
USA	Bellini, Victor				X				
USA	Berkhout, Rudie				X				
USA	Booth, Roberta				X				
USA	Broadbent Consulting				X		X		
USA	Bruck, Richard	X			X		X		
USA	Carl M. Rodia And Associates						X		
USA	Casdin-Silver Holography				X		X		
USA	Center For The Holographic Arts					X		X	
USA	Cherry Optical Holography	X			X				
USA	Cifelli, Dan						X		
USA	Coburn Corporation			X					
USA	Control Module Inc.						X		
USA	CVI Laser Corporation			X					
USA	Deem, Rebecca				X				
USA	DeFreitas Holography Studio				X		X		
USA	Diamond Images	X			X		X		
USA	Dimensional Foods Co.	X		X					
USA	Doris Vila Holographics				X				
USA	Edmund Scientific Company		X	X					
USA	Fantastic Holograms		X						
USA	Feroe Holographic Consulting						X		
USA	Fornari, Arthur				X				
USA	Forth Dimension Holographics	X	X			X	X		
USA	Gorglione, Nancy				X		X		
USA	Holo Art			X					
USA	Holo-Spectra						X		
USA	Hologram Land		X						
USA	Hologram Research	X			X		X	X	
USA	Hologram World	X		X					
USA	Holograms and Lasers International	X	X	X	X		X		
USA	Holograms International	X	X	X					
USA	Holographic Applications						X		
USA	Holographic Design Systems	X							
USA	Holographic Dimensions	X			X		X		
USA	Holographic Images Inc.	X			X				
USA	Holographic Impressions	X		X					
USA	Holographic Industries	X					X		
USA	Holographic Products						X		

TABLE 1 – BUSINESSES THAT SELL OR EXHIBIT HOLOGRAMS

Country	Business Name	Distributor Wholesaler	Retail Shop	Mail Order Catalog	Artist-With Portfolio	Pulsed Portrait Studio	Broker-Sales Rep	Touring exhibit	Gallery/ museum display
USA	Holographic Studios	X	X		X		X		X
USA	Holographics Inc.				X		X		
USA	Holographics North Inc.	X			X				
USA	Holography Institute of San Francisco				X				
USA	Holography Presses On (HPO)			X					
USA	Holographyx Inc.						X		
USA	Holographyx Inc.						X		
USA	Holonix						X		
USA	Holophile						X	X	
USA	Holovision Systems Inc.						X		
USA	Illuminations						X		
USA	Interactive Industries	X	X	X					
USA	Jeffery Murray Custom Holography				X		X		
USA	John, Pearl				X				
USA	Kauffman, John				X				
USA	Krystal Holographics International Inc.	X		X	X		X		
USA	Larry Lieberman Holography	X	X		X				
USA	Lasart Ltd.	X		X	X				
USA	Laser Affiliates				X			X	
USA	Laser Light Designs						X		
USA	Laser Reflections		X		X		X		
USA	Laserworks				X		X		
USA	Lightrix	X		X	X				
USA	LightVision Confections	X					X		
USA	Liti Holographics				X	X			
USA	Lone Star Illusions		X						
USA	M.I.T. Museum		X						X
USA	Man/Environment	X	X	X	X		X		
USA	Marks, Gerald				X		X		
USA	Media Interface						X		
USA	Miller, Peter				X				
USA	Multiplex Moving Holograms	X		X	X				
USA	Museum Of Holography/Chicago		X		X				X
USA	Nakamura, Ikuo				X				
USA	NeoVision Productions						X	X	
USA	New Light Industries						X		
USA	New York Hall Of Science								X
USA	Oregon Laser Consultants						X		
USA	Pacific Holographics Inc.				X		X		
USA	Planet 3-D						X		
USA	Point Source Productions				X		X		
USA	PullTime 3-D Laboratories				X		X		
USA	Rainbow Symphony Inc.	X		X	X		X		
USA	Real Image	X					X		
USA	Red Beam						X		
USA	Regal Press Inc.			X					
USA	Reva's Holographic Illusions		X						
USA	RGB Holographics Inc.	X			X				
USA	Robert Sherwood Holographic Design	X			X		X		
USA	Saint Mary's College				X				
USA	School Of Holography								X
USA	Science Kit & Boreal Labs			X					
USA	Sharon McCormack Holography				X		X		
USA	Smith, Steven L				X				
USA	Sommers Plastic Products		.				X		
USA	Southern Indiana Holographics				X				
USA	Star Magic		X	X					
USA	Three Dimensional Imagery				X				
USA	Three-D Light Gallery		X		X				
USA	Trace Holographic Art & Design						X		
USA	Tyler Group						X		
USA	U.K. Gold Purchasers	X							
USA	University Of Wisconsin/Madison			X					
USA	Unterseher & Associates				X		X		
USA	Van Leer Metallized Products	X							
USA	Visions Unlimited		X						
USA	Visual Visionaries						X		
USA	Wave Mechanics				X				
USA	Wesley, Ed				X		X		
USA	World Holographics						X		
USA	Zero Gravity		X						
USA	Zone Holografix				X		X		

TABLE 2 – ARTWORK ORIGINATION AND HOLOGRAM MASTERING SERVICES

Country	Business Name	3D models - sculpted	3D computer modeling	Master for silver halide	Master for DCG	Master for photo-polymer	Master for em-bossed	Master using stereo-gram printer
Australia	3DIMAGE	X	X	X		X		
Australia	Australian Holographics	X		X		X		X
Australia	Menning, Melinda	X		X	X	X	X	
Australia	Moonbeamers	X	X	X			X	
Austria	Holography Center of Austria		X	X				X
Canada	Dimension 3	X	X	X		X	X	X
Canada	General Holographics, Inc.	X	X	X	X		X	X
Canada	Holocrafts	X	X		X			
Canada	Inspeck, Inc.	X	X					
Canada	Melissa Crenshaw Holography	X	X	X	X	X	X	
Canada	Royal Holographic Art Gallery			X				
Canada	Silverbridge Group	X	X		X			
China	Holography Development Co.Ltd.	X	X	X			X	
Czech Rep.	Lightgate, Ltd	X		X			X	
France	GEHOL sarl			X				
France	Hologram Industries	X	X					
France	Krystal Holographics France SARL	X	X			X		X
Germany	AB Rück Holoart	X	X	X				
Germany	AKS Holographie-Galerie GmbH	X	X	X				
Germany	Alfred Dirksen + Sohn	X						
Germany	Bundesdruckerei	X	X				X	
Germany	Decolux GmbH	X	X				X	
Germany	EPA - Elektro-Physik Aachen	X	X		X			
Germany	ETA-Optik Gmbh				X			
Germany	Highlite	X	X		X			
Germany	Holar Seele KG	X	X	X				
Germany	Holo Time Gericke	X	X	X	X	X		
Germany	Holodesign	X	X	X	X			
Germany	Holografie Studio Nürnberg	X	X	X	X			
Germany	HOLOGRAM INDUSTRIES	X	X	X	X			
Germany	Hologram Company RAKO GmbH	X	X				X	
Germany	Holographic Laserdesign	X	X	X				
Germany	Holographic Studios	X	X	X				
Germany	Holographie Labor	X	X	X		X		
Germany	Holopress Holographic Techn.	X	X	X				
Germany	Holoptics	X	X	X				
Germany	Holotec	X	X	X				
Germany	HSM	X	X				X	
Germany	KomSa Kommunikationstechnik	X	X		X			
Germany	Krystal Holographics Vertriebs-	X	X			X		
Germany	Laserfilm	X	X	X				
Germany	LichtBlicke Classen & Voss	X	X	X				
Germany	Magick Signs Holografie	X	X				X	
Germany	mario liedtke pro design	X	X					
Germany	Oeserwerk Ernst Oeser & Söhne KG	X	X				X	
Germany	Ose Holografie-Design	X	X	X				
Germany	topac GmbH	X	X				X	
Hungary	Artplay Holographika Studio		X	X			X	
Hungary	Hologram Varga Miklos	X	X			X		
Hungary	Optopol Panoramic Metrology Con.	X		X				
India	Holostik India Pvt. Ltd.	X					X	
India	Print-M-Boss	X					X	
India	Shriram Holographics	X					X	
India	Spatial Holodynamics (India)	X	X				X	
Italy	Diavy srl	X	X				X	
Italy	Holo 3D S.p.A.	X	X				X	
Japan	Asahi Glass Co.	X		X				
Japan	Dai Nippon Printing Co., Ltd.	X	X	X	X	X	X	X
Japan	Topcon Inc.	X	X				X	
Japan	Toppan Printing Co., Ltd.	X	X	X	X	X	X	X
Japan	Toyama National College Of Marit	X	X	X				X
Latvia	University of Latvia						X	
Lithuania	Geola	X		X				
Mexico	Corporacion Mexicana De Impresion	X	X				X	
Mexico	Empaques y Envolturas Holograficas,		X				X	
Mexico	HOLOGRAMAS, S.A. DE C.V.	X	X				X	
Netherlands	Dutch Holographic Laboratory BV	X	X	X	X	X	X	X
Poland	Holographic Dimensions, Poland S.A.	X					X	
Portugal	INETI - Institute of Information Tech.	X	X				X	

TABLE 2 - ARTWORK ORIGINATION AND HOLOGRAM MASTERING SERVICES

Country	Business Name	3D models - sculpted	3D computer modeling	Master for silver halide	Master for DCG	Master for photo-polymer	Master for em-bossed	Master using stereo-gram printer
Russia	OPTIC	X	X	X			X	
Spain	Holosco, Ernest Barnes	X	X	X			X	
Sweden	Holovision AB	X	X	X				
Switzerland	3D Ltd.	X	X				X	X
Taiwan	Fong Teng Technology	X	X				X	
Taiwan	Holoart Studio	X		X				
Taiwan	Infox Corporation						X	
Thailand	Electro-Optics Lab, NECTEC	X		X		X	X	
Ukraine	Feofaniya Ltd.	X	X	X			X	
Ukraine	Institute of Applied Optics	X		X				
Ukraine	Tair Hologram Company			X				
UK	3D-4D Holographics	X	X	X				
UK	Anaspec Ltd.,	X	X	X				
UK	API Foils		X				X	X
UK	Applied Holographics, Plc.	X	X				X	
UK	Astor Universal Ltd.	X	X				X	
UK	Colour Holographics	X		X	X	X	X	X
UK	De La Rue Holographics	X	X				X	
UK	Embossing Technology Ltd	X	X				X	
UK	Holocrafts Europe Limited.	X	X	X	X	X	X	
UK	Holographics (Uk) Ltd.	X	X	X				
UK	Jayco Holographics	X	X	X			X	
UK	K.C. Brown Holographics	X		X				
UK	Light Impressions International, Ltd.	X	X				X	
UK	London Holographic Image Studio	X	X	X				X
UK	Op-Graphics (Holography) Ltd.			X				
UK	Optical Security Group - England	X	X				X	
UK	SEAREACH, Plc	X	X				X	
UK	Spatial Imaging Limited	X	X			X	X	X
USA	3M - Safety and Security Systems						X	
USA	AD 2000, Inc.					X	X	X
USA	Amagic Holographics, Inc.		X	X			X	X
USA	American Banknote Holographics, .						X	X
USA	Ana MacArthur			X				
USA	API Foild		X	X				X
USA	Avant-Garde Studio	X						
USA	Blue Ridge Holographics, Inc.	X		X		X	X	
USA	Bridgestone Technologies, Inc.	X	X				X	
USA	Broadbent Consulting			X	X	X	X	
USA	Bruck, Richard Holography	X	X	X	X	X	X	
USA	California Institute of the Arts	X	X					
USA	Casdin-Silver Holography	X		X				
USA	CFC Applied Holographics	X	X				X	X
USA	Cfc Northern Bank Note Co.	X	X				X	X
USA	Cherry Optical Holography	X		X		X		
USA	Chromagem Inc.					X	X	
USA	Crown Roll Leaf, Inc.,	X	X				X	
USA	David Dann Modelmaking Studios	X	X					
USA	Deem, Rebecca	X	X	X	X	X	X	
USA	DeFreitas Holography Studio			X				
USA	Diamond Images, Inc.			X	X		X	X
USA	Dri-Print Foils	X					X	
USA	Engineering Animation, Inc.		X					
USA	Feroe Holographic Consulting			X			X	
USA	FLEXcon	X	X				X	
USA	Foilmark Holographic Images	X					X	
USA	Forth Dimension Holographics			X				
USA	General Design		X					
USA	Holo Art	X		X				X
USA	Holo Sciences, LLC	X	X	X	X	X	X	X
USA	Holo-Source Corporation	X					X	
USA	HoloCom	X	X				X	X
USA	Hologram Research, Inc.	X	X	X				X
USA	Holograms and Lasers International	X	X	X				
USA	Holographic Design Systems	X	X	X	X	X	X	
USA	Holographic Dimensions	X	X				X	X
USA	Holographic Images Inc.	X		X				
USA	Holographic Label Converting (HLC)	X					X	
USA	Holographic Studios	X	X					X

TABLE 2 - ARTWORK ORIGINATION AND HOLOGRAM MASTERING SERVICES

Country	Business Name	3D models - sculpted	3D computer modeling	Master for silver halide	Master for DCG	Master for photo-polymer	Master for em-bossed	Master using stereo-gram printer
USA	Holographics Inc.	X		X				
USA	Holographics North Inc.	X		X				X
USA	Image Engine	X						
USA	Imagination Plantation		X					
USA	Infinity Laser Laboratories			X		X		
USA	International Holographic Paper	X	X				X	X
USA	Intrepid World Communications	X		X				
USA	Jeffery Murray Custom Holography			X	X	X	X	
USA	Krystal Holographics Intl.	X	X	X	X	X		
USA	Krystal Holographics International	X	X	X	X	X		X
USA	Kurz Foils						X	
USA	Label Systems	X	X				X	
USA	Larry Lieberman Holography	X		X				
USA	Lasart Ltd.	X		X	X			
USA	Laser Reflections	X		X	X			
USA	Lightrix, Inc.	X	X	X		X	X	
USA	Linda Law Holographics		X					
USA	Liti Holographics	X	X	X				X
USA	Louis Paul Jonas Studios, Inc.	X						
USA	Man/Environment, Inc.			X		X	X	
USA	Marks, Gerald	X	X	X		X	X	
USA	Merrick, Michael G.			X				
USA	Multiplex Moving Holograms			X				X
USA	NovaVision	X	X				X	
USA	OpSec - USA	X	X				X	
USA	Optical Security Group	X	X				X	X
USA	Optimation Holographics	X	X				X	
USA	Pacific Holographics Inc.			X			X	
USA	Pangaea Design	X						
USA	Polaroid Corporation	X	X	X		X		
USA	PullTime 3-D Laboratories	X	X	X				
USA	Rainbow Symphony Inc.			X		X	X	
USA	Ralcon	X	X	X	X	X		
USA	Red Beam, Inc.	X	X	X	X	X	X	
USA	Regal Press Inc.	X	X				X	
USA	Reynolds Metals Co.	X					X	
USA	RGB Holographics Inc.	X	X			X		X
USA	Robert Sherwood Holographic Design	X		X	X	X	X	
USA	Scharr Industries	X	X				X	
USA	Sharon McCormack Holography	X	X	X				X
USA	Simian Co.	X	X				X	X
USA	Sommers Plastic Products	X	X					
USA	Spectratek Inc.						X	
USA	Three Dimensional Imagery			X				
USA	Trace Holographic Art & Design, Inc.	X					X	
USA	Transfer Print Foils, Inc.						X	
USA	Unterseher & Associates			X	X	X		
USA	Van Leer Metallized Products	X	X				X	
USA	Virtual Image	X	X				X	
USA	Visual Visionaries	X	X	X	X	X	X	
USA	Wavefront Technology	X	X				X	
USA	Wesley, Ed	X		X	X	X	X	
USA	Witchcraft Tape Products, Inc.						X	
USA	Zebra Imaging, Inc	X	X			X		X
USA	Zone Holografix	X	X	X	X	X	X	

TABLE 3 - MASS PRODUCTION / REPLICATION SERVICES

Country	Business Name	silver halide glass	silver halide film	DCG	Photo-polymer	Color Photo-polymer	Security Holo-grams	Embossed substrates
Australia	3DIMAGE	X	X					
Australia	Australian Holographics	X						
Australia	Moonbeamers	X	X					
Austria	Holography Center of Austria	X	X					
Canada	Dimension 3				X			X
Canada	General Holographics, Inc.	X		X				
Canada	Holocrafts			X				
Canada	Melissa Crenshaw Holography Studio	X	X	X				
Canada	Royal Holographic Art Gallery	X	X					
Canada	Silverbridge Group			X				
China	Minjian Laser Holography						X	
China	Zhuhai Xiangzhou Great Wall Laser						X	
France	Hologram Industries						X	X
France	Krystal Holographics France SARL				X	X		
Germany	AB Rück Holoart	X	X					
Germany	AKS Holographie-Galerie GmbH	X	X					
Germany	Bundesdruckerei	X	X				X	X
Germany	Decolux GmbH						X	X
Germany	EPA - Elektro-Physik Aachen GmbH			X				
Germany	ETA-Optik Gmbh			X				
Germany	Highlite			X				
Germany	Holar Seele KG	X	X					
Germany	Holodesign	X	X					
Germany	Holografie Studio Nürnberg	X	X					
Germany	HOLOGRAM INDUSTRIES	X	X					
Germany	Hologram Company RAKO GmbH				X			X
Germany	Holographic Laserdesign	X	X					
Germany	Holographic Studios		X					
Germany	Holographie Labor	X						X
Germany	Holopress Holographic Techn. GmbH	X	X					X
Germany	Holoptics	X	X					
Germany	Holotec	X	X					
Germany	HSM						X	X
Germany	KornSa Kommunikationstechnik GmbH			X				
Germany	Krystal Holographics Vertriebs-GmbH				X	X		
Germany	Laserfilm	X	X					
Germany	LichtBlicke Classen & Voss	X	X					
Germany	Magick Signs Holografie							X
Germany	Oeserwerk Ernst Oeser & Söhne KG							X
Germany	Ose Holografie-Design	X	X					
Germany	topac GmbH							X
Germany	W. Cordes GmbH + Co.							X
Hungary	Artplay Holographika Studio						X	X
Hungary	Optopol Panoramic Metrology Consulting	X						
India	Holographic Security Marking Syst..						X	
India	Holostik India Pvt. Ltd.						X	X
India	Print-M-Boss							X
India	Shriram Holographics							X
India	Spatial Holodynamics (India) Pvt. Ltd.							X
Italy	Diavy srl							X
Italy	Holo 3D S.p.A.						X	X
Japan	Asahi Glass Co.	X						
Japan	Dai Nippon Printing Co., Ltd.				X	X	X	X
Japan	Topcon Inc.							X
Japan	Toppan Printing Co., Ltd.	X	X·		X		X	X
Latvia	University of Latvia							X
Lithuania	Geola	X	X					
Mexico	Corporacion Mexicana De Impresion .						X	X
Mexico	Empaques y Envolturas Holograficas,						X	X
Mexico	HOLOGRAMAS, S.A. DE C.V.						X	X
Netherlands	Dutch Holographic Laboratory BV	X	X		X			X
Panama	HDIPanama						X	
Poland	Holographic Dimensions, Poland S.A.	X						X
Portugal	INETI - Institute of Information Tech.						X	
Russia	OPTIC		X				X	X
Sweden	Holovision AB	X	X					
Switzerland	3D Ltd.						X	X
Taiwan	Ahead Optoelectronics, Inc.						X	
Taiwan	Fong Teng Technology							X
Taiwan	Infox Corporation							X
Thailand	Electro-Optics Lab, NECTEC	X	X				X	X

TABLE 3 - MASS PRODUCTION / REPLICATION SERVICES

Country	Business Name	silver halide glass	silver halide film	DCG	Photo-polymer	Color Photo-polymer	Security Holograms	Embossed substrates
Ukraine	Feofaniya Ltd.							X
Ukraine	Institute of Applied Optics	X	X					
Ukraine	Tair Hologram Company	X						
UK	Action Tapes							X
UK	Anaspec Ltd., Holography and Image Sys.						X	
UK	API Foils							X
UK	Applied Holographics, Plc.							X
UK	Astor Universal Ltd.							X
UK	De La Rue Holographics							X
UK	Embossing Technology Ltd							X
UK	Holocrafts Europe Limited.			X				
UK	Light Impressions International, Ltd.				X			X
UK	London Holographic Image Studio	X						
UK	Op-Graphics (Holography) Ltd.		X		X			
UK	Optical Security Group - England				X		X	X
UK	SEAREACH, Plc						X	X
UK	Spatial Imaging Limited						X	
UK	STI - Europe						X	X
UK	Whiley Foils Limited (API Foils)							X
USA	3M - Safety and Security Systems						X	X
USA	AD 2000, Inc.				X		X	X
USA	Advanced Deposition Tech.						X	X
USA	Amagic Holographics, Inc.				X			X
USA	American Banknote Holographics, Inc.							X
USA	American Paper Optics Inc.							X
USA	Another Dimension Inc. (Spectore/ADI)							X
USA	API Foild						X	X
USA	Bridgestone Technologies, Inc.						X	X
USA	CFC Applied Holographics				X		X	X
USA	Cfc Northern Bank Note Co.						X	X
USA	Cherry Optical Holography				X			
USA	Coburn Corporation							X
USA	Control Module Inc.						X	
USA	Creative Label							X
USA	Crown Roll Leaf, Inc.,						X	X
USA	Datacard Corporation						X	
USA	Dimensional Arts							X
USA	Dri-Print Foils							X
USA	DuPont (E.I. DuPont De Nemours & Co.)					X		
USA	FLEXcon						X	X
USA	Foilmark Holographic Images						X	X
USA	Forth Dimension Holographics	X	X					
USA	Holman Technology, Inc.							X
USA	Holo-Source Corporation						X	X
USA	HoloCom						X	X
USA	Holograms and Lasers International	X						
USA	Holographic Design Systems						X	
USA	Holographic Dimensions						X	X
USA	Holographic Label Converting (HLC)						X	X
USA	Holographic Products						X	
USA	Holographics Inc.	X						
USA	Holographics North Inc.		X					
USA	International Holographic Paper							X
USA	Intrepid World Communications	X	X					
USA	Krystal Holographics International Inc.			X	X	X		
USA	Krystal Holographics International Inc.			X	X	X	X	
USA	Kurz Foils							X
USA	Label Systems						X	X
USA	Larry Lieberman Holography	X						
USA	Lasart Ltd.	X		X				
USA	Laser Reflections	X		X				
USA	Lightrix, Inc.				X			X
USA	LightVision Confections							X
USA	Nimbus Manufacturing, Inc.						X	
USA	NovaVision						X	X
USA	OpSec - USA				X		X	X
USA	Optical Security Group						X	X
USA	Optimation Holographics							X
USA	Parker Foils, Inc.							X
USA	Polaroid Corporation		X		X		X	
USA	Pyramid Foils, Inc.							X

TABLE 3 - MASS PRODUCTION / REPLICATION SERVICES

Country	Business Name	silver halide glass	silver halide film	DCG	Photo-polymer	Color Photo-polymer	Security Holo-grams	Embossed substrates
USA	Rainbow Symphony Inc.							X
USA	Ralcon			X	X			
USA	Regal Press Inc.							X
USA	Reynolds Metals Co.							X
USA	RGB Holographics Inc.				X	X		
USA	Scharr Industries							X
USA	Simian Co.						X	X
USA	Sommers Plastic Products				X			X
USA	Spectratek Inc.				X			X
USA	Three Dimensional Imagery	X	X					
USA	Transfer Print Foils, Inc.						X	X
USA	Tyler Group						X	
USA	Unifoil Corporation							X
USA	Van Leer Metallized Products						X	X
USA	Virtual Image							X
USA	Wavefront Technology							X
USA	Witchcraft Tape Products, Inc.							X
USA	Zebra Imaging, Inc				X	X	X	

TABLE 4 - HOLOGRAM APPLICATION / FINISHING / CONVERTING SERVICES

Country	Business Name	Die cutting-slitting	Over printing with ink	Foil hot stamping	Pressure sensitive / automated	Pressure sensitive / hand applied	Laminating	Textiles
Canada	Dimension 3	X	X	X	X	X		
China	Foshan Holosun Packaging Co.	X	X	X	X	X	X	
China	Shanghai Dahua Printing Factory	X	X	X	X	X		
China	Wuhan Packaging and Printing	X	X	X	X		X	
France	Hologram Industries	X	X	X	X		X	
Germany	Bundesdruckerei	X	X	X	X			
Germany	Decolux GmbH	X	X	X	X			
Germany	Holographie Labor	X	X	X	X			
Germany	Holopress Holographic Techn.	X	X	X	X			
Germany	HSM	X	X	X	X			
Germany	Kolbe-Druck	X	X	X	X	X	X	
Germany	Magick Signs Holografie	X	X	X	X			
Germany	Oeserwerk Ernst Oeser	X	X	X	X			
Germany	topac GmbH		X	X	X			
Hungary	Artplay Holographika Studio	X	X					
India	Holostik India Pvt. Ltd.	X	X	X	X		X	
India	I.S. Gill	X	X	X	X			
India	Shriram Holographics	X	X	X	X			
Italy	Diavy srl	X	X	X				
Italy	Holo 3D S.p.A.	X	X	X	X	X		
Japan	Dai Nippon Printing Co., Ltd.	X	X	X	X			
Japan	Topcon Inc.	X	X	X	X	X		
Japan	Toppan Printing Co., Ltd.	X	X	X	X	X		
Mexico	Empaques y Envolturas Holo.	X	X	X	X	X		
Mexico	HOLOGRAMAS, S.A. DE C.V.	X	X	X	X	X	X	
Poland	Holographic Dimensions, Poland	X	X	X	X	X	X	
Russia	OPTIC	X	X			X		
South Africa	Synchron Pty Ltd.			X				
Switzerland	3D Ltd.	X	X	X	X	X		
Taiwan	Fong Teng Technology	X	X	X	X		X	
Thailand	Electro-Optics Lab, NECTEC	X	X	X	X		X	
Ukraine	Feofaniya Ltd.	X	X	X	X			
UK	API Foils	X	X	X	X	X	X	X
UK	Applied Holographics, Plc.	X	X	X	X	X	X	
UK	Art Foil Graphic Machinery	X	X	X	X			
UK	Astor Universal Ltd.	X	X	X	X	X	X	
UK	Embossing Technology Ltd	X	X					
UK	Light Impressions International,	X	X	X	X	X		
USA	21st Century Finishing Inc.	X	X	X	X	X	X	
USA	American Banknote Holo.	X	X	X	X	X	X	
USA	API Foils	X	X	X	X			

TABLE 4 - HOLOGRAM APPLICATION / FINISHING / CONVERTING SERVICES

Country	Business Name	Die cutting-slitting	Over printing with ink	Foil hot stamping	Pressure sensitive / automated	Pressure sensitive / hand applied	Laminating	Textiles
USA	Bridgestone Technologies, .	X	X	X	X	X	X	
USA	CFC Applied Holographics	X		X			X	
USA	Cfc Northern Bank Note Co.	X	X	X	X	X	X	
USA	Creative Label	X	X	X				
USA	Crown Roll Leaf, Inc.,	X	X	X	X	X	X	
USA	D. Brooker & Associates		X					
USA	FLEXcon	X	X	X	X	X	X	
USA	FlexSystems USA						X	X
USA	Foilmark Holographic Images	X	X	X	X		X	
USA	Grafix Plastics	X			X	X	X	
USA	Holo Art	X				X	X	
USA	Holo-Source Corporation	X	X	X	X	X	X	
USA	Holographic Dimensions	X	X	X	X	X	X	
USA	Holographic Finishing, Inc.	X		X				
USA	Holographic Impressions	X	X	X	X	X	X	
USA	Holographic Label Converting	X	X	X	X	X	X	
USA	Holography Presses On (HPO)				X	X		X
USA	International Holographic Paper	X	X	X	X	X	X	
USA	Krystal Holographics Intl.	X	X		X		X	
USA	Letterhead Press, Inc.	X	X	X	X	X	X	X
USA	Luminer Printing and Converting	X	X		X		X	
USA	MGM Converters Inc.	X	X	X	X	X	X	
USA	OpSec - USA	X	X				X	
USA	Optical Security Group	X	X	X	X		X	
USA	Optimation Holographics	X	X				X	
USA	Polaroid Corporation	X	X		X	X	X	X
USA	Rainbow Symphony Inc.	X	X	X	X	X	X	
USA	Regal Press Inc.	X	X	X	X	X	X	
USA	Reynolds Metals Co.		X					
USA	Simian Co.	X	X	X	X	X	X	
USA	Smith & McKay Printing Co. Inc.	X	X	X	X	X	X	
USA	Sommers Plastic Products	X	X	X	X	X	X	
USA	Textile Graphics, Inc.							X
USA	Transfer Print Foils, Inc.	X		X	X	X	X	
USA	Unifoil Corporation	X			X		X	
USA	Van Leer Metallized Products	X	X	X			X	
USA	Virtual Image	X	X	X	X	X	X	
USA	Witchcraft Tape Products, Inc.	X	X	X	X	X	X	
USA	Zebra Imaging, Inc	X						

TABLE 5 - SUPPLIERS OF LASERS, OPTICS AND RELATED STUDIO EQUIPMENT

Country	Business Name	New CW lasers	New pulsed lasers	Used or surplus lasers	Laser Repair	Laser Equipment, accessories	Optics and related items	Darkroom supplies and chemicals
Austria	Feinmechanische Optische Betriebs						X	
Austria	LST - Laser & Strahl Technik						X	
Canada	Lasiris Inc.						X	
China	Chongqing Yinhe Laser Products	X		X	X	X	X	
France	Quantel	X				X		
France	S.O.P.R.A.					X		
Germany	ACR Laser- und Medientechnik	X			X	X		
Germany	AF Elektronik						X	X
Germany	C. Roth GmH + Co.							X
Germany	Coherent Deutschland GmbH	X			X	X		
Germany	Coherent Lübeck GmbH	X	X		X	X		
Germany	Dilas Diodenlaser GmbH	X						
Germany	Fink Feinoptik						X	
Germany	G. Franck OptroniK GmbH - GFO						X	X
Germany	Gerhard Winopal Forschungsbedarf						X	
Germany	Gigahertz-Optik						X	
Germany	Gresser E. KG					X		

TABLE 5 – SUPPLIERS OF LASERS, OPTICS AND RELATED STUDIO EQUIPMENT

Country	Business Name	New CW lasers	New pulsed lasers	Used or surplus lasers	Laser Repair	Laser Equipment. accessories	Optics and related items	Darkroom supplies and chemicals
Germany	Gsänger Optoelektronik GmbH						X	
Germany	HAM Kristall Technologie						X	X
Germany	HB Laserkomponenten						X	X
Germany	HMS-Elektronik					X		
Germany	Holostudio Beate Krengel						X	
Germany	Kempter	X				X		
Germany	L.O.T. Oriel GmbH & Co. KG						X	
Germany	Laser 2000 GmbH	X			X	X		
Germany	Laser Solution Management	X				X		
Germany	LASOS						X	
Germany	Mefoma Fototechnik GmbH						X	X
Germany	MeiLing - Lasertechnik						X	
Germany	Melles Griot GmbH	X	X		X	X		
Germany	Metrologic Instruments GmbH						X	X
Germany	Micos GmbH & Co KG						X	X
Germany	Microplan GmbH						X	X
Germany	NEC Electronics (Europe) GmbH	X		X		X		
Germany	Newport GmbH					X	X	X
Germany	Nika Präzisionsoptik GmbH						X	
Germany	OMIKRON GmbH							X
Germany	Optical Test Equipment						X	
Germany	Optlectra						X	
Germany	OWIS GmbH						X	X
Germany	Physik Instrumente (PI)				X	X	X	
Germany	PHYWE Systeme GmbH						X	X
Germany	Polytec GmbH	X				X		
Germany	Radiant dyes accessories GmbH						X	
Germany	Rofin-Sinar Laser GmbH	X				X		
Germany	Spectra-Physics GmbH	X			X	X		
Germany	SpectroLas GmbH						X	
Germany	Spindler & Hoyer GmbH & Co.	X				X	X	
Germany	Steeg & Reuter GmbH						X	
Germany	Steinbichler Optotechnik GmbH						X	
Germany	Steiner GmbH							X
Germany	Technolas Laser Technik Gmbh	X						
Germany	Thorlabs Inc.						X	
Germany	Topag GmbH	X				X		
Germany	Uniphase Vertriebs-GmbH	X				X		
Germany	University of Erlangen						X	
Germany	Verwaltungsgesellschaft						X	X
Germany	Waldner Laboreinrichtungen							X
Germany	Werner Jülich						X	
Japan	Fuji Electric Co. Ltd	X				X		
Japan	Kimmon Electric Co., Ltd.	X				X		
Japan	Marubun Corporation	X				X	X	X
Korea	Dan Han Optics						X	
Lithuania	Geola		X		X	X	X	
Mexico	Centro de Invest. en Optica, A.C.						X	
Russia	Technoexan Ltd	X				X		
Spain	Lasing S.A.						X	X
Taiwan	Superbin Co. Ltd	X				X		
Ukraine	TAVEX, Ltd.						X	
UK	Anaspec Ltd., Holography	X						
UK	Customer Service Instrumentation						X	
UK	INNOLAS (UK) Ltd.		X			X		
UK	Lumonics Ltd.		X			X		
USA	Advanced Optics, Inc.						X	
USA	American Laser Corporation	X				X		
USA	Anderson Lasers, Inc.	X	X	X	X	X	X	X
USA	Burleigh Instruments, Inc.						X	
USA	Cambridge Laser Labs			X		X		
USA	City Chemical							X
USA	Coherent, Inc. - Laser Group	X				X		
USA	Continental Optical						X	
USA	Control Optics						X	X
USA	Corion Corp.						X	
USA	Corning Incorporated						X	

TABLE 5 - SUPPLIERS OF LASERS, OPTICS AND RELATED STUDIO EQUIPMENT

Country	Business Name	New CW lasers	New pulsed lasers	Used or surplus lasers	Laser Repair	Laser Equipment, accessories	Optics and related items	Darkroom supplies and chemicals
USA	CVI Laser Corporation						X	
USA	Data Optics, Inc.						X	X
USA	Dell Optics Company, Inc.						X	
USA	Edmund Scientific Company	X				X	X	X
USA	El Don Engineering			X	X	X		
USA	Electro Optical Industries, Inc.						X	
USA	Evergreen Laser Corp.			X	X	X		
USA	Excitek Inc.			X	X	X		
USA	Fisher Scientific	X				X	X	X
USA	G.M. Vacuum Coating Lab, Inc.						X	
USA	Glass Mountain Optics						X	
USA	Holo-Spectra			X		X	X	
USA	Honeywell Technology Center						X	
USA	Images Company	X					X	X
USA	Infinity Laser Laboratories			X				
USA	Infrared Optical Products, Inc.						X	
USA	Inrad, Inc.						X	
USA	Jodon Inc.	X		X			X	X
USA	Ka-Lor Cubicle & Supply Co.							X
USA	Kaiser Optical Systems, Inc.						X	
USA	Kinetic Systems, Inc.						X	
USA	Laser and Motion Development			X		X	X	
USA	Laser Innovations			X	X	X		
USA	Laser Las Vegas			X	X	X		
USA	Laser Optics, Inc.			X	X	X	X	
USA	Laser Reflections		X			X		
USA	Laser Resale Inc.			X	X	X		
USA	Laser Technical Services	X	X	X	X	X		
USA	LaserMax, Inc.	X	X			X	X	
USA	Lasermetrics, Inc.				X	X		
USA	Laserworks			X			X	
USA	Lenox Laser	X					X	
USA	Lexel Laser, Inc.	X				X		
USA	LiCONiX, A Melles Griot	X				X		
USA	Lightwave Electronics	X						
USA	Lumenx Technologies, Inc.	X			X	X		
USA	M.O.M. Inc.						X	
USA	Melles Griot	X		X		X	X	
USA	Melles Griot Laser Group	X				X		
USA	Meredith Instruments			X		X		
USA	Midwest Laser Products	X		X		X		
USA	Mitutoyo Measuring Instruments						X	
USA	MWK Industries			X		X	X	
USA	New Focus, Inc	X				X	X	
USA	Newport Corporation						X	X
USA	Norland Products, Inc.						X	
USA	Odhner Holographics						X	
USA	Optics Plus Inc.						X	
USA	Optimation						X	
USA	Oriel Instruments						X	
USA	Panatron Inc.			X	X	X		
USA	Photo Research, Inc.						X	X
USA	Photo Sciences						X	
USA	Potomac Photonics, Inc.	X						
USA	Rolyn Optics						X	
USA	Science Kit & Boreal Labs						X	X
USA	Spectra-Physics Lasers Inc.	X				X		
USA	Swift Instruments						X	
USA	Thorlabs Inc.						X	
USA	Uniphase Lasers	X				X		
USA	Uvex Safety Inc.							X

TABLE 6 - SUPPLIERS OF HOLOGRAPHIC RECORDING / REPLICATION MATERIALS

Country	Business Name	Silver halide glass-red sensitive	Silver halide film-red sensitive	Silver halide multi-color	DCG	Photo-polymer	Photo-resist	Hot stamp foil	Rec Material OEM	Plastic Carrier
Australia	General Optics Pty. Ltd.	X	X		X					
Austria	Holography Center of Austria	X	X	X					X	
Belgium	Agfa - Gevaert N.V.	X	X							
Canada	Royal Holographic Art Gallery		X							
China	Tianjin Holdor Optics Inc. China	X		X				X		
Denmark	Ibsen Micro Structures A/S						X			
Finland	Starcke, Ky.							X		
Germany	b+g Banse und Grohmann	X	X							
Germany	FilmoTec GmbH	X	X							
Germany	GTO Lasertechnik GmbH	X	X							
Germany	HRT GmbH	X	X	X					X	
Germany	S. C. Ahlemeyer	X	X					X		
Germany	Zanders Feinpapiere AG						X	X		
Latvia	University of Latvia								X	
Lithuania	Slavich Int. Wholesale Office	X	X	X	X					
Russia	Slavich Joint Stock Company	X	X	X						
UK	Anaspec Ltd.,	X	X					X		
UK	API Foils							X		
UK	Astor Universal Ltd.									
USA	3Deep Company	X	X							
USA	Crown Roll Leaf, Inc.,							X		
USA	Dri-Print Foils							X		
USA	DuPont (E.I. DuPont & Co.)					X			X	
USA	Forth Dimension Holographics	X	X							
USA	Hologram Research, Inc.	X								
USA	Images Company	X								
USA	Integraf	X	X	X		X				
USA	Jodon Inc.	X	X							
USA	Keystone Scientific Co.	X								
USA	Kurz Foils							X		
USA	Process Technologies						X			
USA	Shipley Chemical Co.						X		X	
USA	Sillcocks Plastics International									X
USA	Sommers Plastic Products									X
USA	Three Deep Co.	X								
USA	Towne Technologies						X			
USA	VinTeq, Ltd.	X	X							

TABLE 7 - SUPPLIERS OF HOLOGRAM-RELATED PRODUCTION EQUIPMENT

Country	Business Name	Sales - HOPs standard or custom	Emboss web	Electro-plating, shims, etc	Photo-polymer processing	Registration devices	Hot stamp press	Label application equip
Germany	Baier Praegepressen		X			X		
Germany	HAKRO		X			X		
Germany	Hologram Company RAKO		X			X		
Germany	Kremo		X					
Germany	Leonhard Kurz GmbH		X			X		
Germany	Steuer KG GmbH & Co.					X	X	
Netherlands	Dutch Holographic	X	X					
Taiwan	Ahead Optoelectronics, Inc.	X	X					
UK	Art Foil Graphic Machinery						X	
UK	Global Images		X	X				
UK	Molins PLC						X	X
UK	Spatial Imaging Limited	X						
UK	Star Foil Technology Ltd.						X	
UK	Westmead Technology Ltd	X	X					
USA	Anderson Lasers, Inc.	X						
USA	Bobst Group Inc.					X	X	X
USA	Brandtjen & Kluge, Inc.,					X	X	X
USA	Capitol Converting Eq.						X	
USA	Digital Matrix Corp.	X						
USA	Dimensional Arts	X	X					
USA	DuPont (E.I. DuPont)				X			
USA	Frank J. Deutsch Inc.					X	X	X

TABLE 7 - SUPPLIERS OF HOLOGRAM-RELATED PRODUCTION EQUIPMENT

Country	Business Name	Sales - HOPs standard or custom	Emboss Web	Electro-plating, shims, etc	Photo-polymer processing	Registration devices	Hot stamp press	Label application equip
USA	Holman Technology, Inc.	X	X	X				
USA	HoloCom	X	X					
USA	James River Products		X			X		
USA	Man/Environment, Inc.	X						
USA	New Light Industries	X	X					
USA	Peacock Laboratories, Inc.			X				
USA	Standard Paper Box Machine.						X	
USA	Total Register Inc.					X		
USA	Tyler Group	X						
USA	Ultra-Res Corporation	X						
USA	Unifoil Corporation		X			X		

TABLE 8 - INDUSTRIAL HOLOGRAPHY

Country	Business Name	NDT-stress analysis	NDT-fluid analysis	NDT R&D	HOE design / produce	HOE R&D	CGH software	CGH R&D	CGH production	Bio-Medical display	Real Time R&D
Australia	Moonbeamers				X						
Belgium	Center for Applied Res. in Art					X					
Belgium	Free University Of Brussels.					X					
Belgium	Laboratory Vinckiner			X		X					
Canada	Capilano College							X			
Canada	Lasiris Inc.				X						
Canada	Universite Laval					X		X			
China	Guangzhou Inst.of Electronics					X					
China	Nanjing Sanle Laser Tech.				X						
Denmark	Ibsen Micro Structures A/S				X						
Finland	Heptagon Oy				X						
France	Aerospatiale	X		X							
Germany	AKS Holographie-GaleriebH								X		
Germany	Bernhard Halle Nachf.o.				X						
Germany	BIAS			X							
Germany	BIFO				X						
Germany	Carls Zeiss GmbH				X						
Germany	Chemnitzer Werstoffmechanik	X									
Germany	CHIRON Technolas GmbH				X						
Germany	Coburn Europe GmbH				X						
Germany	Daimler Benz Aerospace	X		X		X		X			
Germany	DLR e.V.				X						
Germany	Dornier Medizintechnik GmbH				X	X				X	
Germany	EPA - Elektro-Physik Aachen				X						
Germany	es - Lasersysteme D. Baur				X					X	
Germany	Ettemeyer GmbH & Co	X									
Germany	F & E Labor	X									
Germany	FhG Fraunhofer Gesellschaft							X			
Germany	GOM Gesellschaft	X									
Germany	Humboldt University Berlin					X					
Germany	Inst. für Licht- und Bautechnik	X									
Germany	Institute for Applied Physics		X	X		X					
Germany	Leseberg Dr. Detlef					X		X			
Germany	Mannheim							X			
Germany	Optilas GmbH				X						
Germany	Rottenkolber Holo-System	X									
Germany	Technische Fachhochschule			X		X		X			
Germany	Technische Universität Berlin			X		X		X			X
Germany	University of Bundeswehr		X	X							
Germany	University of Jena										X
Germany	University of Karlsruhe							X			
Germany	University of Magdeburg							X			
Germany	University of Stuttgart			X		X					
Germany	Volkswagen AG	X									
Hungary	Artplay Holographika Studio				X		X		X		
Israel	Holo-Or Ltd				X						
Italy	Universita Di Roma			X		X					
Japan	Canon Inc. R&D Headquarters					X		X			
Japan	Central Glass Co., Ltd.				X	X					

TABLE 8 - INDUSTRIAL HOLOGRAPHY

Country	Business Name	NDT-stress analysis	NDT-fluid analysis	NDT R&D	HOE design / produce	HOE R&D	CGH software	CGH R&D	CGH production	Bio-Medical display	Real Time R&D
Japan	Chiba University					X		X			
Japan	Chiba University			X		X		X			
Japan	Fujitsu Laboratories Ltd.					X		X			
Japan	Keio University	X		X							
Japan	Laboratories of Image Info.			X		X					
Japan	Mazda Motor Corp.	X		X							
Japan	Ministry Of International Trade					X					
Japan	Mitsubishi Heavy Industries	X		X							
Japan	Nihon University			X							
Japan	Nippondenso Co., Ltd.				X						
Japan	Nissan Motor				X						
Japan	Numazu College Of Tech.			X		X		X			
Japan	Sharp Corp.					X		X			
Japan	Shonan Institute Of Tech.					X					
Japan	Sophia University		X	X				X			
Japan	Tama Art University					X					
Japan	Tokai University					X					
Japan	Tokyo Institute Of Technology					X					
Japan	Toshihiro Kubota.					X		X			
Japan	University Of Tokyo					X		X		X	X
Japan	University Of Tsukuba					X		X			
Japan	Waseda University					X		X		X	
Mexico	Centro de Investigaciones A.C.				X						
Portugal	Universidade Do Porto			X							
Russia	Kaliningrad State University	X		X							
Slovakia	Technical University Zvolen	X									
Sweden	Lund Institute Of Tech			X		X		X			
Switzerland	Universite De Neuchatel			X		X		X			
Taiwan	Industrial Tech. Research					X					
Taiwan	Institute Of Optical Science			X		X		X			
Ukraine	Feofaniya Ltd.	X									
Ukraine	TAVEX, Ltd.										X
UK	British Aerospace Plc.	X									
UK	Colour Holographics				X						
UK	Electro Optics Developments .				X						
UK	Imperial College Of Science			X							
UK	Loughborough Univ. Of Tech.			X		X					
UK	Op-Graphics (Holography) .										X
UK	Pilkington Optronics	X			X			X	X		
UK	Richmond Holographic Studios					X					
UK	Turing Institute									X	
USA	3-D Systems				X			X			
USA	Accuwave Corp.				X						
USA	Advanced Precision Tech.				X						
USA	Advanced Tech. Program					X		X			X
USA	Alabama A&M University										X
USA	American Holographic Inc.				X						
USA	American Soc. for NDT			X							
USA	Ann Arbor Optical Co.				X				X		
USA	Applied Optics					X		X			
USA	Batelle Pacific Northwest			X		X				X	
USA	Broadbent Consulting				X						
USA	Chronomotion										X
USA	Control Module Inc.				X						
USA	Datacard Corporation				X						
USA	Diffraction Ltd.				X						
USA	Excalibur Engineering	X			X	X	X		X		
USA	Flight Dynamics				X						
USA	Holographic Optics Inc.				X						
USA	Holographics Inc.			X		X					
USA	Holonix				X						
USA	Holotek					X					
USA	IBM Almaden Research Center										X
USA	Illinois Institute Of Technology					X		X			X
USA	ImEdge Technology				X						
USA	Kaiser Optical Systems, Inc.		X		X					X	
USA	Krystal Holographics				X						
USA	Laboratory for Optical Data					X		X			

TABLE 8 - INDUSTRIAL HOLOGRAPHY

Country	Business Name	NDT-stress analysis	NDT-fluid analysis	NDT R&D	HOE design / produce	HOE R&D	CGH soft-ware	CGH R&D	CGH produc-tion	Bio-Medical display	Real Time R&D
USA	Laboratory for Optical Data					X		X			
USA	Laser Technology, Inc.	X									
USA	Lawrence Berkeley Laboratory	X	X	X	X	X		X			
USA	M.I.T. (Mass. Inst. of Tech)					X		X			X
USA	McMahan Electro-Optic	X									
USA	Metrologic Instruments, Inc.				X						
USA	Nimbus Manufacturing, Inc.				X						
USA	Northern Illinois University	X		X							
USA	Optical Research Associates						X				
USA	Optineering	X	X								
USA	Polaris Research Group	X	X								
USA	Polaroid Corporation				X				X		
USA	Potomac Photonics, Inc.				X						
USA	Ralcon				X						
USA	Rice Systems, Inc.		X								
USA	Richardson Grating Laboratory				X						
USA	Rochester Inst. Of Technology					X		X			
USA	Rochester Photonics				X						
USA	Rowland Institute For Science			X							
USA	Saginaw Valley State Univ					X					
USA	San Jose State University					X		X			
USA	Sandia National Laboratories	X	X	X	X	X	X	X	X		X
USA	Silhouette Technology Inc.				X						
USA	Sinclair Optics, Inc.				X				X		
USA	Siros Technologies, Inc.										X
USA	Smith, Steven L.					X		X			
USA	Stanford University			X		X		X			X
USA	Ultra-Res Corporation	X			X						X
USA	Univ Of Alabama at Huntsville			X		X					
USA	University Of Arizona			X		X		X			
USA	University Of Dayton			X		X		X			
USA	University Of Michigan					X					
USA	University Of Rochester			X		X		X			
USA	University Of Southern Cal					X		X			
USA	Veeco Instruments Inc.	X	X								
USA	Wavefront Research, Inc.				X	X					

TABLE 9 - HOLOGRAPHY RELATED EDUCATION & RESOURCES

Country	Business Name	Assoc commercial /trade	Assoc hobbyist / artist	News-letter	journal or magazine	Published book	Publisher distributor	Multi-media publisher
Germany	AHT 3D-Medien		X		X			
Germany	Design + Kunst e.V. Chemnitz		X					
Germany	Deutsche Gesellschaft fur Holografie	X	X	X				
Germany	Deutscher Drucker Gerlags				X			
Germany	Dorra, Bodo					X		
Germany	Dt. Gesellschaft für Elektro.	X						
Germany	Dt. Gesellschaft für Angewandte	X						
Germany	Dt. Gesellschaft für Stereografie		X					
Germany	Dt. Physikalische Gesellschaft e.V.	X						
Germany	GESA - Arbeitskreis für optische	X						
Germany	Holopublic Unbehaun			X				
Germany	Meisinger Verlag					X		
Germany	Rita Wittig Fachbuchverlag					X		
Germany	Springer Verlag Berlin					X		
Germany	Verlag 3D Magazin				X			
Japan	Holographic Display Artists & Eng.	X	X	X				
Spain	Karas Studios Holografia		X					
Sweden	Hologram Center Holmby		X					
Sweden	Holography Group TEM		X					
Sweden	Royal Institute of Technology					X		
UK	Creative Holography Index, The						X	
UK	Dr. Hans Bjelkhagen					X		
UK	International Hologram Manuf.	X						
UK	Reconnaissance International Ltd.			X				
UK	Saxby, Graham					X		
USA	American Society for NDT	X			X			
USA	Center For The Holographic Arts		X					X
USA	DeFreitas Holography Studio							
USA	Foil Stamping and Embossing Ass.	X						
USA	Holo Art		X					
USA	Holography Marketplace				X	X	X	X
USA	Holoworld.com							X
USA	Images Company							X
USA	Inside Finishing Magazine	X			X			
USA	Laser Arts Society For Education		X					
USA	Laser Focus World				X			
USA	Laser Institute Of America	X			X			
USA	Optical Society of America (OSA)	X		X	X			
USA	Photonics Spectra				X			
USA	Reconnaissance International Ltd.			X				
USA	Ross Books				X	X	X	X
USA	SPIE	X		X	X			
USA	SPIE's Holography Working Group	X		X	X			
USA	Springer-Verlag New York						X	
USA	Unifoil Corporation	X						

Index

DID YOU BORROW THIS COPY?
GET YOUR OWN COPY DELIVERED TO YOUR DOOR
ANYWHERE IN THE WORLD!

YOUR COST
(INCLUDES ALL SHIPPING AND HANDLING FEES*)

$30 - CONTINENTAL U.S.
$35 - ALASKA, HAWAII, CANADA, MEXICO
$45 - ALL OTHER COUNTRIES

ALL BOOKS SHIPPED PROMPTLY UPON RECEIPT OF PAYMENT BY UPS OR U.S. POSTAL AIR MAIL.

* We are not responsible for any additional international entry fees that may be imposed.

ROSS BOOKS
TOLL-FREE ORDER PHONE (USA):
1-800-367-0930
VOICE PHONE:
1- 510-841-2474
FAX:
1-510-841-2695
INTERNET WEB SITE:
www.holoinfo.com
EMAIL:
sales@rossbooks.com
MAILING ADDRESS:
Ross Books P.O. Box 4340
Berkeley, CA 94704 USA

To Order HMP 8th Edition
Fill out and send us the accompanying order form by fax or by mail.
Include a check or money order (U.S.$), or your credit card information.
Or phone, email or place your order directly on our Internet Web site.

Payment Information:
Check # _____   Money Order # _____   American Express ___ Visa ___ Mastercard __
Credit Card Number _____
Credit Card Expiration Date (month) _____ (year) _____
Name on Card _____ Authorized Signature _____

"Ship to" Information:
Name: _____
Company: _____
Address: _____
City/State/Province: _____
Country: _____
Postal Code: _____
Phone, Fax, or Email: _____

_____Please send me information about advertising rates.
_____Please contact me when additional or related publications become available.
_____Please contact me so my company can be listed in the HMP International Business Directory.